DYNAMICS OF STOCHASTIC SYSTEMS

Translated from Russian
by A. Vinogradov

DYNAMICS OF STOCHASTIC SYSTEMS

V.I. Klyatskin

Institute of Atmospheric Physics
Russian Academy of Sciences
Moscow 119017
Russia

2005

ELSEVIER

Amsterdam – Boston – Heidelberg – London – New York – Oxford – Paris
San Diego – San Francisco – Singapore – Sydney – Tokyo

ELSEVIER B.V.   ELSEVIER Inc.   ELSEVIER Ltd   ELSEVIER Ltd
Radarweg 29   525 B Street, Suite 1900   The Boulevard, Langford Lane   84 Theobalds Road
P.O. Box 211, 1000 AE   San Diego, CA 92101-4495   Kidlington, Oxford OX5 1GB   London WC1X 8RR
Amsterdam, The Netherlands   USA   UK   UK

First edition 2005

Library of Congress Cataloging in Publication Data
A catalog record is available from the Library of Congress.

British Library Cataloguing in Publication Data
A catalogue record is available from the British Library.

ISBN: 0-444-51796-0

Printed in The Netherlands.

Preface

Writing this book, I issued from the course that I gave to scientific associates at the Institute of Calculus Mathematics, Russian Academy of Sciences. In the book, I use the functional approach to uniformly formulate general methods of statistical description and analysis of dynamic systems described in terms of different types of equations with fluctuating parameters, such as ordinary differential equations, partial differential equations, boundary-value problems, and integral equations. Asymptotic methods of analyzing stochastic dynamic systems — the delta-correlated random process (field) approximation and the diffusion approximation — are also considered. General ideas are illustrated by the examples of coherent phenomena in stochastic dynamic systems, such as clustering of particles and passive tracer in random velocity field and dynamic localization of plane waves in randomly layered media.

The book consists of three parts.

The first part may be viewed as an introductory text. It takes up a few typical physical problems to discuss their solutions obtained under random perturbations of parameters affecting the system behavior. More detailed formulations of these problems and relevant statistical analysis may be found in other parts of the book.

The second part is devoted to the general theory of statistical analysis of dynamic systems with fluctuating parameters described by differential and integral equations. This theory is illustrated by analyzing specific dynamic systems. In addition, this part considers asymptotic methods of dynamic system statistical analysis, such as the delta-correlated random process (field) approximation and the diffusion approximation.

The third part deals with analysis of specific physical problems associated with coherent phenomena. These are clustering and diffusion of particles and passive tracer in a random velocity field, dynamic localization of plane waves propagating in layered random media. These phenomena are described by ordinary differential equations and partial differential equations.

Each chapter is appended with problems the reader to solve by himself (herself), which will be a good training for independent investigations.

The book is intended primarily for scientific workers; however, it may be useful also for senior and postgraduate students specialized in mathematics and physics and dealing with stochastic dynamic systems.

Valery I. Klyatskin

Contents

Introduction

Different areas of physics pose statistical problems in ever-greater numbers. Apart from issues traditionally obtained in statistical physics, many applications call for including fluctuation effects into consideration. While fluctuations may stem from different sources (such as thermal noise, instability, and turbulence), methods used to treat them are very similar. In many cases, the statistical nature of fluctuations may be deemed known (either from physical considerations or from problem formulation) and the physical processes may be modeled by differential, integro-differential or integral equations.

We will consider a statistical theory of dynamic and wave systems with fluctuating parameters. These systems can be described by ordinary differential equations, partial differential equations, integro-differential equations and integral equations. A popular way to solve such systems is by obtaining a closed system of equations for statistical characteristics of such systems to study their solutions as comprehensively as possible.

We note that often wave problems are boundary-value problems. When this is the case, one may resort to the imbedding method to reformulate the equations at hand to initial value problems, thus considerably simplifying the statistical analysis [1], [2].

The purpose of this book is to demonstrate how different physical problems described by stochastic equations may be solved on the base of a general approach.

In stochastic problems with fluctuating parameters, the variables are functions. It would be natural therefore to resort to functional methods for their analysis. We will use a *functional method* devised by Novikov [3] for Gaussian fluctuations of parameters in a turbulence theory and developed by the author of this book [1], [4] [6] for the general case of dynamic systems and fluctuating parameters of arbitrary nature.

However, only a few dynamic systems lend themselves to analysis yielding solutions in a general form. It proved to be more efficient to use an asymptotic method where the statistical characteristics of dynamic problem solutions are expanded in powers of a small parameter which is essentially a ratio of the random impact's correlation time to the time of observation or to other characteristic time scale of the problem (in some cases, these may be spatial rather than temporal scales). This method is essentially a generalization of the theory of Brownian motion. It is termed the *delta-correlated random process (field) approximation*.

For dynamic systems described by ordinary differential stochastic equations with Gaussian fluctuations of parameters, this method leads to a Markovian problem solving model, and the respective equation for transition probability density has the form of the *Fokker–Planck equation*. In this book, we will consider in depth the methods of analysis available for this equation and its boundary conditions. We will analyze solutions and validity conditions by way of integral transformations. In more complicated problems described by partial differential equations, this method leads to a generalized equation of Fokker–Planck type in which variables are the derivatives of the solution's characteristic functional. For dynamic problems with non-Gaussian fluctuations of parameters, this method also yields Markovian type solutions. Under the circumstances, the probability density of respective dynamic stochastic equations satisfies a closed operator equation.

In physical investigations, Fokker–Planck and similar equations are usually set up from rule of thumb considerations, and dynamic equations are invoked only to calculate the coefficients of these equations. This approach is inconsistent, generally speaking. Indeed, the statistical problem is completely defined by dynamic equations and assumptions on the statistics of random impacts. For example, the Fokker–Planck equation must be a logical sequence of the dynamic equations and some assumptions on the character of random impacts. It is clear that not all problems lend themselves for reducing to a Fokker–Planck equation. The functional approach allows one to derive a Fokker–Planck equation from the problem's dynamic equation along with its applicability conditions.

For a certain class of random processes (Markovian telegrapher's processes, Gaussian Markovian process and the like), the developed functional approach also yields closed equations for the solution probability density with allowance for a finite correlation time of random interactions.

For processes with Gaussian fluctuations of parameters, one may construct a better physical approximation than the delta-correlated random process (field) approximation, — the *diffusion approximation* that allows for finiteness of correlation time radius. In this approximation, the solution is Markovian and its applicability condition has transparent physical meaning, namely, the statistical effects should be small within the correlation time of fluctuating parameters. This book treats these issues in depth from a general standpoint and for some specific physical applications.

In recent time, the interest of both theoreticians and experimenters has been attracted to relation of the behavior of average statistical characteristics of a problem solution with the behavior of the solution in certain happenings (realizations). This is especially important for geophysical problems related to the atmosphere and ocean where, generally speaking, a respective averaging ensemble is absent and experimenters, as a rule, have to do with individual observations.

Seeking solutions to dynamic problems for these specific realizations of medium parameters is almost hopeless due to extreme mathematical complexity of these problems. At the same time, researchers are interested in main characteristics of these phenomena without much need to know specific details. Therefore, the idea to use a well developed approach to random processes and fields based on ensemble averages rather than separate observations proved to be very fruitful. By way of example, almost all physical problems of atmosphere and ocean to some extent are treated by statistical analysis.

Randomness in medium parameters gives rise to a stochastic behavior of physical fields. Individual samples of scalar two-dimensional fields $\rho\left(\mathbf{R}, t\right)$, $\mathbf{R} = (x, y)$, say, recall a rough mountainous terrain with randomly scattered peaks, troughs, ridges and saddles. Common methods of statistical averaging (computing mean-type averages — $\langle \rho\left(\mathbf{R}, t\right) \rangle$, space-time correlation function — $\langle \rho\left(\mathbf{R}, t\right) \rho\left(\mathbf{R}', t'\right) \rangle$ etc., where $\langle ... \rangle$ implies averaging over an ensemble of random parameter samples) smooth the qualitative features of specific samples. Frequently, these statistical characteristics have nothing in common with the behavior of specific samples, and at first glance may even seem to be at variance with them. For example, the statistical averaging over all observations makes the field of average concentration of a passive tracer in a random velocity field ever more smooth, whereas each its realization sample tends to be more irregular in space due to mixture of areas with substantially different concentrations.

Thus, these types of statistical average usually characterize 'global' space-time dimensions of the area with stochastic processes but tell no details about the process behavior inside the area. For this case, details heavily depend on the velocity field pattern, specifically, on whether it is divergent or solenoidal. Thus, the first case will show with the total probability that *clusters* will be formed, i.e. compact areas of enhanced concentration of tracer surrounded by vast areas of low-concentration tracer. In the circumstances, all statistical moments of the distance between the particles will grow with time exponentially; that is, on average, a statistical recession of particles will take place [7].

In a similar way, in case of waves propagating in random media, an exponential spread of the rays will take place on average; but simultaneously, with the total probability, *caustics* will form at finite distances. One more example to illustrate this point is the *dynamic localization* of plane waves in layered randomly inhomogeneous media. In this phenomenon, the wave field intensity exponentially decays inward the medium with the probability equal to unity when the wave is incident on the half-space of such a medium, while all statistical moments increase exponentially with distance from the boundary of the medium [1, 8].

These physical processes and phenomena occurring with the probability equal to unity will be referred to as *coherent* processes and phenomena [9]. This type of *statistical coherence* may be viewed as some organization of the complex dynamic system, and retrieval of its *statistically stable characteristics* is similar to the concept of *coherence* as *self-organization* of multicomponent systems that evolve from the random interactions of their elements [10]. In the general case, it is rather difficult to say whether or not the phenomenon occurs with the probability equal to unity. However, for a number of applications amenable to treatment with the simple models of fluctuating parameters, this may be handled by analytical means. In other cases, one may verify this by performing numerical modeling experiments or analyzing experimental findings.

The complete statistic (say, the whole body of all n-point space-time moment functions), would undoubtedly contain all the information about the investigated dynamic system. In practice, however, one may succeed only in studying the simplest statistical characteristics associated mainly with simultaneous and one-point probability distributions. It would be reasonable to ask how with these statistics on hand one would look into the quantitative and qualitative behavior of some system happenings?

This question is answered by *methods of statistical topography*. These methods were highlighted by [11], who seems to had coined this term. Statistical topography yields a different philosophy of statistical analysis of dynamic stochastic systems, which may prove useful for experimenters planning a statistical processing of experimental data. These issues are treated in depths in this book.

More details about the material of this book and more exhaustive references can be found in mentioned textbooks [1], [4]–[6], recent reviews [2, 9, 12, 13], and recently published textbook [14].

Part I

Dynamical description of stochastic systems

Chapter 1

Examples, basic problems, peculiar features of solutions

In this chapter, we consider several dynamic systems described by differential equations of different types and discuss the features in the behaviors of solutions to these equations under random disturbances of parameters. Here, we content ourselves with the problems in the simplest formulation. More complete formulations will be discussed below in the sections dealing with statistical analysis of corresponding systems.

1.1 Ordinary differential equations: initial value problems

1.1.1 Particles under the random velocity field

In the simplest case, a particle under the random velocity field is described by the system of ordinary differential equations of the first order

$$\frac{d}{dt}\mathbf{r}(t) = \mathbf{U}(\mathbf{r}, t), \quad \mathbf{r}(t_0) = \mathbf{r}_0, \tag{1.1}$$

where $\mathbf{U}(\mathbf{r}, t) = \mathbf{u}_0(\mathbf{r}, t) + \mathbf{u}(\mathbf{r}, t)$, $\mathbf{u}_0(\mathbf{r}, t)$ is the deterministic component of the velocity field (mean flow), and $\mathbf{u}(\mathbf{r}, t)$ is the random component. In the general case, field $\mathbf{u}(\mathbf{r}, t)$ can have both divergence-free (solenoidal, for which $\operatorname{div} \mathbf{u}(\mathbf{r}, t) = 0$) and divergent (for which $\operatorname{div} \mathbf{u}(\mathbf{r}, t) \neq 0$) components.

We dwell on stochastic features of the solution to problem (1.1) for a system of particles in the absence of mean flow ($\mathbf{u}_0(\mathbf{r}, t) = 0$). From Eq. (1.1) formally follows that every particle moves independently of other particles. However, if random field $\mathbf{u}(\mathbf{r}, t)$ has a finite spatial correlation radius l_{cor}, particles spaced by a distance shorter than l_{cor} appear in the common zone of infection of random field $\mathbf{u}(\mathbf{r}, t)$, and the behavior of such a system can show new collective features.

For steady-state velocity field $\mathbf{u}(\mathbf{r}, t) \equiv \mathbf{u}(\mathbf{r})$, Eq. (1.1) reduces to

$$\frac{d}{dt}\mathbf{r}(t) = \mathbf{u}(\mathbf{r}), \quad \mathbf{r}(0) = \mathbf{r}_0. \tag{1.2}$$

This equation clearly shows that stationary points $\tilde{\mathbf{r}}$ (at which $\mathbf{u}(\tilde{\mathbf{r}}) = 0$) remain the fixed points. Depending on whether these points are stable or unstable, they will attract or repel nearby particles. In view of randomness of function $\mathbf{u}(\mathbf{r})$, points $\tilde{\mathbf{r}}$ are random too.

It is expected that the similar behavior will also be characteristic of the general case of the space-time random field of velocities $\mathbf{u}(\mathbf{r},t)$.

If some points $\tilde{\mathbf{r}}$ remain stable during sufficiently long time, then clusters of particles (i.e., compact regions with enhanced particle concentration, which occur merely in rarefied zones) must arise around these points in separate realizations of random field $\mathbf{u}(\mathbf{r},t)$. On the contrary, if the stability of these points alternates with instability sufficiently rapidly and particles have no time for significant rearrangement, no clusters of particles will occur.

Simulations [15, 16] show that the behavior of a system of particles essentially depends on whether the random field of velocities is divergence-free or divergent. By way of example, Fig.1.1a shows a schematic of evolution of the two-dimensional system of particles uniformly distributed in the circle for a particular realization of the divergence-free steady-state field $\mathbf{u}(\mathbf{r})$.

Here, we use the dimensionless time related to statistical parameters of field $\mathbf{u}(\mathbf{r})$. In this case, the area of surface patch within the contour remains intact and particles relatively uniformly fill the region within the deformed contour. The only feature consists in the fractal-type irregularity of the deformed contour. On the contrary, in the case of the divergent velocity field $\mathbf{u}(\mathbf{r})$, particles uniformly distributed in the square at the initial instance will form clusters during the temporal evolution. Results simulated for this case are shown in Fig. 1.1b. We emphasize that the formation of clusters is purely kinematic effect. This feature of particle dynamics disappears on averaging over an ensemble of realizations of random velocity field .

To demonstrate the process of particle clustering, we consider the simplest problem [13], in which the random velocity field $\mathbf{u}(\mathbf{r},t)$ has the form

$$\mathbf{u}(\mathbf{r},t) = \mathbf{v}(t)f(\mathbf{r}), \tag{1.3}$$

where $\mathbf{v}(t)$ is the random vector process and the deterministic function

$$f(\mathbf{r}) = \sin 2(\mathbf{kr}) \tag{1.4}$$

is a function of one variable. Note that this form of function $f(\mathbf{r})$ corresponds to the first term of the expansion in harmonic components and is commonly used in numerical simulations.

In this case, Eq. (1.1) can be written in the form

$$\frac{d}{dt}\mathbf{r}(t) = \mathbf{v}(t)\sin 2(\mathbf{kr}), \quad \mathbf{r}(0) = \mathbf{r}_0.$$

In the context of this model, motions of a particle along vector $\mathbf{k}$ and in the plane perpendicular to vector $\mathbf{k}$ are independent and can be separated. If we direct the x-axis along vector $\mathbf{k}$, then the equations assume the form

$$\begin{aligned} \frac{d}{dt}x(t) &= v_x(t)\sin(2kx), \quad x(0) = x_0, \\ \frac{d}{dt}\mathbf{R}(t) &= \mathbf{v_R}(t)\sin(2kx), \quad \mathbf{R}(0) = \mathbf{R}_0. \end{aligned} \tag{1.5}$$

The solution of the first equation in (1.5) is

$$x(t) = \frac{1}{k}\arctan\left[e^{T(t)}\tan(kx_0)\right], \tag{1.6}$$

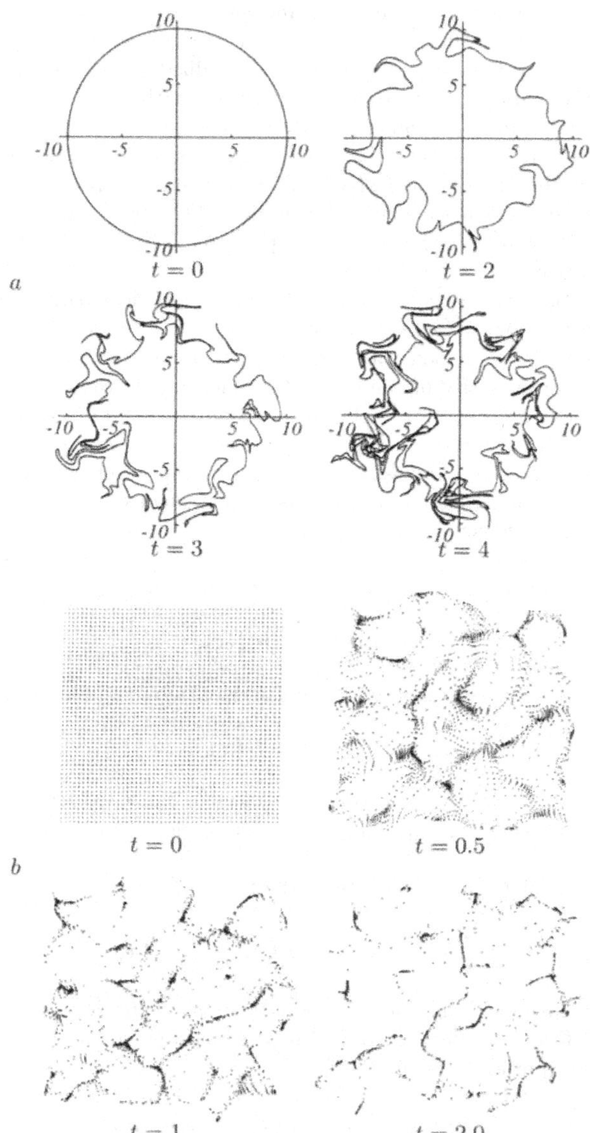

Figure 1.1: Diffusion of a system of particles described by Eqs. (1.2) numerically simulated for (a) solenoidal and (b) divergence-free random velocity field $\mathbf{u}(\mathbf{r})$.

where

$$T(t) = 2k \int\limits_0^t d\tau\, v_x(\tau). \tag{1.7}$$

and we can write the second equation in (1.5) in the form

$$\frac{d}{dt}\mathbf{R}(t|\mathbf{r}_0) = \sin(2kx_0)\frac{\mathbf{v_R}(t)}{e^{-T(t)}\cos^2(kx_0) + e^{T(t)}\sin^2(kx_0)}.$$

As a result, we have

$$\mathbf{R}(t|\mathbf{r}_0) = \mathbf{R}_0 + \sin(2kx_0) \int\limits_0^t d\tau\frac{\mathbf{v_R}(\tau)}{e^{-T(\tau)}\cos^2(kx_0) + e^{T(\tau)}\sin^2(kx_0)}. \tag{1.8}$$

Consequently, if the initial particle position x_0 is such that

$$kx_0 = n\frac{\pi}{2}, \tag{1.9}$$

where $n = 0, \pm 1, ...$, then the particle will be the fixed particle and $\mathbf{r}(t) \equiv \mathbf{r}_0$.

Equalities (1.9) define planes in the general case and points in the one-dimensional case. They correspond to zeros of the velocity field. Stability of these points depends on the sign of function $\mathbf{v}(t)$, and this sign changes during the evolution process. As a result, we can expect that particles will be concentrated around these points if $v_x(t) \neq 0$, which just corresponds to clustering of particles.

In the case of the divergence-free velocity field, $v_x(t) = 0$ and, consequently, $T(t) \equiv 0$; as a result, we have

$$x(t|x_0) \equiv x_0, \quad \mathbf{R}(t|\mathbf{r}_0) = \mathbf{R}_0 + \sin 2(kx_0) \int\limits_0^t d\tau\mathbf{v_R}(\tau),$$

which means that no clustering occurs.

Figure 1.2a shows a fragment of the realization of random process $T(t)$ obtained by numerical integration of Eq. (1.7) for a realization of random process $v_x(t)$; we used this fragment for simulating the temporal evolution of coordinates of four particles $x(t)$, $x \in (0, \pi/2)$ initially located at coordinates $x_0(i) = \frac{\pi}{2}\frac{i}{5}$ ($i = 1, 2, 3, 4$) (see Fig. 1.2b). Figure 1.2b shows that particles form a cluster in the vicinity of point $x = 0$ at the dimensionless time $t \approx 4$ (see [13]). Further, at time $t \approx 16$ the initial cluster disappears and new one appears in the vicinity of point $x = \pi/2$. At moment $t \approx 40$, the cluster appears again in the vicinity of point $x = 0$, and so on. In this process, particles in clusters remember their past history and significantly diverge during intermediate temporal segments (see Fig. 1.2c).

Thus, we see that, in this example, the cluster does not move from one region to another; instead, it first collapses and then a new cluster is formed. Here, the lifetime of clusters significantly exceeds the duration of intermediate segments. It seems that this feature is characteristic of the particular model of the velocity field and follows from stationary property of points (1.9).

As regards the particle diffusion along the y-direction, no cluster occurs in this direction.

Note that such clustering in a system of particles was found to all appearance for the first time in papers [17, 18] as a result of simulating the so-called Eole experiment

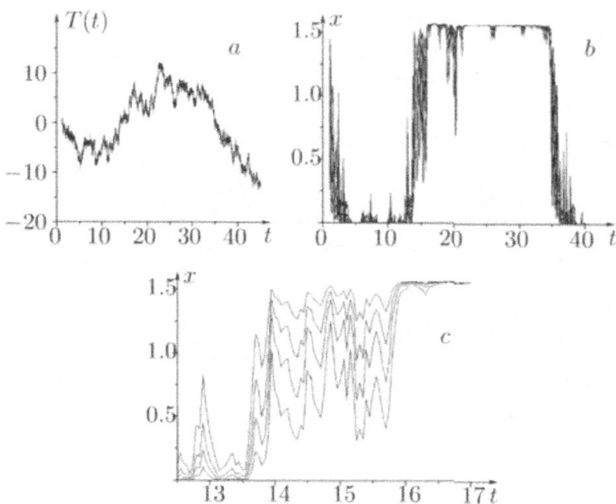

Figure 1.2: (a) Segment of a realization of random process $T(t)$ obtained by numerically integrating Eq. (1.7) for a realization of random process $v_x(t)$; (b), (c) x-coordinates simulated with this segment for four particles versus time.

with the use of simplest equations of atmospheric dynamics. In the context of this global experiment, 500 constant-density balloons were launched in Argentina in 1970-1971; these balloons traveled at a height of about 12 km and spread along the whole of the southern hemisphere. Figure 1.3 shows the balloon distribution over the southern hemisphere for day 105 from the beginning of this process simulation [17]; this distribution clearly shows that balloons are concentrated in groups, which just corresponds to clustering .

Now, we dwell on another stochastic aspect related to dynamic equations of type (1.1); namely, we consider the *phenomenon of transfer* caused by random fluctuations.

Consider the one-dimensional nonlinear equation

$$\frac{d}{dt}x(t) = x\left(1 - x^2\right) + f(t), \quad x(0) = x_0; \; \lambda > 0, \tag{1.10}$$

where $f(t)$ is the random function of time. In the absence of randomness ($f(t) \equiv 0$), the solution of Eq. (1.10) has two stable steady-state states $x = \pm 1$ and one instable state $x = 0$. Depending on the initial condition, solution of Eq. (1.10) arrives at one of the stable states. However, in the presence of small random disturbances $f(t)$, dynamic system (1.10) will first approach the vicinity of one of the stable states and then, after the lapse of certain time, it will be transferred into the vicinity of another stable state.

It is clear that the similar behavior can occur in more complicated situations.

The system of equations (1.1) describes also the behavior of a particle under the field of random external forces $\mathbf{f}(\mathbf{r}, t)$. In the simplest case, the behavior of a particle in the presence of linear friction is described by the system of the first-order differential equations

$$\begin{aligned}
\frac{d}{dt}\mathbf{r}(t) &= \mathbf{v}(t), \quad \frac{d}{dt}\mathbf{v}(t) = -\lambda\mathbf{v}(t) + \mathbf{f}(\mathbf{r}, t), \\
\mathbf{r}(0) &= \mathbf{r}_0, \quad \mathbf{v}(0) = \mathbf{v}_0.
\end{aligned} \tag{1.11}$$

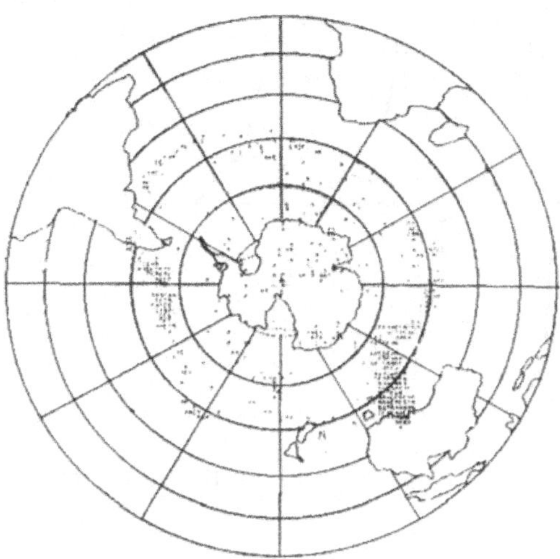

Figure 1.3: Balloon distribution in the atmosphere for day 105 from the beginning of process simulation.

The behavior of a particle under the deterministic potential field in the presence of linear friction and random forces is described by the system of equations

$$\frac{d}{dt}\mathbf{r}(t) = \mathbf{v}(t), \quad \frac{d}{dt}\mathbf{v}(t) = -\lambda\mathbf{v}(t) - \frac{\partial U(\mathbf{r},t)}{\partial \mathbf{r}} + \mathbf{f}(\mathbf{r},t),$$
$$\mathbf{r}(0) = \mathbf{r}_0, \quad \mathbf{v}(0) = \mathbf{v}_0, \tag{1.12}$$

which is the simplest example of *Hamiltonian systems*. In statistical problems, equations of type (1.11), (1.12) are widely used for describing *Brownian motion* of particles.

1.1.2 Systems with blow-up singularities

The simplest stochastic system showing singular behavior in time is described by the following equation commonly used in the statistical theory of waves

$$\frac{d}{dt}x(t) = -\lambda x^2(t) + f(t), \quad x(0) = x_0, \quad \lambda > 0, \tag{1.13}$$

where $f(t)$ is the random function of time.

In the absence of randomness ($f(t) = 0$), the solution to Eq.(1.13) has the form

$$x(t) = \frac{1}{\lambda(t - t_0)}, \quad t_0 = -\frac{1}{\lambda x_0}.$$

For $x_0 > 0$, we have $t_0 < 0$, and solution $x(t)$ monotonically tends to zero with increasing the time. On the contrary, for $x_0 < 0$, solution $x(t)$ reaches a value of $-\infty$ within a finite time $t_0 = -1/\lambda x_0$, which means that the solution becomes *singular* and shows the *blow-up behavior*. In this case, random force $f(t)$ has insignificant effect on the behavior of the system. The effect becomes significant only for positive parameter x_0.

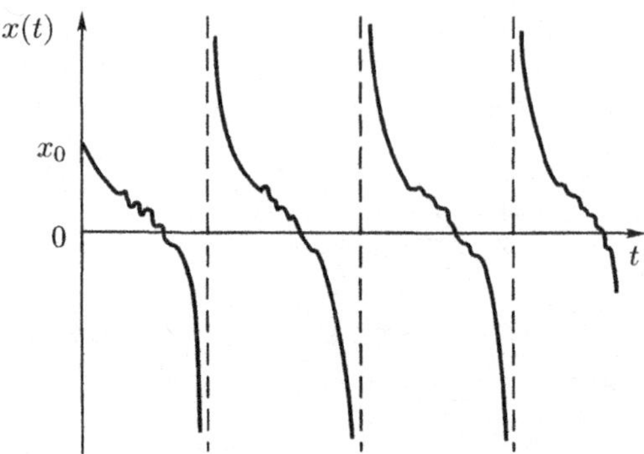

Figure 1.4: Typical realization of the solution to Eq. (1.13).

In this case, the solution, slightly fluctuating, decreases with time as long as it remains positive. On reaching sufficiently small value $x(t)$, the force $f(t)$ *transfers* the solution into the region of negative values of x, where it will reach a value of $-\infty$ within certain finite time.

Thus, in the stochastic case, the solution to problem (1.13) shows the blow-up behavior for arbitrary values of parameter x_0 and always reaches $-\infty$ within a finite time t_0. Figure 1.4 schematically shows the temporal realization of the solution $x(t)$ to problem(1.13) for $t > t_0$; its behavior resembles the *quasi-periodic* structure.

1.1.3   Oscillator with randomly varying frequency (stochastic parametric resonance)

In the above stochastic examples, we considered the effect of additive random actions (forces) on the behavior of systems. The simplest nontrivial system with multiplicative (parametric) action can be illustrated using the *stochastic parametric resonance* as an example. Such a system is described by the second-order equation

$$\frac{d^2}{dt^2}x(t) + \omega_0^2[1 + z(t)]x(t) = 0,$$

$$x(0) = x_0, \quad \frac{d}{dt}x(0) = v_0, \tag{1.14}$$

where $z(t)$ is the random function of time. This equation is characteristic of almost all areas of physics. Physically, it is obvious that dynamic system (1.14) is capable of parametric excitation, because random process $z(t)$ has harmonic components of all frequencies, including frequencies $2\omega_0/n$ $(n = 1, 2, ...)$ that exactly correspond to the frequencies of parametric resonance in the system with periodic function $z(t)$, as it is the case, for example, in the *Mathieu equation*.

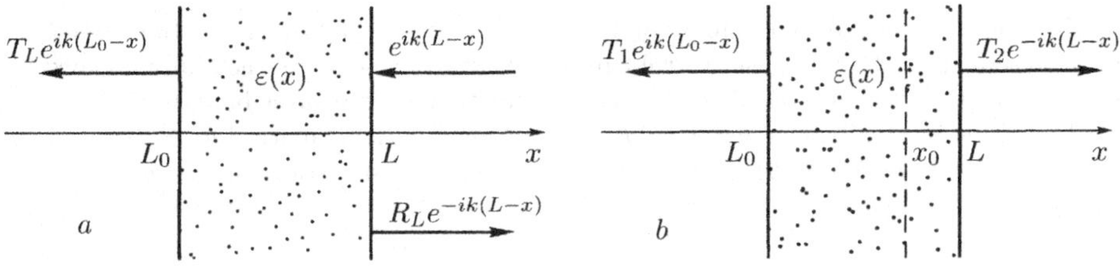

Figure 1.5: (*a*) Plane wave incident on medium layer and (*b*) source inside the medium layer.

1.2 Boundary-value problems for linear ordinary differential equations (plane waves in layered media)

In the previous section, we considered several dynamic systems described by a system of ordinary differential equations with given initial conditions. Now, we consider the simplest linear boundary-value problem, namely, the stationary one-dimensional wave problem.

Let the layer of inhomogeneous medium occupies the segment of space $L_0 < x < L$ and let the unit-amplitude plane wave $u_0(x) = e^{-ik(x-L)}$ is incident on this layer from the region $x > L$ (Fig. 1.5a). The wavefield satisfies the *Helmholtz equation*,

$$\frac{d^2}{dx^2}u(x) + k^2(x)u(x) = 0, \tag{1.15}$$

where

$$k^2(x) = k^2[1 + \varepsilon(x)]$$

and function $\varepsilon(x)$ describes medium inhomogeneities. We assume that $\varepsilon(x) = 0$, i.e., $k(x) = k$ outside the layer; inside the layer, we set $\varepsilon(x) = \varepsilon_1(x) + i\gamma$, where $\varepsilon_1(x)$ is the real part responsible for wave scattering in the medium and the imaginary part $\gamma \ll 1$ describes the absorption of the wave in the medium.

In region $x > L$, the wavefield has the following structure

$$u(x) = e^{-ik(x-L)} + R_L e^{ik(x-L)},$$

where R_L is the complex reflection coefficient. In region $x < L_0$, the structure of the wavefield is

$$u(x) = T_L e^{ik(L_0-x)},$$

where T_L is the complex transmission coefficient. Boundary conditions for Eq. (1.15) are the continuity of the field and the field derivative at the layer boundaries; mathematically, they can be written as follows

$$u(L) + \frac{i}{k}\frac{du(x)}{dx}\bigg|_{x=L} = 2, \qquad u(L_0) - \frac{i}{k}\frac{du(x)}{dx}\bigg|_{x=L_0} = 0. \tag{1.16}$$

Thus, the wavefield in the layer of an inhomogeneous medium is described by the boundary-value problem (1.15), (1.16). Dynamic equation (1.15) coincides in form with Eq. (1.14). Note that, in the problem under consideration, function $\varepsilon(x)$ is discontinuous at the boundaries. We will call boundary-value problem (1.15), (1.16) the *unmatched boundary-value problem*. In such problems, wave scattering occurs not only on medium inhomogeneities, but also on discontinuities of function $\varepsilon(x)$ at layer boundaries.

If medium parameters (function $\varepsilon_1(x)$) are specified in the statistical form, then solving the stochastic problem (1.15), (1.16) consists in obtaining statistical characteristics of the reflection and transmission coefficients, which are related to the wavefield values at the layer boundaries by the relationships

$$R_L = u(L) - 1, \quad T_L = u(L_0),$$

and the wavefield intensity

$$I(x) = |u(x)|^2$$

inside the inhomogeneous medium. Determination of these characteristics constitutes the subject of the *statistical theory of radiative transfer*.

Consider some features characteristic of solutions to the stochastic boundary-value problem (1.15), (1.16). On the assumption that medium inhomogeneities are absent ($\varepsilon_1(x) = 0$) and absorption γ is sufficiently small, the intensity of the wavefield in the medium slowly decays with distance according to the exponential law

$$I(x) = |u(x)|^2 = e^{-k\gamma(L-x)}. \tag{1.17}$$

Figure 1.6 shows two realizations of the intensity of a wave in a sufficiently thick layer of medium. These realizations were simulated for two realizations of medium inhomogeneities [19]. Omitting the detailed description of problem parameters, we mention only that this figure clearly shows the prominent tendency of a sharp exponential decay (accompanied by significant spikes toward both higher and nearly zero-valued intensity values), which is caused by multiple reflections of the wave in the chaotically inhomogeneous random medium (the phenomenon of *dynamic localization*). Recall that absorption is small ($\gamma \ll 1$), so that it cannot significantly affect the dynamic localization.

The *imbedding method* offers a possibility of reformulating boundary-value problem (1.15), (1.16) to the dynamic initial value problem with respect to parameter L (this parameter is the geometrical position of the layer right-hand boundary) by considering the solution to the boundary-value problem as a function of parameter L (see the next Chapter). On such reformulation, the reflection coefficient R_L satisfies the Riccati equation

$$\frac{d}{dL}R_L = 2ikR_L + \frac{ik}{2}\varepsilon(L)\left(1 + R_L\right)^2, \quad R_{L_0} = 0 \tag{1.18}$$

and the wavefield in the medium layer $u(x) \equiv u(x; L)$ satisfies the linear equation

$$\begin{aligned}
\frac{\partial}{\partial L}u(x; L) &= iku(x; L) + \frac{ik}{2}\varepsilon(L)\left(1 + R_L\right)u(x; L), \\
u(x; x) &= 1 + R_x.
\end{aligned} \tag{1.19}$$

The equation for the reflection coefficient squared modulus $W_L = |R_L|^2$ for absent attenuation (i.e., at $\gamma = 0$) follows from Eq. (1.18):

$$\frac{d}{dL}W_L = -\frac{ik}{2}\varepsilon_1(L)\left(R_L - R_L^*\right)\left(1 - W_L\right), \quad W_{L_0} = 0. \tag{1.20}$$

Note that condition $W_{L_0} = 1$ will be the initial condition to Eq. (1.20) in the case of totally reflecting boundary at L_0. In this case, the wave incident on the layer of a non-absorptive medium ($\gamma = 0$) is totally reflected from the layer, i.e., $W_L = 1$.

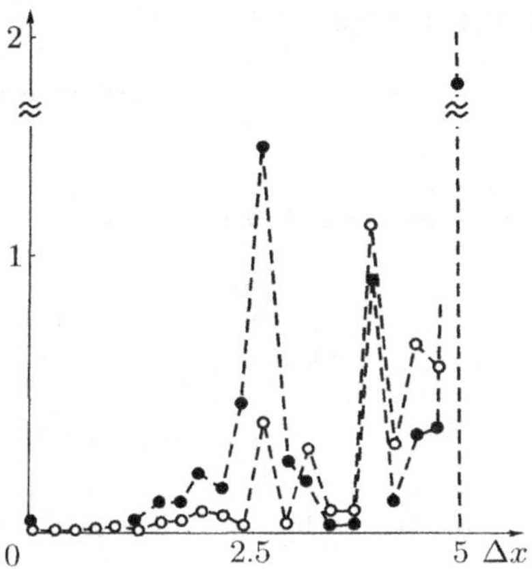

Figure 1.6: Dynamic localization phenomenon simulated for two realizations of medium inhomogeneities.

In the general case of arbitrarily reflecting boundary L_0, the steady-state (independent of L) solution $W_L = 1$ corresponding to the total reflection of incident wave formally exists for a half-space ($L_0 \to -\infty$) filled with non-absorptive random medium, too. This solution, as it will be shown later, is actually realized in the statistical problem with a probability equal to unity.

If, in contrast to the above problem, we assume that function $k(x)$ is continuous at boundary $x = L$, i.e., if we assume that the wave number in the free half-space $x > L$ is equal to $k(L)$, then boundary conditions (1.16) of problem (1.15) will be replaced with the conditions

$$u(L) + \frac{i}{k(L)}\frac{du(x)}{dx}\bigg|_{x=L} = 2, \quad u(L_0) - \frac{i}{k(L_0)}\frac{du(x)}{dx}\bigg|_{x=L_0} = 0. \qquad (1.21)$$

We will call the boundary-value problem (1.15), (1.21) the *matched boundary problem*.

The field of a point source located in the layer of random medium is described by the similar boundary-value problem for Green's function of the Helmholtz equation:

$$\frac{d^2}{dx^2}G(x; x_0) + k^2[1 + \varepsilon(x)]G(x; x_0) = 2ik\delta(x - x_0),$$

$$G(L; x_0) + \frac{i}{k}\frac{dG(x; x_0)}{dx}\bigg|_{x=L} = 0, \quad G(L_0; x_0) - \frac{i}{k}\frac{dG(x; x_0)}{dx}\bigg|_{x=L_0} = 0.$$

Outside the layer, the solution has here the form of outgoing waves (Fig. 1.5*b*)

$$G(x; x_0) = T_1 e^{ik(x-L)} \quad (x \geq L), \qquad G(x; x_0) = T_2 e^{-ik(x-L_0)} \quad (x \leq L_0).$$

Note that, for the source located at the layer boundary $x_0 = L$, this problem coincides with the boundary-value problem (1.15), (1.16) on the wave incident on the layer, which yields

$$G(x; L) = u(x; L).$$

1.3 Partial differential equations

Consider now several dynamic systems (dynamic fields) described by partial differential equations.

1.3.1 Passive tracer in random velocity field

In the context of linear first-order partial differential equations, the simplest problems concern the equation of continuity for the concentration of a conservative tracer and the equation of transfer of a nonconservative passive tracer by random velocity field $\mathbf{U}(\mathbf{r}, t)$:

$$\left(\frac{\partial}{\partial t} + \frac{\partial}{\partial \mathbf{r}} \mathbf{U}(\mathbf{r}, t)\right) \rho(\mathbf{r}, t) = 0, \quad \rho(\mathbf{r}, 0) = \rho_0(\mathbf{r}), \tag{1.22}$$

$$\left(\frac{\partial}{\partial t} + \mathbf{U}(\mathbf{r}, t) \frac{\partial}{\partial \mathbf{r}}\right) q(\mathbf{r}, t) = 0, \quad q(\mathbf{r}, 0) = q_0(\mathbf{r}). \tag{1.23}$$

The conservative tracer is a tracer whose total mass remains intact

$$M_0 = \int d\mathbf{r} \rho(\mathbf{r}, t) = \int d\mathbf{r} \rho_0(\mathbf{r}) \tag{1.24}$$

We can use the method of characteristics to solve the linear first-order partial differential equations (1.22), (1.23). Introducing *characteristic curves* (particles)

$$\frac{d}{dt} \mathbf{r}(t) = \mathbf{U}(\mathbf{r}, t), \quad \mathbf{r}(0) = \mathbf{r}_0, \tag{1.25}$$

we can write these equations in the form

$$\frac{d}{dt} \rho(t) = -\frac{\partial \mathbf{U}(\mathbf{r}, t)}{\partial \mathbf{r}} \rho(t), \quad \rho(0) = \rho_0(\mathbf{r}_0),$$

$$\frac{d}{dt} q(t) = 0, \quad q(0) = q_0(\mathbf{r}_0). \tag{1.26}$$

This formulation of the problem corresponds to the *Lagrangian description*, while the initial dynamic equations (1.22), (1.23) correspond to the *Eulerian description*.

Here, we introduced the characteristic vector parameter $\mathbf{r}_0$ in the system of equations (1.25), (1.26). With this parameter, Eq. (1.25) coincides with Eq. (1.1) that describes particle dynamics in random velocity field.

The solution of system of equations (1.25), (1.26) depends on initial value $\mathbf{r}_0$,

$$\mathbf{r}(t) = \mathbf{r}(t|\mathbf{r}_0), \quad \rho(t) = \rho(t|\mathbf{r}_0), \tag{1.27}$$

which we will isolate in the argument list by the vertical bar symbol.

The first equality in Eq. (1.27) can be considered as the algebraic equation in characteristic parameter; the solution of this equation

$$\mathbf{r}_0 = \mathbf{r}_0(\mathbf{r}, t)$$

exists because *divergence* $j(t|\mathbf{r}_0) = \det \|\partial r_i(t|\mathbf{r}_0)/\partial r_{0k}\|$ is different from zero. Consequently, we can write the solution of the initial equation (1.22) in the form

$$\rho(\mathbf{r}, t) = \rho(t|\mathbf{r}_0(\mathbf{r}, t)) = \int d\mathbf{r}_0 \rho(t|\mathbf{r}_0) j(t|\mathbf{r}_0) \delta\left(\mathbf{r}(t|\mathbf{r}_0) - \mathbf{r}\right).$$

Integrating this expression over $\mathbf{r}$, we obtain, in view of Eq. (1.24), the relationship between functions $\rho(t|\mathbf{r}_0)$ and $j(t|\mathbf{r}_0)$

$$\rho(t|\mathbf{r}_0) = \frac{\rho_0(\mathbf{r}_0)}{j(t|\mathbf{r}_0)}, \tag{1.28}$$

and, consequently, the density field can be rewritten in the form of equality

$$\rho(\mathbf{r}, t) = \int d\mathbf{r}_0 \rho(t|\mathbf{r}_0) j(t|\mathbf{r}_0) \delta\left(\mathbf{r}(t|\mathbf{r}_0) - \mathbf{r}\right) = \int d\mathbf{r}_0 \rho_0(\mathbf{r}_0) \delta\left(\mathbf{r}(t|\mathbf{r}_0) - \mathbf{r}\right) \tag{1.29}$$

that states the relationship between the Lagrangian and Eulerian characteristics. For the position of the Lagrangian particle, the delta-function appeared in the right-hand side of this equality is the *indicator function* (see Chapter 3).

For a divergence-free velocity field ($\operatorname{div} \mathbf{U}(\mathbf{r}, t) = 0$), both particle divergence and particle density are conserved, i.e.,

$$j(t|\mathbf{r}_0) = 1, \quad \rho(t|\mathbf{r}_0) = \rho_0(\mathbf{r}_0), \quad q(t|\mathbf{r}_0) = q_0(\mathbf{r}_0).$$

Consider now the stochastic features of solutions to problem (1.22). A convenient way of analyzing random field dynamics consists in using *topographic* concepts. Indeed, in the case of the divergence-free velocity field, temporal evolution of the contour of constant concentration $\rho = \text{const}$ coincides with the dynamics of particles in this velocity field and, consequently, coincides with the dynamics shown in Fig. 1.1a. In this case, the area within the contour remains constant and, as it is seen from Fig. 1.1a, the pattern becomes highly indented, which is manifested in gradient sharpening and the appearance of contour dynamics for progressively shorter scales. In the other limiting case of a divergent velocity field, the area within the contour tends to zero, and the concentration field condenses in clusters. One can find examples simulated for this case in papers [15, 16]. These features of particle dynamics disappear on averaging over an ensemble of realizations. Cluster formation in the Eulerian description can be traced using the random velocity field of form (1.3), (1.4) [13].

For the divergence-free velocity field $v_x(t) = 0$, $T(t) \equiv 0$, and we have the expression

$$\rho(\mathbf{r}, t) = \rho_0 \left(\mathbf{r} - \sin(2kx) \int_0^t d\tau \mathbf{v}(\tau) \right)$$

from which follows that no clustering occurs.

If $v_x(t) \neq 0$, then concentration field $\rho(\mathbf{r}, t)$ in the particular case of uniform (independent of $\mathbf{r}$) initial distribution $\rho_0(\mathbf{r}) = \rho_0$ can be represented by the following expression

$$\rho(\mathbf{r}, t)/\rho_0 = \frac{1}{e^{T(t)} \cos^2(kx) + e^{-T(t)} \sin^2(kx)}, \tag{1.30}$$

where function $T(t)$ is given by Eq. (1.7).

Figure 1.7 shows the Eulerian concentration field $1 + \rho(\mathbf{r}, t)/\rho_0$ and its space-time evolution calculated by Eq. (1.30) in the dimensionless space-time variables (the density field is added with a unity to avoid the difficulties of dealing with nearly zero-valued concentrations in the logarithmic scale). This figure shows successive patterns of concentration field rearrangement toward narrow neighborhoods of points $x \approx 0$ and $x \approx \pi/2$, i.e., the cluster formation. Figures 1.7a and 1.7b show the temporal pattern ($t = 1 \div 10$) of cluster formation around point $x \approx 0$. Figures 1.7c and 1.7d show the temporal pattern ($t = 16 \div 25$)

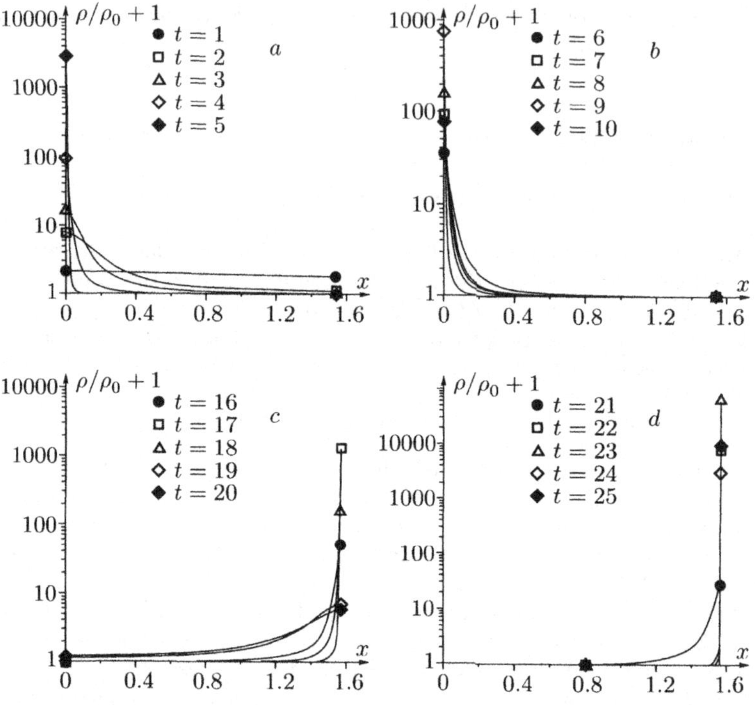

Figure 1.7: Space-time evolution of the Eulerian density field given by Eq. (1.30).

of rearranging the concentration field from the neighborhood of point $x \approx 0$ toward the neighborhood of point $x \approx \pi/2$, i.e., they show the removal of the cluster near $x \approx 0$ and the birth of a new cluster near $x \approx \pi/2$. This process is then repeated in time. As is seen from figures, the lifetimes of such clusters coincide on the order of magnitude with the time of cluster formation.

This model provides an insight into the difference between the diffusion processes in divergent and divergence-free velocity fields. In divergence-free (incompressible) velocity fields, particles (and, consequently, the concentration field) have no time for attracting to stable centers of attraction during the lifetime of these centers, and particles slightly fluctuate relative their initial location. On the contrary, in the divergent (compressible) velocity field, lifetime of stable centers of attraction is sufficient for particles attracted to them, because the speed of attraction increases exponentially, which is clearly seen from Eq (1.30).

From the above description, it becomes obvious that dynamic equation (1.22) considered as the model equation describing actual physical phenomena can be used only on finite temporal intervals. A more complete analysis assumes the consideration of the field of tracer concentration gradient $\mathbf{p}(\mathbf{r},t) = \boldsymbol{\nabla}\rho(\mathbf{r},t)$ that satisfies the equation

$$
\begin{aligned}
\left(\frac{\partial}{\partial t} + \frac{\partial}{\partial \mathbf{r}}\mathbf{U}(\mathbf{r},t)\right) p_i(\mathbf{r},t) &= -p_k\,(\mathbf{r},t)\frac{\partial U_k(\mathbf{r},t)}{\partial r_i} - \rho(\mathbf{r},t)\frac{\partial^2 \mathbf{U}(\mathbf{r},t)}{\partial r_i \partial \mathbf{r}},\\
\mathbf{p}(\mathbf{r},0) &= \mathbf{p}_0(\mathbf{r}) = \boldsymbol{\nabla}\rho_0(\mathbf{r}).
\end{aligned}
\tag{1.31}
$$

In addition, one should also include the effect of molecular diffusion (with molecular diffusion coefficient μ) that smooths the mentioned gradient sharpness; this effect is described by the linear second-order partial differential equation

$$
\left(\frac{\partial}{\partial t} + \frac{\partial}{\partial \mathbf{r}}\mathbf{U}(\mathbf{r},t)\right)\rho(\mathbf{r},t) = \mu\Delta\rho(\mathbf{r},t), \quad \rho(\mathbf{r},0) = \rho_0(\mathbf{r}).
\tag{1.32}
$$

1.3.2 Quasilinear and nonlinear first-order partial differential equations

Consider now the simplest quasilinear equation for scalar quantity $q(\mathbf{r},t)$, which we write in the form

$$
\left(\frac{\partial}{\partial t} + \mathbf{U}(t,q)\frac{\partial}{\partial \mathbf{r}}\right)q(\mathbf{r},t) = Q\,(t,q)\,, \quad q(\mathbf{r},0) = q_0(\mathbf{r}),
\tag{1.33}
$$

where we assume for simplicity that functions $\mathbf{U}(t,q)$ and $Q(t,q)$ have no explicit dependence on spatial variable $\mathbf{r}$.

Supplement Eq. (1.33) with the equation for the gradient $\mathbf{p}(\mathbf{r},t) = \boldsymbol{\nabla}q(\mathbf{r},t)$, which follows from Eq. (1.33), and the equation of continuity for conserved quantity $I(\mathbf{r},t)$:

$$
\begin{aligned}
\left(\frac{\partial}{\partial t} + \mathbf{U}(t,q)\frac{\partial}{\partial \mathbf{r}}\right)\mathbf{p}(\mathbf{r},t) + \frac{\partial\{\mathbf{U}(t,q)\mathbf{p}(\mathbf{r},t)\}}{\partial q}\mathbf{p}(\mathbf{r},t) &= \frac{\partial Q(t,q)}{\partial q}\mathbf{p}(\mathbf{r},t),\\
\frac{\partial}{\partial t}I(\mathbf{r},t) + \frac{\partial}{\partial \mathbf{r}}\{\mathbf{U}(t,q)I(\mathbf{r},t)\} &= 0.
\end{aligned}
\tag{1.34}
$$

From Eqs. (1.34) follows that

$$
\int d\mathbf{r}I(\mathbf{r},t) = \int d\mathbf{r}I_0(\mathbf{r}).
\tag{1.35}
$$

In terms of characteristic curves determined by the system of ordinary differential equations, Eqs. (1.33) and (1.34) can be written in the form

$$
\begin{aligned}
\frac{d}{dt}\mathbf{r}(t) &= \mathbf{U}\,(t,q)\,, \quad \frac{d}{dt}q(t) = Q\,(t,q)\,, \quad \mathbf{r}\,(0) = \mathbf{r}_0, \quad q\,(0) = q_0\,(\mathbf{r}_0)\,,\\
\frac{d}{dt}\mathbf{p}(t) &= -\frac{\partial\{\mathbf{U}\,(t,q)\,\mathbf{p}(t)\}}{\partial q}\mathbf{p}(t) + \frac{\partial Q(t,q)}{\partial q}\mathbf{p}(t), \quad \mathbf{p}\,(0) = \frac{\partial q_0(\mathbf{r}_0)}{\partial \mathbf{r}_0},\\
\frac{d}{dt}I(t) &= -\frac{\partial\{\mathbf{U}\,(t,q)\,\mathbf{p}(t)\}}{\partial q}I(t), \quad I\,(0) = I_0\,(\mathbf{r}_0)\,.
\end{aligned}
\tag{1.36}
$$

Thus, the Lagrangian description considers the system (1.36) as the initial value problem. In this description, the two first equations form the closed system that determines characteristic curves.

Expressing now characteristic parameter $\mathbf{r}_0$ in terms of t and $\mathbf{r}$, one can write the solution to Eqs. (1.33) and (1.34) in the Eulerian description as

$$
\begin{aligned}
q\,(\mathbf{r},t) &= \int d\mathbf{r}_0 q\,(t|\mathbf{r}_0)\,j\,(t|\mathbf{r}_0)\,\delta\,(\mathbf{r}\,(t|\mathbf{r}_0) - \mathbf{r})\,,\\
I\,(\mathbf{r},t) &= \int d\mathbf{r}_0 I\,(t|\mathbf{r}_0)\,j\,(t|\mathbf{r}_0)\,\delta\,(\mathbf{r}\,(t|\mathbf{r}_0) - \mathbf{r})\,.
\end{aligned}
\tag{1.37}
$$

The feature of the transition from the Lagrangian description (1.36) to the Eulerian description (1.37) consists in the general appearance of ambiguities, which results in discontinuous solutions. These ambiguities are related to the fact that the divergence — Jacobian

$$j\left(t|\mathbf{r}_0\right) = \det\left\|\frac{\partial}{\partial r_{0k}} r_i\left(t|\mathbf{r}_0\right)\right\|$$

— can vanish at certain moments.

Quantities $I(t|\mathbf{r}_0)$ and $j(t|\mathbf{r}_0)$ are not independent. Indeed, integrating $I(\mathbf{r},t)$ in Eq. (1.37) over $\mathbf{r}$ and taking into account Eq. (1.35), we see that there is the evolution integral

$$j\left(t|\mathbf{r}_0\right) = \frac{I_0\left(\mathbf{r}_0\right)}{I\left(t|\mathbf{r}_0\right)}, \qquad (1.38)$$

from which follows that zero-valued divergence $j(t|\mathbf{r}_0)$ is accompanied by the infinite value of conservative quantity $I(t|\mathbf{r}_0)$.

It is obvious that all these results can be easily extended to the case in which functions $\mathbf{U}(\mathbf{r},t,q)$ and $Q(\mathbf{r},t,q)$ explicitly depend on spatial variable $\mathbf{r}$ and Eq. (1.33) itself is the vector equation. As a particular physical example, we consider the equation for the velocity field $\mathbf{V}(\mathbf{r},t)$ of low-inertia particles moving in the hydrodynamic flow whose velocity field is $\mathbf{u}(\mathbf{r},t)$ (see, e.g., [20])

$$\left(\frac{\partial}{\partial t} + \mathbf{V}(\mathbf{r},t)\frac{\partial}{\partial \mathbf{r}}\right)\mathbf{V}(\mathbf{r},t) = -\lambda\left[\mathbf{V}(\mathbf{r},t) - \mathbf{u}(\mathbf{r},t)\right]. \qquad (1.39)$$

We will assume this equation the phenomenological equation.

In the general case, the solution to Eq. (1.39) can be nonunique, it can have discontinuities, etc. However, in the case of asymptotically small inertia property of particles (parameter $\lambda \to \infty$), which is of our concern here, the solution will be unique during reasonable temporal intervals. Note that, in the right-hand side of Eq. (1.39), term $\mathbf{F}(\mathbf{r},t) = \lambda\mathbf{V}(\mathbf{r},t)$ linear in the velocity field $\mathbf{V}(\mathbf{r},t)$ is, according to the known *Stokes formula*, the resistance force acting on a slowly moving particle. If we approximate the particle by a sphere of radius a, parameter λ will be $\lambda = 6\pi a\eta/m_\mathrm{p}$, where η is the dynamic viscosity coefficient and m_p is the mass of the particle (see, e.g., [21]).

From Eq. (1.39) follows that velocity field $\mathbf{V}(\mathbf{r},t)$ is the divergent field (div $\mathbf{V}(\mathbf{r},t) \neq 0$) even if hydrodynamic flow $\mathbf{u}(\mathbf{r},t)$ is free of divergence field (div $\mathbf{u}(\mathbf{r},t) = 0$). As a consequence, particle number density $n(\mathbf{r},t)$ in divergence-free hydrodynamic flows, which satisfies the linear equation of continuity

$$\left(\frac{\partial}{\partial t} + \frac{\partial}{\partial \mathbf{r}}\mathbf{V}(\mathbf{r},t)\right)n(\mathbf{r},t) = 0, \quad n(\mathbf{r},0) = n_0(\mathbf{r}) \qquad (1.40)$$

similar to Eq. (1.22), shows the cluster behavior [22].

For large parameters $\lambda \to \infty$ (inertialess particles), we have

$$\mathbf{V}(\mathbf{r},t) \approx \mathbf{u}(\mathbf{r},t), \qquad (1.41)$$

and particle number density $n(\mathbf{r},t)$ in divergence-free hydrodynamic flows shows no cluster behavior.

The first-order partial differential equation (1.39) (the Eulerian description) is equivalent to the system of ordinary differential characteristic equations (the Lagrangian description)

$$\frac{d}{dt}\mathbf{r}(t) = \mathbf{V}\left(\mathbf{r}(t), t\right), \quad \mathbf{r}(0) = \mathbf{r}_0,$$

$$\frac{d}{dt}\mathbf{V}(t) = -\lambda\left[\mathbf{V}(t) - \mathbf{u}\left(\mathbf{r}(t), t\right)\right], \quad \mathbf{V}(0) = \mathbf{V}_0(\mathbf{r}_0) \tag{1.42}$$

that describe diffusion of a particle under random external force and linear friction and coincide with Eq. (1.11). In the simplest case of random force independent of spatial coordinates, we have the system

$$\frac{d}{dt}\mathbf{r}(t) = \mathbf{v}(t), \quad \frac{d}{dt}\mathbf{v}(t) = -\lambda\left[\mathbf{v}(t) - \mathbf{f}(t)\right],$$

$$\mathbf{r}(0) = \mathbf{r}_0, \quad \mathbf{v}(0) = \mathbf{v}_0. \tag{1.43}$$

In the general case, a nonlinear scalar first-order partial differential equation can be written in the form

$$\frac{\partial}{\partial t}q(\mathbf{r}, t) + H\left(\mathbf{r}, t, q, \mathbf{p}\right) = 0, \quad q\left(\mathbf{r}, 0\right) = q_0\left(\mathbf{r}\right), \tag{1.44}$$

where $\mathbf{p}(\mathbf{r}, t) = \nabla q(\mathbf{r}, t)$.

In terms of the Lagrangian description, this equation can be rewritten in the form of the system of characteristic equations:

$$\frac{d}{dt}\mathbf{r}(t|\mathbf{r}_0) = \frac{\partial}{\partial \mathbf{p}}H\left(\mathbf{r}, t, q, \mathbf{p}\right), \quad \mathbf{r}(0|\mathbf{r}_0) = \mathbf{r}_0;$$

$$\frac{d}{dt}\mathbf{p}(t|\mathbf{r}_0) = -\left(\frac{\partial}{\partial \mathbf{r}} + \mathbf{p}\frac{\partial}{\partial q}\right)H\left(\mathbf{r}, t, q, \mathbf{p}\right), \quad \mathbf{p}(0|\mathbf{r}_0) = \mathbf{p}_0(\mathbf{r}_0);$$

$$\frac{d}{dt}q(t|\mathbf{r}_0) = \left(\mathbf{p}\frac{\partial}{\partial \mathbf{p}} - 1\right)H\left(\mathbf{r}, t, q, \mathbf{p}\right), \quad q(0|\mathbf{r}_0) = \mathbf{q}_0(\mathbf{r}_0). \tag{1.45}$$

Now, we supplement Eq. (1.44) with the equation for the conservative quantity $I(\mathbf{r}, t)$

$$\frac{\partial}{\partial t}I(\mathbf{r}, t) + \frac{\partial}{\partial \mathbf{r}}\left\{\frac{\partial H\left(\mathbf{r}, t, q, \mathbf{p}\right)}{\partial \mathbf{p}}I(\mathbf{r}, t)\right\} = 0, \quad I\left(\mathbf{r}, 0\right) = I_0\left(\mathbf{r}\right). \tag{1.46}$$

From Eq. (1.46) follows that

$$\int d\mathbf{r}I(\mathbf{r}, t) = \int d\mathbf{r}I_0(\mathbf{r}). \tag{1.47}$$

Then, in the Lagrangian description, the corresponding quantity satisfies the equation

$$\frac{d}{dt}I(t|\mathbf{r}_0) = -\frac{\partial^2 H\left(\mathbf{r}, t, q, \mathbf{p}\right)}{\partial \mathbf{r}\partial \mathbf{p}}I(\mathbf{r}, t), \quad I\left(0|\mathbf{r}_0\right) = I_0\left(\mathbf{r}_0\right),$$

so that the solution to Eq. (1.46) has the form

$$I\left(\mathbf{r}, t\right) = I\left(t|\mathbf{r}_0\left(t, \mathbf{r}\right)\right) = \int d\mathbf{r}_0 I\left(t|\mathbf{r}_0\right) j\left(t|\mathbf{r}_0\right) \delta\left(\mathbf{r}\left(t|\mathbf{r}_0\right) - \mathbf{r}\right), \tag{1.48}$$

where $j\left(t|\mathbf{r}_0\right) = \det\left\|\partial r_i\left(t|\mathbf{r}_0\right)/\partial r_{0j}\right\|$ is the divergence (Jacobian).

Quantities $I(t|\mathbf{r}_0)$ and $j(t|\mathbf{r}_0)$ are related to each other. Indeed, substituting Eq. (1.48) for $I(\mathbf{r}, t)$ in Eq. (1.47), we see that there is the evolution integral

$$j(t|\mathbf{r}_0) = \frac{I_0(\mathbf{r}_0)}{I(t|\mathbf{r}_0)},$$

and Eq. (1.48) assumes the form

$$I(\mathbf{r}, t) = \int d\mathbf{r}_0 I_0(\mathbf{r}_0)\, \delta(\mathbf{r}(t|\mathbf{r}_0) - \mathbf{r}).$$

1.3.3   Parabolic equation of quasioptics (waves in randomly inhomogeneous media)

We will describe wave propagation in media with large-scale three-dimensional inhomogeneities responsible for small-angle scattering on the base of the *parabolic equation of quasioptics*,

$$\frac{\partial}{\partial x} u(x, \mathbf{R}) = \frac{i}{2k} \Delta_\mathbf{R} u(x, \mathbf{R}) + \frac{ik}{2} \varepsilon(x, \mathbf{R}) u(x, \mathbf{R}), \quad u(0, \mathbf{R}) = u_0(\mathbf{R}), \qquad (1.49)$$

where k is the wave number, $\Delta_\mathbf{R} = \partial^2/\partial \mathbf{R}^2$, and $\varepsilon_1(\mathbf{r}) = \varepsilon(x, \mathbf{R})$ is the deviation of the refractive index (or dielectric permittivity) from unity. It was successfully used in many problems on wave propagation in Earth's atmosphere and ocean [23, 24].

Introducing the amplitude–phase representation of the wavefield in Eq. (1.49) by the formula

$$u(x, \mathbf{R}) = A(x, \mathbf{R}) e^{iS(x, \mathbf{R})},$$

we can write the equation for the wavefield intensity $I(x, \mathbf{R}) = u(x, \mathbf{R}) u^*(x, \mathbf{R})$ in the form

$$\frac{\partial}{\partial x} I(x, \mathbf{R}) + \frac{1}{k} \nabla_\mathbf{R} \{ \nabla_\mathbf{R} S(x, \mathbf{R}) I(x, \mathbf{R}) \} = 0, \quad I(0, \mathbf{R}) = I_0(\mathbf{R}). \qquad (1.50)$$

From this equation follows that the power of a wave in plane $x = $ const is conserved in the general case of arbitrary incident wave beam:

$$E_0 = \int I(x, \mathbf{R}) d\mathbf{R} = \int I_0(\mathbf{R}) d\mathbf{R}.$$

Equation (1.50) coincides in form with Eq. (1.22). Consequently, we can treat it as the transport equation for conservative tracer in the potential velocity field. However, this tracer can be considered the passive tracer only in the *geometrical optics approximation*, in which case the phase of the wave, the transverse gradient of the phase $\mathbf{p}(x, \mathbf{R}) = \frac{1}{k} \nabla_\mathbf{R} S(x, \mathbf{R})$, and the matrix of the phase second derivatives $u_{ij}(x, \mathbf{R}) = \frac{1}{k} \frac{\partial^2}{\partial R_i \partial R_j} S(x, \mathbf{R})$ characterizing the *curvature of the phase front* $S(x, \mathbf{R}) = $ const satisfy the closed system of equations

$$\frac{\partial}{\partial x} S(x, \mathbf{R}) + \frac{k}{2} \mathbf{p}^2(x, \mathbf{R}) = \frac{k}{2} \varepsilon(x, \mathbf{R}),$$

$$\left(\frac{\partial}{\partial x} + \mathbf{p}(x, \mathbf{R}) \nabla_\mathbf{R} \right) \mathbf{p}(x, \mathbf{R}) = \frac{1}{2} \nabla_\mathbf{R} \varepsilon(x, \mathbf{R}),$$

$$\left(\frac{\partial}{\partial x} + \mathbf{p}(x, \mathbf{R}) \nabla_\mathbf{R} \right) u_{ij}(x, \mathbf{R}) + u_{ik}(x, \mathbf{R}) u_{kj}(x, \mathbf{R}) = \frac{1}{2} \frac{\partial^2}{\partial R_i \partial R_j} \varepsilon(x, \mathbf{R}), \quad (1.51)$$

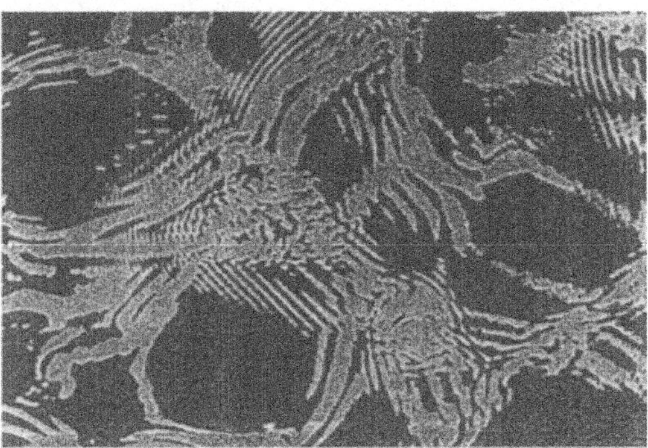

Figure 1.8: Transverse section of a laser beam in turbulent medium.

Figure 1.9: Caustics in a pool.

In the general case, i.e., with the inclusion of diffraction effects, this tracer is the active tracer.

According to the material of Sect. 1.3.1, realizations of intensity must show the cluster behavior, which manifests itself in the appearance of the *caustic* structures. An example demonstrating the appearance of the wave field caustic structures is given in Fig. 1.8, which is a fragment of the photo on the back of the cover — the flyleaf — of book [23] that shows the transverse section of a laser beam in turbulent medium (laboratory measurements). A photo of the pool in Fig. 1.9 also shows the prominent caustic structure of the wave field on the pool bottom. Such structures appear due to light refraction and reflection by rough water surface, which corresponds to scattering by the so-called *phase screen*.

Consider now geometrical optics approximation (1.51) for parabolic equation (1.49). In this approximation, the equation for the wave phase is the Hamilton–Jacobi equation and the equation for the transverse gradient of the phase (1.51) is the closed quasilinear first-order partial differential equation, and we can solve it by the method of characteristics.

Equations for the characteristic curves (*rays*) have the form

$$\frac{d}{dx}\mathbf{R}(x) = \mathbf{p}(x), \qquad \frac{d}{dx}\mathbf{p}(x) = \frac{1}{2}\nabla_{\mathbf{R}}\varepsilon(x, \mathbf{R}), \tag{1.52}$$

and the wave field intensity and matrix of the phase second derivatives along the characteristic curves will satisfy the equations

$$\frac{d}{dx}I(x) = -I(x)u_{ii}(x),$$
$$\frac{d}{dx}u_{ij}(x) + u_{ik}(x)u_{kj}(x) = \frac{1}{2}\frac{\partial^2}{\partial R_i \partial R_j}\varepsilon(x, \mathbf{R}). \tag{1.53}$$

Equations (1.52) coincide in appearance with the equations for a particle under random external forces in the absence of friction (1.11) and form the system of the Hamilton equations.

In the two-dimensional case ($R = y$), Eqs. (1.52), (1.53) become significantly simpler and assume the form

$$\frac{d}{dx}y(x) = p(x), \qquad \frac{d}{dx}p(x) = \frac{1}{2}\frac{\partial}{\partial y}\varepsilon(x, y),$$
$$\frac{d}{dx}I(x) = -I(x)u(x), \qquad \frac{d}{dx}u(x) + u^2(x) = \frac{1}{2}\frac{\partial^2}{\partial y^2}\varepsilon(x, y). \tag{1.54}$$

The last equation for $u(x)$ in (1.54) is similar to Eq. (1.13) whose solution shows the singular behavior. The only difference between these equations consists in the random term that has now a more complicated structure. Nevertheless, it is quite clear that solutions to stochastic problem (1.54) will show the blow-up behavior; namely, function $u(x)$ will reach minus infinity and intensity will reach plus infinity at a finite distance. Such a behavior of a wave field in randomly inhomogeneous media corresponds to *random focusing*, i.e., to the formation of caustics, which means the appearance of points of multivaluedness (and discontinuity) in the solutions of quasilinear equation (1.51) for the transverse gradient of the wave field phase.

1.3.4   Navier–Stokes equation: random forces in hydrodynamic theory of turbulence

Consider now the turbulent motion model that assumes the presence of external forces $\mathbf{f}(\mathbf{r}, t)$ acting on the liquid. Such a model is evidently only imaginary, because there is no actual analogues for these forces. However, assuming that forces $\mathbf{f}(\mathbf{r}, t)$ on average ensure an appreciable energy income only to large-scale velocity components, we can expect that, within the concepts of theory of local isotropic turbulence, the imaginary nature of field $\mathbf{f}(\mathbf{r}, t)$ will only slightly affect statistical properties of small-scale turbulent components [25]. Consequently, this model is quite appropriate for describing small-scale properties of turbulence.

Motion of an incompressible liquid under external forces is governed by the *Navier–Stokes equation*

$$\left(\frac{\partial}{\partial t} + \mathbf{u}(\mathbf{r}, t)\frac{\partial}{\partial \mathbf{r}}\right)\mathbf{u}(\mathbf{r}, t) = -\frac{1}{\rho_0}\frac{\partial}{\partial \mathbf{r}}p(\mathbf{r}, t) + \nu\Delta\mathbf{u}(\mathbf{r}, t) + \mathbf{f}(\mathbf{r}, t),$$
$$\frac{\partial}{\partial \mathbf{r}}\mathbf{u}(\mathbf{r}, t) = 0, \qquad \frac{\partial}{\partial \mathbf{r}}\mathbf{f}(\mathbf{r}, t) = 0. \tag{1.55}$$

Here, ρ_0 is the density of the liquid, ν is the kinematic viscosity, and pressure field $p(\mathbf{x}, t)$ is expressed in terms of the velocity field at the same instant by the relationship

$$p(\mathbf{r}, t) = -\rho_0 \int \Delta^{-1}(\mathbf{r}, \mathbf{r}') \frac{\partial^2 (u_i(\mathbf{r}', t) u_j(\mathbf{r}', t))}{\partial r_i' \partial r_j'} \, d\mathbf{r}', \tag{1.56}$$

where $\Delta^{-1}(\mathbf{r}, \mathbf{r}')$ is the integral operator inverse to the Laplace operator (repeated indexes assume summation).

If we substitute Eq. (1.56) in Eq. (1.55) to exclude the pressure field, then we obtain in the three-dimensional case that the Fourier transform of the velocity field with respect to spatial coordinates

$$\hat{u}_i(\mathbf{k}, t) = \int d\mathbf{r} u_i(\mathbf{r}, t) e^{-i\mathbf{k}\mathbf{r}}, \quad u_i(\mathbf{r}, t) = \frac{1}{(2\pi)^3} \int d\mathbf{k} \hat{u}_i(\mathbf{k}, t) e^{i\mathbf{k}\mathbf{r}},$$

$(\hat{u}_i^*(\mathbf{k}, t) = \hat{u}_i(-\mathbf{k}, t))$ satisfies the nonlinear integro-differential equation

$$\frac{\partial}{\partial t} \hat{u}_i(\mathbf{k}, t) + \frac{i}{2} \int d\mathbf{k}_1 \int d\mathbf{k}_2 \Lambda_i^{\alpha\beta}(\mathbf{k}_1, \mathbf{k}_2, \mathbf{k}) \hat{u}_\alpha(\mathbf{k}_1, t) \hat{u}_\beta(\mathbf{k}_2, t) - \nu k^2 \hat{u}_i(\mathbf{k}, t) =$$
$$= \hat{f}_i(\mathbf{k}, t), \tag{1.57}$$

where

$$\Lambda_i^{\alpha\beta}(\mathbf{k}_1, \mathbf{k}_2, \mathbf{k}) = \frac{1}{(2\pi)^3} \{k_\alpha \Delta_{i\beta}(\mathbf{k}) + k_\beta \Delta_{i\alpha}(\mathbf{k})\} \delta(\mathbf{k}_1 + \mathbf{k}_2 - \mathbf{k}),$$

$$\Delta_{ij}(\mathbf{k}) = \delta_{ij} - \frac{k_i k_j}{\mathbf{k}^2} \quad (i, \, \alpha, \, \beta = 1, 2, 3), \tag{1.58}$$

and $\widehat{\mathbf{f}}(\mathbf{k}, t)$ is the spatial Fourier harmonics of external forces,

$$\widehat{\mathbf{f}}(\mathbf{k}, t) = \int d\mathbf{r} \mathbf{f}(\mathbf{r}, t) e^{-i\mathbf{k}\mathbf{r}}, \quad \mathbf{f}(\mathbf{r}, t) = \frac{1}{(2\pi)^3} \int d\mathbf{k} \widehat{\mathbf{f}}(\mathbf{k}, t) e^{i\mathbf{k}\mathbf{r}}.$$

A specific feature of the three-dimensional hydrodynamic motions consists in the fact that the absence of external forces and viscosity-driven effects is sufficient for energy conservation.

It appears convenient to describe the stationary turbulence in terms of the space-time Fourier harmonics of the velocity field

$$\hat{u}_i(\mathbf{K}) = \int d\mathbf{x} \int_{-\infty}^{\infty} dt u_i(\mathbf{x}, t) e^{-i(\mathbf{k}\mathbf{x} + \omega t)}, \quad u_i(\mathbf{x}, t) = \frac{1}{(2\pi)^4} \int d\mathbf{k} \int_{-\infty}^{\infty} d\omega \hat{u}_i(\mathbf{K}) e^{i(\mathbf{k}\mathbf{x} + \omega t)},$$

where $\mathbf{K}$ is the four-dimensional wave vector $\{\mathbf{k}, \omega\}$ and field $\hat{u}_i^*(\mathbf{K}) = \hat{u}_i(-\mathbf{K})$ because field $u_i(\mathbf{r}, t)$ is real. In this case, we obtain the equation for component $\hat{u}_i(\mathbf{K})$ by performing the Fourier transformation of Eq. (1.57) with respect to time:

$$(i\omega + \nu \mathbf{k}^2) \hat{u}_i(\mathbf{K}) +$$
$$+ \frac{i}{2} \int d^4 \mathbf{K}_1 \int d^4 \mathbf{K}_2 \Lambda_i^{\alpha\beta}(\mathbf{K}_1, \mathbf{K}_2, \mathbf{K}) \hat{u}_\alpha(\mathbf{K}_1) \hat{u}_\beta(\mathbf{K}) = \hat{f}_i(\mathbf{K}), \tag{1.59}$$

where

$$\Lambda_i^{\alpha\beta}(\mathbf{K}_1, \mathbf{K}_2, \mathbf{K}) = \frac{1}{2\pi} \Lambda_i^{\alpha\beta}(\mathbf{k}_1, \mathbf{k}_2, \mathbf{k}) \delta(\omega_1 + \omega_2 - \omega),$$

and $\hat{f}_i(\mathbf{K})$ are the space-time Fourier harmonics of external forces. The obtained Eq. (1.59) is now the integral (and not integro-differential) nonlinear equation.

Chapter 2

Solution dependence on problem type, medium parameters, and initial data

Below, we considered a number of dynamic systems described by both ordinary and partial differential equations. Many applications concerned with studying statistical characteristics of the solutions to these equations require the knowledge of the solution dependence (generally, in the functional form) on the medium parameters appeared in the equation as coefficients and the initial conditions. Some properties appear common of all such dependencies, and two of them are of special interest in the context of statistical descriptions. We illustrate these dependencies by the example of the simplest problem, namely, the system of ordinary differential equations (1.1) that describes particle dynamics in the random velocity field, which we reformulate in the form of the nonlinear integral equation

$$\mathbf{r}(t) = \mathbf{r}_0 + \int_{t_0}^{t} \mathbf{U}(\mathbf{r}(\tau), \tau) d\tau. \tag{2.1}$$

The solution to Eq. (2.1) functionally depends on the vector field $\mathbf{U}(\mathbf{r}', \tau)$ and initial values $\mathbf{r}_0$, t_0.

2.1 Functional representation of problem solution

2.1.1 Variational (functional) derivatives

Recall first the general definition of a functional. One says that a functional is given if a rule is fixed that associates a number to every function from certain function family. Below, we give some examples of functionals.

$$\textbf{(a)} \qquad F[\varphi(\tau)] = \int_{t_1}^{t_2} d\tau a(\tau)\varphi(\tau),$$

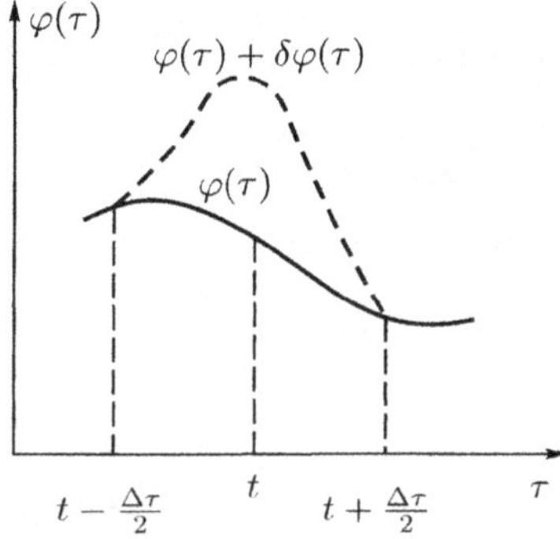

Figure 2.1: To definition of variational derivative.

where $a(t)$ is the given (fixed) function and limits t_1 and t_2 can be both finite and infinite. This is the linear functional.

$$(\mathbf{b}) \qquad F[\varphi(\tau)] = \int\limits_{t_1}^{t_2} \int\limits_{t_1}^{t_2} d\tau_1 d\tau_2 B(\tau_1, \tau_2) \varphi(\tau_1) \varphi(\tau_2),$$

where $B(t_1, t_2)$ is the given (fixed) function. This is the quadratic functional.

$$(\mathbf{c}) \qquad F[\varphi(\tau)] = f\left(\Phi[\varphi(\tau)]\right),$$

where $f(x)$ is the given function and quantity $\Phi[\varphi(\tau)]$ is by itself the functional.

Evaluate the difference between the values of a functional calculated for functions $\varphi(\tau)$ and $\varphi(\tau) + \delta\varphi(\tau)$, where $\delta\varphi(\tau) \neq 0$ for $t - \frac{1}{2}\Delta t < \tau < t + \frac{1}{2}\Delta t$ (see Fig. 2.1).

The variation of a functional is defined as the linear (in $\delta\varphi(\tau)$) portion of the difference

$$\delta F[\varphi(\tau)] = \{F[\varphi(\tau) + \delta\varphi(\tau)] - F[\varphi(\tau)]\}.$$

The limit

$$\frac{\delta F[\varphi(\tau)]}{\delta\varphi(t)dt} = \lim_{\Delta t \to 0} \frac{\delta F[\varphi(\tau)]}{\int\limits_{\Delta t} d\tau \delta\varphi(\tau)}. \tag{2.2}$$

is called the *variational* (or *functional*) *derivative* (see, e.g., [25]).

For short, we will use notation $\delta F[\varphi(\tau)]/\delta\varphi(t)$ instead of $\delta F[\varphi(\tau)]/\delta\varphi(t)dt$.

Note that, if we use function $\delta\varphi(\tau) = \alpha\delta(\tau)$, where $\delta(\tau)$ is the Dirac delta-function in Eq. (2.2), then Eq. (2.2) can be represented in the form of the ordinary derivative

$$\frac{\delta F[\varphi(\tau)]}{\delta\varphi(t)} = \lim_{\alpha \to 0} \frac{d}{d\alpha} F[\varphi(\tau) + \alpha\delta(\tau - t)].$$

The variational derivative of functional $F[\varphi(\tau)]$ is again the functional of $\varphi(\tau)$, which depends additionally on point t as a parameter. As a result, this variational derivative will have two types of derivatives; one can differentiate it in the ordinary sense with respect to parameter t and in the functional sense with respect to $\varphi(\tau)$ at point $\tau = t'$, thus obtaining the second variational derivative of the initial functional

$$\frac{\delta^2 F[\varphi(\tau)]}{\delta\varphi(t')\delta\varphi(t)} = \frac{\delta}{\delta\varphi(t')}\left[\frac{\delta F[\varphi(\tau)]}{\delta\varphi(t)}\right].$$

The second variational derivative will now be the functional of $\varphi(\tau)$ dependent on two points t and t', and so forth.

Determine the variational derivatives of functionals (**a**), (**b**), and (**c**).

In the case (**a**), we have

$$\delta F[\varphi(\tau)] = F[\varphi(\tau) + \delta\varphi(\tau)] - F[\varphi(\tau)] = \int\limits_{t-\frac{1}{2}\Delta t}^{t+\frac{1}{2}\Delta t} d\tau a(\tau)\delta\varphi(\tau).$$

If function $a(t)$ is continuous on segment Δt, then, by the average theorem,

$$\delta F[\varphi(\tau)] = a(t') \int\limits_{\Delta t} d\tau \delta\varphi(\tau)$$

where point t' belongs to segment $\left[t - \frac{1}{2}\Delta t, t + \frac{1}{2}\Delta t\right]$. Consequently,

$$\frac{\delta F[\varphi(\tau)]}{\delta\varphi(t)} = \lim_{\Delta t \to 0} a(t') = a(t). \tag{2.3}$$

In the case (**b**), we obtain similarly

$$\frac{\delta F[\varphi(\tau)]}{\delta\varphi(t)} = \int\limits_{t_1}^{t_2} d\tau\,[B(\tau,t)+B(t,\tau)]\,\varphi(\tau) \quad (t_1 < t < t_2).$$

Note that function $B(\tau_1, \tau_2)$ can always be assumed a symmetric function of its arguments here.

In the case (**c**), we have

$$F[\varphi(\tau) + \delta\varphi(\tau)] = f\left(\Phi[\varphi(\tau)]\right) + \frac{\partial f(\Phi[\varphi(\tau)])}{\partial\Phi}\delta\Phi[\varphi(\tau)] + \ldots$$
$$= F[\varphi(\tau)] + \frac{\partial f(\Phi[\varphi(\tau)])}{\partial\Phi}\delta\Phi[\varphi(\tau)] + \ldots$$

and, consequently,

$$\frac{\delta}{\delta\varphi(t)}f\left(\Phi[\varphi(\tau)]\right) = \frac{\partial f(\Phi[\varphi(\tau)])}{\partial\Phi}\frac{\delta}{\delta\varphi(t)}\Phi[\varphi(\tau)]. \tag{2.4}$$

Consider now functional $\Phi[\varphi(\tau)] = F_1[\varphi(\tau)]F_2[\varphi(\tau)]$. We have

$$\delta\Phi[\varphi(\tau)] = \{F_1[\varphi(\tau) + \delta\varphi(\tau)]F_2[\varphi(\tau) + \delta\varphi(\tau)] - F_1[\varphi(\tau)]F_2[\varphi(\tau)]\}$$
$$= F_1[\varphi(\tau)]\delta F_2[\varphi(\tau)] + F_2[\varphi(\tau)]\delta F_1[\varphi(\tau)]$$

and, consequently,

$$\frac{\delta}{\delta\varphi(t)}F_1[\varphi(\tau)]F_2[\varphi(\tau)] = F_1[\varphi(\tau)]\frac{\delta}{\delta\varphi(t)}F_2[\varphi(\tau)] + F_2[\varphi(\tau)]\frac{\delta}{\delta\varphi(t)}F_1[\varphi(\tau)]. \tag{2.5}$$

We can define the variational derivative of functional $\varphi(\tau_0)$ with respect to function $\varphi(t)$ by the formal relationship

$$\frac{\delta\varphi(\tau_0)}{\delta\varphi(t)} = \delta(\tau_0 - t). \tag{2.6}$$

Formula (2.6) can be proved, for example, by considering the linear functional of the form

$$F[\varphi(\tau)] = \frac{1}{\sqrt{2\pi}\sigma} \int_{-\infty}^{\infty} d\tau\, \varphi(\tau) \exp\left\{ -\frac{(\tau - \tau_0)^2}{2\sigma^2} \right\}. \tag{2.7}$$

According to Eq. (2.3), the variational derivative of this functional has the form

$$\frac{\delta}{\delta\varphi(t)} F[\varphi(\tau)] = \frac{1}{\sqrt{2\pi}\sigma} \exp\left\{ -\frac{(t - \tau_0)^2}{2\sigma^2} \right\}. \tag{2.8}$$

Performing now formal limit process $\sigma \to 0$ in Eqs. (2.7) and (2.8), we obtain the desired formula (2.6).

Formula (2.6) is very convenient for functional differentiation of functionals explicitly dependent on $\varphi(\tau)$. Indeed, for the quadratic functional **(b)**, we have

$$\frac{\delta}{\delta\varphi(t)} \int_{t_1}^{t_2}\int_{t_1}^{t_2} d\tau_1 d\tau_2 B(\tau_1, \tau_2)\varphi(\tau_1)\varphi(\tau_2)$$

$$\stackrel{(2.5)}{=} \int_{t_1}^{t_2}\int_{t_1}^{t_2} d\tau_1 d\tau_2 B(\tau_1, \tau_2) \left[\frac{\delta\varphi(\tau_1)}{\delta\varphi(t)}\varphi(\tau_2) + \varphi(\tau_1)\frac{\delta\varphi(\tau_2)}{\delta\varphi(t)} \right]$$

$$\stackrel{(2.6)}{=} \int_{t_1}^{t_2} d\tau\, [B(t,\tau)+B(\tau,t)]\,\varphi(\tau) \quad (t_1 < t < t_2).$$

Consider the functional

$$F[\varphi(\tau)] = \int_{t_1}^{t_2} d\tau\, L\left(\tau, \varphi(\tau), \frac{d\varphi(\tau)}{d\tau} \right)$$

as another example. In this case,

$$\frac{\delta}{\delta\varphi(t)} F[\varphi(\tau)]$$

$$\stackrel{(2.4)}{=} \int_{t_1}^{t_2} d\tau \left[\frac{\delta L\left(\tau, \varphi(\tau), \frac{d\varphi(\tau)}{d\tau} \right)}{\delta\varphi(\tau)} + \frac{\delta L\left(\tau, \varphi(\tau), \frac{d\varphi(\tau)}{d\tau} \right)}{\delta\dot\varphi(\tau)} \frac{d}{d\tau} \right] \frac{\delta\varphi(\tau)}{\delta\varphi(t)}$$

$$\stackrel{(2.6)}{=} \left(-\frac{d}{dt}\frac{\partial}{\partial\dot\varphi(t)} + \frac{\partial}{\partial\varphi(t)} \right) L\left(t, \varphi(t), \frac{d\varphi(t)}{dt} \right),$$

where $\dot\varphi(t) = \frac{d}{dt}\varphi(t)$ if point t belongs to interval (t_1, t_2).

Just as a function can be expanded in the Taylor series, a functional $F[\varphi(\tau) + \eta(\tau)]$ can be expanded in the functional Taylor series in function $\eta(\tau)$ for $\eta(\tau) \sim 0$

$$
\begin{aligned}
F[\varphi(\tau) + \eta(\tau)] &= F[\varphi(\tau)] + \int\limits_{-\infty}^{\infty} dt \frac{\delta F[\varphi(\tau)]}{\delta\varphi(t)} \eta(t) \\
&\quad + \frac{1}{2!} \int\limits_{-\infty}^{\infty} \int\limits_{-\infty}^{\infty} dt_1 dt_2 \frac{\delta^2 F[\varphi(\tau)]}{\delta\varphi(t_1)\delta\varphi(t_2)} \eta(t_1)\eta(t_2) + \dots .
\end{aligned}
\tag{2.9}
$$

Note that the operator expression

$$
\begin{aligned}
1 + \int\limits_{-\infty}^{\infty} dt\eta(t) \frac{\delta}{\delta\varphi(t)} &+ \frac{1}{2!} \int\limits_{-\infty}^{\infty} \int\limits_{-\infty}^{\infty} dt_1 dt_2 \eta(t_1)\eta(t_2) \frac{\delta^2}{\delta\varphi(t_1)\delta\varphi(t_2)} + \dots \\
&= 1 + \int\limits_{-\infty}^{\infty} dt\eta(t) \frac{\delta}{\delta\varphi(t)} + \frac{1}{2!} \left[\int\limits_{-\infty}^{\infty} dt\eta(t) \frac{\delta}{\delta\varphi(t)} \right]^2 + \dots
\end{aligned}
\tag{2.10}
$$

can be written shortly as the operator

$$
\exp\left\{ \int\limits_{-\infty}^{\infty} dt\eta(t) \frac{\delta}{\delta\varphi(t)} \right\},
\tag{2.11}
$$

the action of which should be treated precisely in the sense of expansion (2.10). Using this operator, we can rewrite Eq. (2.9) in the form

$$
F[\varphi(\tau) + \eta(\tau)] = e^{\int\limits_{-\infty}^{\infty} dt\eta(t) \frac{\delta}{\delta\varphi(t)}} F[\varphi(\tau)],
\tag{2.12}
$$

which enables us to interpret operator (2.11) as the functional shift operator.

Consider now functional $F[t; \varphi(\tau)]$ dependent on parameter t. We can differentiate this functional with respect to t and determine its variational derivative with respect to $\varphi(t')$, as well. One can easily see that these operations commute, i.e., the equality

$$
\frac{\partial}{\partial t} \frac{\delta F[t; \varphi(\tau)]}{\delta\varphi(t')} = \frac{\delta}{\delta\varphi(t')} \frac{\partial F[t; \varphi(\tau)]}{\partial t}
\tag{2.13}
$$

holds. If the domain of τ is independent of t, the validity of Eq. (2.13) is obvious. Otherwise, for example, for functionals $F[t; \varphi(\tau)]$ with $0 \leq \tau \leq t$, the validity of Eq. (2.13) can be checked on by expanding functional $F[t; \varphi(\tau)]$ in the functional Taylor series.

2.1.2 Principle of dynamic causality

Vary Eq. (2.1) with respect to field $\mathbf{U}(\mathbf{r}, t)$. Assuming that the initial position $\mathbf{r}_0$ is independent of field $\mathbf{U}$, we obtain the equation linear in variational derivative

$$
\frac{\delta r_i(t)}{\delta U_j(\mathbf{r}, t')} = \delta_{ij}\delta\left(\mathbf{r} - \mathbf{r}(t')\right) \theta\left(t' - t_0\right) \theta\left(t - t'\right) + \int\limits_{t_0}^{t} \frac{\partial U_i(\mathbf{r}(\tau), \tau)}{\partial r_k} \frac{\delta r_k(\tau)}{\delta U_j(\mathbf{r}, t')} d\tau,
\tag{2.14}
$$

where $\delta\left(\mathbf{r}-\mathbf{r}'\right)$ is the Dirac delta function, and $\theta\left(z\right)$ is the Heaviside step function. From Eq. (2.14) follows that

$$\frac{\delta r_i\left(t\right)}{\delta U_j\left(\mathbf{r},t'\right)}=0 \quad \text{for } t'>t \text{ or } t'<t_0, \tag{2.15}$$

which means that solution to the dynamic problem (2.1) $\mathbf{r}(t)$ as a functional of field $\mathbf{U}\left(\mathbf{r},t'\right)$ depends only on $\mathbf{U}\left(\mathbf{r},t'\right)$ for $t_0<t'<t$. Consequently, function $\mathbf{r}(t)$ will remain unchanged if field $\mathbf{U}\left(\mathbf{r},t'\right)$ varies outside interval (t_0,t), i.e., for $t'<t_0$ or $t'>t$. We will call condition (2.15) the *dynamic causality condition*.

Taking this condition into consideration, we can rewrite Eq. (2.14) in the form

$$\frac{\delta r_i\left(t\right)}{\delta U_j\left(\mathbf{r},t'\right)}=\delta_{ij}\delta\left(\mathbf{r}-\mathbf{r}\left(t'\right)\right)\theta\left(t'-t_0\right)\theta\left(t-t'\right)+\int_{t'}^{t}\frac{\partial U_i\left(\mathbf{r}\left(\tau\right),\tau\right)}{\partial r_k}\frac{\delta r_k\left(\tau\right)}{\delta U_j\left(\mathbf{r},t'\right)}d\tau. \tag{2.16}$$

As a consequence, proceeding to limit $t\rightarrow t'+0$, we obtain the equality

$$\left.\frac{\delta r_i\left(t\right)}{\delta U_j\left(\mathbf{r},t'\right)}\right|_{t=t'+0}=\delta_{ij}\delta\left(\mathbf{r}-\mathbf{r}\left(t'\right)\right). \tag{2.17}$$

Integral equation (2.16) in variational derivative is obviously equivalent to the linear differential equation with the initial condition

$$\frac{\partial}{\partial t}\left(\frac{\delta r_i\left(t\right)}{\delta U_j\left(\mathbf{r},t'\right)}\right)=\frac{\partial U_i\left(\mathbf{r}\left(t\right),t\right)}{\partial r_k}\left(\frac{\delta r_k\left(t\right)}{\delta U_j\left(\mathbf{r},t'\right)}\right),\quad \left.\frac{\delta r_i\left(t\right)}{\delta U_j\left(\mathbf{r},t'\right)}\right|_{t=t'}=\delta_{ij}\delta\left(\mathbf{r}-\mathbf{r}\left(t'\right)\right).$$
$$\tag{2.18}$$

The dynamic causality condition is the general property of initial value problems, i.e., problems described by differential equations with initial conditions. The boundary-value problems possess no such property. Indeed, in the case of problem (1.15), (1.16) that describes propagation of a plane wave in a layer of inhomogeneous medium, wave field $u(x)$ at point x and reflection and transmission coefficients depend functionally on function $\varepsilon(x)$ for all x of layer (L_0,L). However, using the imbedding method, we can convert this problem into the initial value problem with respect to auxiliary parameter L and make use of the causality property in terms of imbedding equations.

2.2 Solution dependence on problem's parameters

2.2.1 Solution dependence on initial data

We will use now the vertical bar symbol to isolate the dependence of the solution $\mathbf{r}(t)$ to Eq. (2.1) on the initial parameters $\mathbf{r}_0$ and t_0:

$$\mathbf{r}\left(t\right)=\mathbf{r}\left(t|\mathbf{r}_0,t_0\right),\quad \mathbf{r}_0=\mathbf{r}\left(t_0|\mathbf{r}_0,t_0\right).$$

Let us differentiate Eq. (2.1) with respect to parameters r_{0k} and t_0. As a result, we obtain linear equations for Jacobi's matrix $\frac{\partial}{\partial r_{0k}}r_i\left(t|\mathbf{r}_0,t_0\right)$ and quantity $\frac{\partial}{\partial t_0}r_i\left(t|\mathbf{r}_0,t_0\right)$

$$\frac{\partial r_i\left(t|\mathbf{r}_0,t_0\right)}{\partial r_{0k}} = \delta_{ik}+\int_{t_0}^{t}d\tau\frac{\partial U_i\left(\mathbf{r}\left(\tau\right),\tau\right)}{\partial r_j}\frac{\partial r_j\left(\tau|\mathbf{r}_0,t_0\right)}{\partial r_{0k}},$$

$$\frac{\partial r_i\left(t|\mathbf{r}_0,t_0\right)}{\partial t_0} = -U_i\left(\mathbf{r}_0(t_0),t_0\right)+\int_{t_0}^{t}d\tau\frac{\partial U_i\left(\mathbf{r}\left(\tau\right),\tau\right)}{\partial r_j}\frac{\partial r_j\left(\tau|\mathbf{r}_0,t_0\right)}{\partial t_0}. \tag{2.19}$$

Multiplying the first of these equations by $U_k \left(\mathbf{r}_0 \left(t \right), t \right)$, summing over index k, adding the result to the second equation, and introducing the vector function

$$F_i \left(t | \mathbf{r}_0, t_0 \right) = \left(\frac{\partial}{\partial t_0} + \mathbf{U} \left(\mathbf{r}_0, t_0 \right) \frac{\partial}{\partial \mathbf{r}_0} \right) r_i \left(t | \mathbf{r}_0, t_0 \right),$$

we obtain that this function satisfies the linear homogeneous equation

$$F_i \left(t | \mathbf{r}_0, t_0 \right) = \int\limits_{t_0}^{t} d\tau \frac{\partial U_i \left(\mathbf{r} \left(\tau \right), \tau \right)}{\partial r_k} F_k \left(\tau | \mathbf{r}_0, t_0 \right),$$

whose solution is obviously $F_i \left(t | \mathbf{r}_0, t_0 \right) = 0$. In this way, we obtain the equality

$$\left(\frac{\partial}{\partial t_0} + \mathbf{U} \left(\mathbf{r}_0, t_0 \right) \frac{\partial}{\partial \mathbf{r}_0} \right) r_i \left(t | \mathbf{r}_0, t_0 \right) = 0, \tag{2.20}$$

which can be considered as the linear partial differential equation in terms of derivatives with respect to variables $\mathbf{r}_0$ and t_0 satisfying at $t_0 = t$ the initial condition

$$\mathbf{r} \left(t | \mathbf{r}_0, t \right) = \mathbf{r}_0. \tag{2.21}$$

The variable t appears in problem (2.20), (2.21) as a parameter.

Equation (2.20) is solved using the time direction inverse to that used in solving problem (1.1); for this reason, we will call it the *backward equation*.

Equation (2.20) with initial condition (2.21) obviously possess the property of dynamic causality with respect to parameter t_0. This means that

$$\frac{\delta \mathbf{r} \left(t | \mathbf{r}_0, t_0 \right)}{\delta U_j \left(\mathbf{y}, t' \right)} = 0, \quad \text{if } t' > t \quad \text{or} \quad t' < t_0$$

and, as follows from Eq. (2.20),

$$\left. \frac{\delta \mathbf{r} \left(t | \mathbf{r}_0, t_0 \right)}{\delta U_j \left(\mathbf{r}, t' \right)} \right|_{t'=t_0+0} = \delta \left(\mathbf{r}_0 - \mathbf{r} \right) \frac{\partial \mathbf{r} \left(t | \mathbf{r}_0, t_0 \right)}{\partial r_{0j}}. \tag{2.22}$$

2.2.2　Imbedding method for boundary-value problems

Consider first boundary-value problems formulated in terms of ordinary differential equations. The *imbedding method* (or *invariant imbedding method*, as it is usually called in mathematical literature) offers a possibility of reducing problems at hand to the evolution type initial value problems possessing the property of dynamic causality with respect to an auxiliary parameter.

The idea of this method was suggested by V. A. Ambartsumyan (the so-called *Ambart-sumyan invariance principle*) [26]–[28] for solving equations of the linear theory of radiative transfer. Further, mathematicians grasped this idea and used it to convert boundary-value (nonlinear, in the general case) problems into evolution type initial value problems that are more convenient for simulations. Several monographs (see, e.g., [29]–[31]) deal with this method and consider both physical and computational aspects.

Consider the dynamic system described in terms of the system of ordinary differential equations

$$\frac{d}{dt} \mathbf{x}(t) = \mathbf{F} \left(t, \mathbf{x}(t) \right) \tag{2.23}$$

defined on segment $t \in [0, T]$ with the boundary conditions

$$g\mathbf{x}(0) + h\mathbf{x}(T) = \mathbf{v}, \tag{2.24}$$

where g and h are the constant matrixes.

Dynamic problem (2.23), (2.24) possesses no dynamic causality property, which means that the solution to this problem $\mathbf{x}(t)$ at instant t functionally depends on external forces $\mathbf{F}(\tau, \mathbf{x}(\tau))$ for all $0 \leq \tau \leq T$. Moreover, even boundary values $\mathbf{x}(0)$ and $\mathbf{x}(T)$ are functionals of field $\mathbf{F}(\tau, \mathbf{x}(\tau))$. The absence of dynamic causality in problem (2.23), (2.24) prevents us from using the known statistical methods of analyzing statistical characteristics of the solution to Eq. (2.23) if external force functional $\mathbf{F}(t, \mathbf{x})$ is the random space- and time-domain field. Introducing the one-time probability density $P(t; \mathbf{x})$ of the solution to Eq. (2.23), we can easily see that condition (2.24) is insufficient for determining the value of this probability at any point. The boundary condition imposes only certain functional restriction.

Note that the solution to problem (2.23), (2.24) parametrically depends on T and $\mathbf{v}$, i.e., $\mathbf{x}(t) = \mathbf{x}(t; T, \mathbf{v})$. Adhering to paper [32], we introduce functions

$$\mathbf{R}(T, \mathbf{v}) = \mathbf{x}(T; T, \mathbf{v}), \quad \mathbf{S}(T, \mathbf{v}) = \mathbf{x}(0; T, \mathbf{v})$$

that describe the boundary values of the solution to Eq. (2.23).

Differentiate Eq. (2.23) with respect to T and $\mathbf{v}$. We obtain two linear equations in the corresponding derivatives

$$\frac{d}{dt} \frac{\partial x_i(t; T, \mathbf{v})}{\partial T} = \frac{\partial F_i(t, \mathbf{x})}{\partial x_l} \frac{\partial x_l(t; T, \mathbf{v})}{\partial T},$$

$$\frac{d}{dt} \frac{\partial x_i(t; T, \mathbf{v})}{\partial v_k} = \frac{\partial F_i(t, \mathbf{x})}{\partial x_l} \frac{\partial x_l(t; T, \mathbf{v})}{\partial v_k}. \tag{2.25}$$

These equations are identical in form; consequently, we can expect that their solutions are related by the linear expression

$$\frac{\partial x_i(t; T, \mathbf{v})}{\partial T} = \lambda_k(T, \mathbf{v}) \frac{\partial x_i(t; T, \mathbf{v})}{\partial v_k} \tag{2.26}$$

if vector quantity $\boldsymbol{\lambda}(T, \mathbf{v})$ is such that boundary conditions (2.24) are satisfied and the solution is unique. To determine vector quantity $\boldsymbol{\lambda}(T, \mathbf{v})$, we first set $t = 0$ in Eq. (2.26) and multiply the result by matrix g; then, we set $t = T$ and multiply the result by matrix h; and, finally, we combine the obtained expressions. Taking into account Eq. (2.24), we obtain

$$g \frac{\partial \mathbf{x}(0; T, \mathbf{v})}{\partial T} + h \left. \frac{\partial \mathbf{x}(t; T, \mathbf{v})}{\partial T} \right|_{t=T} = \boldsymbol{\lambda}(T, \mathbf{v}).$$

In view of the fact that

$$\left. \frac{\partial \mathbf{x}(t; T, \mathbf{v})}{\partial T} \right|_{t=T} = \frac{\partial \mathbf{x}(T; T, \mathbf{v})}{\partial T} - \left. \frac{\partial \mathbf{x}(t; T, \mathbf{v})}{\partial t} \right|_{t=T} = \frac{\partial \mathbf{R}(T, \mathbf{v})}{\partial T} - \mathbf{F}(T, \mathbf{R}(T, \mathbf{v}))$$

(with allowance for Eq. (2.23)), we obtain the desired expression for quantity $\boldsymbol{\lambda}(T, \mathbf{v})$,

$$\boldsymbol{\lambda}(T, \mathbf{v}) = -h\mathbf{F}(T, \mathbf{R}(T, \mathbf{v})). \tag{2.27}$$

Expression (2.26) with parameter $\boldsymbol{\lambda}(T, \mathbf{v})$ defined by Eq. (2.27), i.e., the expression

$$\frac{\partial x_i(t; T, \mathbf{v})}{\partial T} = -h_{kl}F_l\left(T, \mathbf{R}(T, \mathbf{v})\right)\frac{\partial x_i(t; T, \mathbf{v})}{\partial v_k} \tag{2.28}$$

can be considered as the linear differential equation; one needs only to supplement it with the corresponding initial condition

$$\mathbf{x}(t; T, \mathbf{v})|_{T=t} = \mathbf{R}(t, \mathbf{v})$$

assuming that function $\mathbf{R}(T, \mathbf{v})$ is known.

The equation for this function can be obtained from the equality

$$\frac{\partial \mathbf{R}(T, \mathbf{v})}{\partial T} = \left.\frac{\partial \mathbf{x}(t; T, \mathbf{v})}{\partial t}\right|_{t=T} + \left.\frac{\partial \mathbf{x}(t; T, \mathbf{v})}{\partial T}\right|_{t=T}. \tag{2.29}$$

The right-hand side of Eq. (2.29) is the sum of the right-hand sides of Eqs. (2.23) and (2.26) at $t = T$. As a result, we obtain the closed nonlinear (quasi-linear) equation

$$\frac{\partial \mathbf{R}(T, \mathbf{v})}{\partial T} = -h_{kl}F_l\left(T, \mathbf{R}(T, \mathbf{v})\right)\frac{\partial \mathbf{R}(T, \mathbf{v})}{\partial v_k} + \mathbf{F}\left(T, \mathbf{R}(T, \mathbf{v})\right). \tag{2.30}$$

The initial condition for Eq. (2.30) follows from Eq. (2.24) for $T \to 0$

$$\mathbf{R}(T, \mathbf{v})|_{T=0} = (g + h)^{-1}\mathbf{v}. \tag{2.31}$$

Setting now $t = 0$ in Eq. (2.25), we obtain for the secondary boundary quantity $\mathbf{S}(T, \mathbf{v}) = \mathbf{x}(0; T, \mathbf{v})$ the equation

$$\frac{\partial \mathbf{S}(T, \mathbf{v})}{\partial T} = -h_{kl}F_l\left(T, \mathbf{R}(T, \mathbf{v})\right)\frac{\partial \mathbf{S}(T, \mathbf{v})}{\partial v_k} \tag{2.32}$$

with the initial condition

$$\mathbf{S}(T, \mathbf{v})|_{T=0} = (g + h)^{-1}\mathbf{v}$$

following from Eq. (2.31).

Thus, the problem reduces to the closed quasi-linear equation (2.30) with initial value (2.31) and linear equation (2.26) whose coefficients and initial value are determined by the solution of Eq. (2.30).

In the problem under consideration, input 0 and output T are symmetric. For this reason, one can solve it not only from T to 0, but also from 0 to T. In the latter case, functions $\mathbf{R}(T, \mathbf{v})$ and $\mathbf{S}(T, \mathbf{v})$ switch the places.

An important point consists in the fact that, despite the initial value problem (2.23) is nonlinear, Eq. (2.26) is the linear equation, because it is essentially the equation in variations. It is Eq. (2.30) that is responsible for nonlinearity.

Note that the above technique of deriving the imbedding equations for Eq. (2.23) can be easily extended to the boundary condition of the form [32]

$$\mathbf{g}\left(\mathbf{x}(0)\right) + \mathbf{h}\left(\mathbf{x}(T)\right) + \int\limits_0^T d\tau \mathbf{K}\left(\tau, \mathbf{x}(\tau)\right) = \mathbf{v},$$

where $\mathbf{g}(\mathbf{x})$, $\mathbf{h}(\mathbf{x})$ and $\mathbf{K}(T, \mathbf{x})$ are arbitrary given vector functions.

If function $\mathbf{F}(t, \mathbf{x})$ is linear in $\mathbf{x}$, $F_i(t, \mathbf{x}) = A_{ij}(t)x_j(t)$, then boundary-value problem (2.23), (2.24) assumes the simpler form

$$\frac{d}{dt}\mathbf{x}(t) = A(t)\,\mathbf{x}(t), \quad g\mathbf{x}(0) + h\mathbf{x}(T) = \mathbf{v},$$

and the solution of Eqs. (2.26), (2.30) and (2.32) will be the function linear in $\mathbf{v}$

$$\mathbf{x}(t; T, \mathbf{v}) = X(t; T)\mathbf{v}. \tag{2.33}$$

As a result, we arrive at the closed matrix Riccati equation for matrix $R(T) = X(T; T)$

$$\frac{d}{dT}R(T) = A(T)R(T) - R(T)hA(T)R(T), \quad R(0) = (g + h)^{-1}. \tag{2.34}$$

As regards matrix $X(t, T)$, it satisfies the linear matrix equation with the initial condition

$$\frac{\partial}{\partial T}X(t; T) = -X(t; T)hA(T)R(T), \quad X(t; T)_{T=t} = R(t). \tag{2.35}$$

Problems

Problem 1 *Helmholtz equation with unmatched boundary. Derive the imbedding equations for the stationary wave boundary-value problem*

$$\left(\frac{d^2}{dx^2} + k^2(x)\right)u(x) = 0,$$

$$\left(\frac{d}{dx} + ik_1\right)u(x)\bigg|_{x=L_0} = 0, \quad \left(\frac{d}{dx} - ik_0\right)u(x)\bigg|_{x=L} = -2ik_0.$$

Instruction. Reformulate this boundary-value problem as the initial-value in terms of functions $u(x) = u(x; L)$ and $v(x; L) = \frac{\partial}{\partial x}u(x; L)$

$$\frac{d}{dx}u(x; L) = v(x; L), \quad \frac{d}{dx}v(x; L) = -k^2(x)u(x; L),$$

$$v(L_0; L) + ik_1 u(L_0; L) = 0, \quad v(L; L) - ik_0 u(L; L) = -2ik_0.$$

Solution.

$$\frac{\partial}{\partial L}u(x; L) = \left\{ik_0 + \frac{i}{2k_0}\left[k^2(L) - k_0^2\right]u(L; L)\right\}u(x; L),$$

$$u(x; L)|_{L=x} = u(x; x),$$

$$\frac{d}{dL}u(L; L) = 2ik_0\left[u(L; L) - 1\right] + \frac{i}{2k_0}\left[k^2(L) - k_0^2\right]u(L; L)^2,$$

$$u(L_0; L_0) = \frac{2k_0}{k_0 + k_1}.$$

Problem 2 *Helmholtz equation with matched boundary. Derive the imbedding equations for the stationary wave boundary-value problem*

$$\left(\frac{d^2}{dx^2} + k^2(x)\right)u(x; L) = 0,$$

$$\left(\frac{d}{dx} + ik_1\right)u(x; L)\bigg|_{x=L_0} = 0, \quad \left(\frac{d}{dx} - ik(L)\right)u(x; L)\bigg|_{x=L} = -2ik(L).$$

Solution.

$$\frac{\partial}{\partial L}u(x;L) = \left\{ik(L) + \frac{1}{2}\frac{k'(L)}{k(L)}\left[2 - u(L;L)\right]\right\}u(x;L),$$

$$u(x;L)|_{L=x} = u(x;x),$$

$$\frac{d}{dL}u(L;L) = ik(L)\left[u(L;L) - 1\right] + \frac{1}{2}\frac{k'(L)}{k(L)}\left[2 - u(L;L)\right]u(L;L),$$

$$u(L_0;L_0) = \frac{2k(L_0)}{k(L_0) + k_1}.$$

Problem 3 *Derive the imbedding equations for the boundary-value problem*

$$\left(\frac{d^2}{dx^2} - \frac{\rho'(x)}{\rho(x)}\frac{d}{dx} + p^2\left[1 + \varepsilon(x)\right]\right)u(x) = 0,$$

$$\left(\frac{1}{\rho(x)}\frac{d}{dx} + ip\right)u(x)\bigg|_{x=L_0} = 0, \qquad \left(\frac{1}{\rho(x)}\frac{d}{dx} - ip\right)u(x)\bigg|_{x=L} = -2ip,$$

where $\rho'(x) = \frac{d\rho(x)}{dx}$

Solution. The imbedding equations with respect to parameter L have the form

$$\frac{\partial}{\partial L}u(x;L) = \left\{ip\rho(L) + \varphi(L)u(L;L)\right\}u(x;L),$$

$$u(x;L)|_{L=x} = u(x;x);$$

$$\frac{d}{dL}u(L;L) = 2ip\rho(L)\left[u(L;L) - 1\right] + \varphi(L)u^2(L;L),$$

$$u(L;L)|_{L=L_0} = 1,$$

where

$$\varphi(x) = \frac{ip}{2\rho(x)}\left[1 + \varepsilon(x) - \rho^2(x)\right].$$

Remark. The above boundary-value problem describes many physical processes, such as acoustic waves in a medium with variable density and some types of electromagnetic waves. In these cases, function $\varepsilon(x)$ describes inhomogeneities of the velocity of wave propagation (refractive index or dielectric permittivity) and function $\rho(x)$ appears only in the case of acoustic waves and characterize medium density.

Problem 4 *Derive the imbedding equations for the matrix Helmholtz equation*

$$\left(\frac{d^2}{dx^2} + \gamma(x)\frac{d}{dx} + K(x)\right)U(x) = 0,$$

$$\left(\frac{d}{dx} + B\right)U(x)\bigg|_{x=L} = D, \qquad \left(\frac{d}{dx} + C\right)U(x)\bigg|_{x=0} = 0,$$

where $\gamma(x)$, $K(x)$, *and* $U(x)$ *are the variable matrixes, while* B, C, *and* D *are the constant matrixes.*

Solution. The imbedding equations with respect to parameter L have the form

$$\frac{\partial}{\partial L} U(x; L) = U(x; L)\Lambda(L), \quad U(x; L)|_{L=x} = U(x; x),$$

$$\frac{d}{dL} U(L; L) = D - \left\{ BU(L; L) + U(L; L)D^{-1}BD \right\}$$

$$+ U(L; L)D^{-1}\gamma(L)D + U(L; L)D^{-1} \left\{ K(L) - \gamma(L)B + B^2 \right\} U(L; L),$$

$$U(0; 0) = (B - C)^{-1}D,$$

where matrix $\Lambda(L)$ is given by the formula

$$\Lambda(L) = D^{-1}\left[\gamma(L) - B\right]D + D^{-1}\left[K(L) + B^2 - \gamma(L)B\right]U(L; L).$$

Chapter 3

Indicator function and Liouville equation

Modern apparatus of the theory of random processes is able of constructing closed descriptions of dynamic systems if these systems meet the condition of dynamic causality and are formulated in terms of linear partial differential equations or certain types of integral equations. The use indicator functions makes it possible to reformulate the initial, generally nonlinear system in terms of the equivalent linear partial differential equations. However, this approach increases the dimension of space of variables. Consider such a conversion using the dynamic systems described in the Chapter 1.

3.1 Ordinary differential equations

Assume that a stochastic problem is described by the system of equations (1.1)

$$\frac{d}{dt}\mathbf{r}(t) = \mathbf{U}\left(\mathbf{r}(t), t\right), \quad \mathbf{r}(t_0) = \mathbf{r}_0. \tag{3.1}$$

We introduce the scalar function

$$\varphi(t; \mathbf{r}) = \delta(\mathbf{r}(t) - \mathbf{r}), \tag{3.2}$$

which is localized on instants at which random process $\mathbf{r}(t)$ crosses a given plane $\mathbf{r}(t) = $ const and is usually called the *indicator function*.

Differentiating Eq. (3.2) with respect to time t and using Eq. (3.1), we obtain the equality

$$\frac{\partial}{\partial t}\varphi(t; \mathbf{r}) = -\frac{\partial}{\partial \mathbf{r}}\delta(\mathbf{r}(t) - \mathbf{r})\frac{d\mathbf{r}(t)}{dt} = -\frac{\partial}{\partial \mathbf{r}}\delta(\mathbf{r}(t) - \mathbf{r})\mathbf{U}\left(\mathbf{r}(t), t\right).$$

Using then the puting-out property of the delta function

$$\delta(\mathbf{r}(t) - \mathbf{r})\mathbf{U}\left(\mathbf{r}(t), t\right) = \delta(\mathbf{r}(t) - \mathbf{r})\mathbf{U}\left(\mathbf{r}, t\right),$$

we obtain the linear partial differential equation

$$\left(\frac{\partial}{\partial t} + \frac{\partial}{\partial \mathbf{r}}\mathbf{U}(\mathbf{r}, t)\right)\varphi(t; \mathbf{r}) = 0, \quad \varphi(t_0; \mathbf{r}) = \delta(\mathbf{r}_0 - \mathbf{r}) \tag{3.3}$$

equivalent to the initial system. This equation is called the *Liouville equation*.

The transition from system (3.1) to Liouville equation (3.3) entails enlargement of the phase space $(t, \mathbf{r})$, whose dimension remains nevertheless finite. Note that Eq. (3.3) coincides in form with the equation for tracer transfer in velocity field $\mathbf{U}(\mathbf{r}, t)$ (1.22); the only difference consists in the initial conditions.

The solution to Eq. (3.1) and, consequently, function (3.2) depend on initial values $t_0, \mathbf{r}_0$. Indeed, function $\mathbf{r}(t) = \mathbf{r}(t|\mathbf{r}_0, t_0)$ as a function of variables $\mathbf{r}_0$ and t_0 satisfies linear first-order partial differential equation (2.20). The equations of such type also allow conversion into the equations in indicator function $\varphi(t; \mathbf{r}|t_0, \mathbf{r}_0)$ (see the next section); in the case under consideration, this equation is again the linear first-order partial differential equation in terms of the derivatives with respect to variables $\mathbf{r}_0$ and t_0

$$\left(\frac{\partial}{\partial t_0} + \mathbf{U}(\mathbf{r}_0, t_0) \frac{\partial}{\partial \mathbf{r}_0} \right) \varphi(t; \mathbf{r}|t_0, \mathbf{r}_0) = 0, \quad \varphi(t; \mathbf{r}|t, \mathbf{r}_0) = \delta(\mathbf{r}_0 - \mathbf{r}). \tag{3.4}$$

Equation (3.4) can be called the *backward Liouville equation*.

3.2 First-order partial differential equations

If the initial value problem is formulated in terms of partial differential equations, we always can convert it to the equivalent description in terms of the linear variational differential equation in the infinite-dimensional space (the *Hopf Equation*)). This conversion appears significantly simpler for some special types of problems. Indeed, if the initial dynamic system is described by the first-order partial differential equation (either *linear* as Eq. (1.22), or *quasilinear* as Eq. (1.33), or, in the general case, *nonlinear* as Eq. (1.44)), then the phase space of the corresponding indicator function will be the finite-dimension space, which follows from the fact that first-order partial differential equations are equivalent to systems of ordinary (characteristic) differential equations. Consider these cases in more details.

3.2.1 Linear equations

Consider the problem on tracer transfer by random velocity field in more details. The problem is formulated in terms of Eq. (1.22) that we rewrite here in the form

$$\left(\frac{\partial}{\partial t} + \mathbf{U}(\mathbf{r}, t) \frac{\partial}{\partial \mathbf{r}} \right) \rho(\mathbf{r}, t) + \frac{\partial \mathbf{U}(\mathbf{r}, t)}{\partial \mathbf{r}} \rho(\mathbf{r}, t) = 0, \quad \rho(\mathbf{r}, 0) = \rho_0(\mathbf{r}). \tag{3.5}$$

To describe the concentration field in the Eulerian description, we introduce the indicator function

$$\varphi(t, \mathbf{r}; \rho) = \delta(\rho(t, \mathbf{r}) - \rho), \tag{3.6}$$

which is similar to function (3.2) and is localized on surface $\rho(\mathbf{r}, t) = \rho = \mathrm{const}$ in the three-dimensional case or on a contour in the two-dimensional case. An equation for this function can be easily obtained either immediately from Eq. (3.5), or from the Liouville equation in the Lagrangian description. Indeed, differentiating Eq. (3.6) with respect to time and using dynamic equation (3.5) and putting-out property of the delta function, we obtain the equation

$$\frac{\partial}{\partial t} \varphi(t, \mathbf{r}; \rho) = -\frac{\partial}{\partial \rho} \frac{\partial \rho(\mathbf{r}, t)}{\partial t} \varphi(t, \mathbf{r}; \rho) =$$

$$= \frac{\partial \mathbf{U}(\mathbf{r}, t)}{\partial \mathbf{r}} \frac{\partial}{\partial \rho} \rho \varphi(t, \mathbf{r}; \rho) + \mathbf{U}(\mathbf{r}, t) \frac{\partial \rho(\mathbf{r}, t)}{\partial \mathbf{r}} \frac{\partial}{\partial \rho} \varphi(t, \mathbf{r}; \rho), \tag{3.7}$$

However, this equation is not closed because the right-hand side includes term $\partial\rho(\mathbf{r},t)/\partial\mathbf{r}$ that cannot be explicitly expressed through $\rho(\mathbf{r},t)$.

On the other hand, differentiating function (3.6) with respect to $\mathbf{r}$, we obtain the equality

$$\frac{\partial}{\partial\mathbf{r}}\varphi(t,\mathbf{r};\rho) = -\frac{\partial\rho(\mathbf{r},t)}{\partial\mathbf{r}}\frac{\partial}{\partial\rho}\varphi(t,\mathbf{r};\rho). \tag{3.8}$$

Eliminating the last term in Eq. (3.7) with the use of (3.8), we obtain the closed Liouville equation in the Eulerian description

$$\left(\frac{\partial}{\partial t} + \mathbf{U}(\mathbf{r},t)\frac{\partial}{\partial\mathbf{r}}\right)\varphi(t,\mathbf{r};\rho) = \frac{\partial\mathbf{U}(\mathbf{r},t)}{\partial\mathbf{r}}\frac{\partial}{\partial\rho}\left[\rho\varphi(t,\mathbf{r};\rho)\right],$$
$$\varphi(0,\mathbf{r};\rho) = \delta(\rho_0(\mathbf{r}) - \rho). \tag{3.9}$$

To obtain a more complete description, we consider the extended indicator function including both density field $\rho(\mathbf{r},t)$ and its spatial gradient $\mathbf{p}(\mathbf{r},t) = \boldsymbol{\nabla}\rho(\mathbf{r},t)$

$$\varphi(t,\mathbf{r};\rho,\mathbf{p}) = \delta\left(\rho(\mathbf{r},t) - \rho\right)\delta\left(\mathbf{p}(\mathbf{r},t) - \mathbf{p}\right). \tag{3.10}$$

Differentiating Eq. (3.10) with respect to time and using dynamic equations (3.5) for concentration and Eq. (1.31) for the concentration spatial gradient, we obtain the closed Liouville equation for the extended indicator function in the case of divergence-free velocity field

$$\left(\frac{\partial}{\partial t} + \mathbf{U}(\mathbf{r},t)\frac{\partial}{\partial\mathbf{r}}\right)\varphi(t,\mathbf{r};\rho,\mathbf{p}) = \frac{\partial U_k(\mathbf{r},t)}{\partial r_i}\frac{\partial}{\partial p_i}p_k\varphi(t,\mathbf{r};\rho,\mathbf{p}),$$
$$\varphi(0,\mathbf{r};\rho,\mathbf{p}) = \delta\left(\rho_0(\mathbf{r}) - \rho\right)\delta\left(\mathbf{p}_0(\mathbf{r}) - \mathbf{p}\right). \tag{3.11}$$

3.2.2   Quasilinear equations

Consider now the simplest quasilinear equation for scalar quantity $q(\mathbf{r},t)$ (1.33)

$$\left(\frac{\partial}{\partial t} + \mathbf{U}(t,q)\frac{\partial}{\partial\mathbf{r}}\right)q(\mathbf{r},t) = Q(t,q), \qquad q(\mathbf{r},0) = q_0(\mathbf{r}). \tag{3.12}$$

In this case, an attempt of deriving a closed equation for the indicator function $\varphi(t,\mathbf{r};q) = \delta(q(\mathbf{r},t) - q)$ on the analogy of the linear problem will fail. Here, we must supplement Eq. (3.12) with Eq. (1.34) for the gradient field $\mathbf{p}(\mathbf{r},t) = \boldsymbol{\nabla}q(\mathbf{r},t)$

$$\left(\frac{\partial}{\partial t} + \mathbf{U}(t,q)\frac{\partial}{\partial\mathbf{r}}\right)\mathbf{p}(\mathbf{r},t) + \frac{\partial\left\{\mathbf{U}(t,q)\mathbf{p}(\mathbf{r},t)\right\}}{\partial q}\mathbf{p}(\mathbf{r},t) = \frac{\partial Q(t,q)}{\partial q}\mathbf{p}(\mathbf{r},t),$$
$$\mathbf{p}(\mathbf{r},0) = \boldsymbol{\nabla}q_0(\mathbf{r}), \tag{3.13}$$

and consider the extended indicator function

$$\varphi(t,\mathbf{r};q,\mathbf{p}) = \delta(q(\mathbf{r},t) - q)\delta(\mathbf{p}(\mathbf{r},t) - \mathbf{p}). \tag{3.14}$$

Differentiating Eq. (3.14) with respect to time and using Eqs. (3.12) and (3.13), we obtain the equation

$$\frac{\partial}{\partial t}\varphi(t,\mathbf{r};q,\mathbf{p}) = -\left[\frac{\partial}{\partial q}\frac{\partial q(\mathbf{r},t)}{\partial t} + \frac{\partial}{\partial p_k}\frac{\partial p_k(\mathbf{r},t)}{\partial t}\right]\varphi(t,\mathbf{r};q,\mathbf{p})$$
$$= \left\{\frac{\partial}{\partial q}\left[\mathbf{p}\mathbf{U}(t,q) - Q(t,q)\right] + \frac{\partial}{\partial p_k}\mathbf{U}(t,q)\frac{\partial p_k(\mathbf{r},t)}{\partial\mathbf{r}}\right\}\varphi(t,\mathbf{r};q,\mathbf{p})$$
$$+ \frac{\partial}{\partial p_k}\left[p_k\left(\mathbf{p}\frac{\partial\mathbf{U}(t,q)}{\partial q} - \frac{\partial Q(t,q)}{\partial q}\right)\right]\varphi(t,\mathbf{r};q,\mathbf{p}), \tag{3.15}$$

which is not closed, however, because of the term $\partial p_k(\mathbf{r}, t)/\partial \mathbf{r}$ in the right-hand side.

Differentiating function (3.14) with respect to $\mathbf{r}$, we obtain the equality

$$\frac{\partial}{\partial \mathbf{r}}\varphi(t, \mathbf{r}; q, \mathbf{p}) = -\left[\mathbf{p}\frac{\partial}{\partial q} + \frac{\partial p_k(\mathbf{r}, t)}{\partial \mathbf{r}}\frac{\partial}{\partial p_k}\right]\varphi(t, \mathbf{r}; q, \mathbf{p}), \tag{3.16}$$

from which follows that

$$\frac{\partial}{\partial p_k}\frac{\partial p_k(\mathbf{r}, t)}{\partial \mathbf{r}}\varphi(t, \mathbf{r}; q, \mathbf{p}) = -\left[\frac{\partial}{\partial \mathbf{r}} + \mathbf{p}\frac{\partial}{\partial q}\right]\varphi(t, \mathbf{r}; q, \mathbf{p}).$$

Consequently, Eq. (3.15) can be rewritten in the closed form

$$\left(\frac{\partial}{\partial t} + \mathbf{U}(t, q)\frac{\partial}{\partial \mathbf{r}}\right)\varphi(t, \mathbf{r}; q, \mathbf{p}) = \frac{\partial}{\partial q}\left\{[\mathbf{p}\mathbf{U}(t, q) - Q(t, q)]\varphi(t, \mathbf{r}; q, \mathbf{p})\right\}$$
$$+ \frac{\partial}{\partial \mathbf{p}}\left\{\mathbf{p}\left(\frac{\partial \mathbf{p}\mathbf{U}(t, q)}{\partial q} - \frac{\partial Q(t, q)}{\partial q}\right)\varphi(t, \mathbf{r}; q, \mathbf{p})\right\},$$

which is just the desired Liouville equation equivalent to the quasilinear equation (3.12) in the extended phase space $\{q, \mathbf{p}\}$.

Note that the equation of continuity for conserved quantity $I(\mathbf{r}, t)$

$$\frac{\partial}{\partial t}I(\mathbf{r}, t) + \frac{\partial}{\partial \mathbf{r}}\left\{\mathbf{U}(t, q)I(\mathbf{r}, t)\right\} = 0, \qquad I(\mathbf{r}, 0) = I_0(\mathbf{r}) \tag{3.17}$$

can be combined with Eqs. (3.12), (3.13). In this case, the indicator function has the form

$$\varphi(t, \mathbf{r}; q, \mathbf{p}, I) = \delta(q(\mathbf{r}, t) - q)\delta(\mathbf{p}(\mathbf{r}, t) - \mathbf{p})\delta(I(\mathbf{r}, t) - I)$$

and derivation of the closed equation for this function appears possible in space $\{q, \mathbf{p}, I\}$, which follows from the fact that, in the Lagrangian description, quantity inverse to $I(\mathbf{r}, t)$ corresponds to divergence.

3.2.3 General-form nonlinear equations

Consider now the scalar nonlinear first-order partial differential equation in the general form (1.44)

$$\frac{\partial}{\partial t}q(\mathbf{r}, t) + H(\mathbf{r}, t, q, \mathbf{p}) = 0, \qquad q(\mathbf{r}, 0) = q_0(\mathbf{r}), \tag{3.18}$$

where $\mathbf{p}(\mathbf{r}, t) = \partial q(\mathbf{r}, t)/\partial \mathbf{r}$. In order to derive the closed Liouville equation in this case, we must supplement Eq. (3.18) with equations for vector $\mathbf{p}(\mathbf{r}, t)$ and second-derivative matrix $U_{ik}(\mathbf{r}, t) = \partial^2 q(\mathbf{r}, t)/\partial r_i \partial r_k$.

Introduce the extended indicator function

$$\varphi(t, \mathbf{r}; q, \mathbf{p}, U, I) = \delta(q(\mathbf{r}, t) - q)\delta(\mathbf{p}(\mathbf{r}, t) - \mathbf{p})\delta(U(\mathbf{r}, t) - U)\delta(I(\mathbf{r}, t) - I),$$

where we included for generality an additional conserved variable $I(\mathbf{r}, t)$ satisfying the equation of continuity (1.46)

$$\frac{\partial}{\partial t}I(\mathbf{r}, t) + \frac{\partial}{\partial \mathbf{r}}\left\{\frac{\partial H(\mathbf{r}, t, q, \mathbf{p})}{\partial \mathbf{p}}I(\mathbf{r}, t)\right\} = 0, \qquad I(\mathbf{r}, 0) = I_0(\mathbf{r}). \tag{3.19}$$

Equations (3.18), (3.19) describe, for example, wave propagation in inhomogeneous media within the frames of the geometrical-optics approximation of the parabolic equation of quasioptics (Eq.(1.51)). Differentiating indicator function with respect to time and using dynamic equations for functions $q(\mathbf{r}, t)$, $\mathbf{p}(\mathbf{r}, t)$, $U(\mathbf{r}, t)$ and $I(\mathbf{r}, t)$, we obtain an unclosed equation containing third-order derivatives of function $q(\mathbf{r}, t)$ with respect to spatial variable $\mathbf{r}$. However, the combination

$$\left(\frac{\partial}{\partial t} + \frac{\partial H(\mathbf{r}, t, q, \mathbf{p})}{\partial \mathbf{p}} \frac{\partial}{\partial \mathbf{r}} \right) \varphi(t, \mathbf{r}; q, \mathbf{p}, U, I)$$

will not include the third-order derivatives; as a result, we obtain the closed Liouville equation in space $\{q, \mathbf{r}, U, I\}$.

3.3 Higher-order partial differential equations

If the initial dynamic system includes higher-order derivatives (e.g., the Laplace operator), derivation of a closed equation for the corresponding indicator function becomes impossible. In this case, only a variational differential equation (the *Hopf equation*) can be derived in the closed form. The variable of this equation is the functional whose average over the ensemble of realizations coincides with the *characteristic functional* of the solution to the corresponding dynamic equation. Consider such a transition using the partial differential equations considered in Chapter 1 as examples.

3.3.1 Parabolic equation of quasioptics

The first example concerns wave propagation in a random medium within the frames of the linear parabolic equation (1.49)

$$\frac{\partial}{\partial x} u(x, \mathbf{R}) = \frac{i}{2k} \Delta_{\mathbf{R}} u(x, \mathbf{R}) + \frac{ik}{2} \varepsilon(x, \mathbf{R}) u(x, \mathbf{R}), \quad u(0, \mathbf{R}) = u_0(\mathbf{R}). \tag{3.20}$$

Consider the functional

$$\varphi[x; v(\mathbf{R}'), v^*(\mathbf{R}')] = \exp \left\{ i \int d\mathbf{R}' \left[u(x, \mathbf{R}')v(\mathbf{R}') + u^*(x, \mathbf{R}')v^*(\mathbf{R}') \right] \right\}, \tag{3.21}$$

where wave field $u(x, \mathbf{R})$ satisfies Eq. (3.20) and $u^*(x, \mathbf{R})$ is the complex conjugated function. Differentiating (3.21) with respect to x and using dynamic equation (3.20) and its complex conjugate, we obtain the equality

$$\frac{\partial}{\partial x} \varphi[x; v(\mathbf{R}'), v^*(\mathbf{R}')]$$

$$= -\frac{1}{2k} \int d\mathbf{R} \left[v(\mathbf{R}) \Delta_{\mathbf{R}} u(x, \mathbf{R}) - v^*(\mathbf{R}) \Delta_{\mathbf{R}} u^*(x, \mathbf{R}) \right] \varphi[x; v(\mathbf{R}'), v^*(\mathbf{R}')]$$

$$- \frac{k}{2} \int d\mathbf{R} \varepsilon(x, \mathbf{R}) \left[v(\mathbf{R}) u(x, \mathbf{R}) - v^*(\mathbf{R}) u^*(x, \mathbf{R}) \right] \varphi[x; v(\mathbf{R}'), v^*(\mathbf{R}')],$$

which can be written as the variational differential equation

$$\frac{\partial}{\partial x} \varphi[x; v(\mathbf{R}'), v^*(\mathbf{R}')] = \frac{ik}{2} \int d\mathbf{R} \varepsilon(x, \mathbf{R}) \widehat{M}(\mathbf{R}) \varphi[x; v(\mathbf{R}'), v^*(\mathbf{R}')]$$

$$+ \frac{i}{2k} \int d\mathbf{R} \left[v(\mathbf{R}) \Delta_{\mathbf{R}} \frac{\delta}{\delta v(\mathbf{R})} - v^*(\mathbf{R}) \Delta_{\mathbf{R}} \frac{\delta}{\delta v^*(\mathbf{R})} \right] \varphi[x; v(\mathbf{R}'), v^*(\mathbf{R}')] \tag{3.22}$$

with the Hermitian operator

$$\widehat{M}(\mathbf{R}) = v(\mathbf{R})\frac{\delta}{\delta v(\mathbf{R})} - v^*(\mathbf{R})\frac{\delta}{\delta v^*(\mathbf{R})},$$

equivalent to the initial Eq. (3.20). A consequence of Eq (3.22) is the equality

$$\frac{\delta}{\delta\varepsilon(x-0,\mathbf{R})}\varphi[x;v(\mathbf{R}'),v^*(\mathbf{R}')] = \frac{ik}{2}\widehat{M}(\mathbf{R})\varphi[x;v(\mathbf{R}'),v^*(\mathbf{R}')]. \tag{3.23}$$

3.3.2 Random forces in hydrodynamic theory of turbulence

Consider now integro-differential equation (1.57) for the Fourier harmonics $\widehat{\mathbf{u}}(\mathbf{k},t)$ of the solution to the Navier–Stokes equation (1.55)

$$\frac{\partial}{\partial t}\hat{u}_i(\mathbf{k},t) + \frac{i}{2}\int d\mathbf{k}_1\int d\mathbf{k}_2\Lambda_i^{\alpha\beta}(\mathbf{k}_1,\mathbf{k}_2,\mathbf{k})\,\hat{u}_\alpha(\mathbf{k}_1,t)\,\hat{u}_\beta(\mathbf{k}_2,t)$$

$$-\nu k^2\hat{u}_i(\mathbf{k},t) = \hat{f}_i(\mathbf{k},t),\qquad \widehat{\mathbf{u}}(\mathbf{k},0) = \widehat{\mathbf{u}}_0(\mathbf{k}), \tag{3.24}$$

$$\Lambda_i^{\alpha\beta}(\mathbf{k}_1,\mathbf{k}_2,\mathbf{k}) = \frac{1}{(2\pi)^3}\{k_\alpha\Delta_{i\beta}(\mathbf{k}) + k_\beta\Delta_{i\alpha}(\mathbf{k})\}\,\delta(\mathbf{k}_1+\mathbf{k}_2-\mathbf{k}),$$

$$\Delta_{ij}(\mathbf{k}) = \delta_{ij} - \frac{k_ik_j}{\mathbf{k}^2}\quad (i,\,\alpha,\,\beta = 1,2,3), \tag{3.25}$$

where complex conjugated field $\widehat{\mathbf{u}}^*(\mathbf{k},t) = \widehat{\mathbf{u}}(-\mathbf{k},t)$ and $\widehat{\mathbf{f}}(\mathbf{k},t)$ is the spatial Fourier harmonics of external forces.

Introduce the functional

$$\varphi[t;\mathbf{z}] = \varphi[t;\mathbf{z}(\mathbf{k}')] = \exp\left\{i\int d\mathbf{k}'\widehat{\mathbf{u}}(\mathbf{k}',t)\mathbf{z}(\mathbf{k}')\right\}. \tag{3.26}$$

Differentiating this functional with respect to time t and using dynamic equation (3.24), we obtain the equality

$$\frac{\partial}{\partial t}\varphi[t;\mathbf{z}(\mathbf{k}')] = i\int d\mathbf{k}\mathbf{z}(\mathbf{k})\frac{\partial\widehat{\mathbf{u}}(\mathbf{k},t)}{\partial t}\varphi[t;\mathbf{z}(\mathbf{k}')]$$

$$= \frac{1}{2}\int d\mathbf{k}z_i(\mathbf{k})\int d\mathbf{k}_1\int d\mathbf{k}_2\Lambda_i^{\alpha\beta}(\mathbf{k}_1,\mathbf{k}_2,\mathbf{k})\,\hat{u}_\alpha(\mathbf{k}_1,t)\,\hat{u}_\beta(\mathbf{k}_2,t)\,\varphi[t;\mathbf{z}(\mathbf{k}')]$$

$$-i\int d\mathbf{k}z_i(\mathbf{k})\left\{\nu k^2\hat{u}_i(\mathbf{k},t) - \hat{f}_i(\mathbf{k},t)\right\}\varphi[t;\mathbf{z}(\mathbf{k}')],$$

which can be rewritten in the functional space as the linear *Hopf equation* containing variational derivatives

$$\frac{\partial}{\partial t}\varphi[t;\mathbf{z}] = -\frac{1}{2}\int d\mathbf{k}z_i(\mathbf{k})\int d\mathbf{k}_1\int d\mathbf{k}_2\Lambda_i^{\alpha\beta}(\mathbf{k}_1,\mathbf{k}_2,\mathbf{k})\frac{\delta^2\varphi[t;\mathbf{z}]}{\delta z_\alpha(\mathbf{k}_1)\,\delta z_\beta(\mathbf{k}_2)}$$

$$-\int d\mathbf{k}z_i(\mathbf{k})\left\{\nu k^2\frac{\delta}{\delta z_i(\mathbf{k})} - i\hat{f}_i(\mathbf{k},t)\right\}\varphi[t;\mathbf{z}]. \tag{3.27}$$

A consequence of Eq. (3.27) is the equality

$$\frac{\delta}{\delta\widehat{\mathbf{f}}(\mathbf{k},t-0)}\varphi[t;\mathbf{z}] = i\mathbf{z}(\mathbf{k})\varphi[t;\mathbf{z}]. \tag{3.28}$$

Problems

Problem 5 *Determine the relationship between the Eulerian and Lagrangian indicator functions for linear problem (3.5), page 43.*

Solution.

$$\varphi(t,\mathbf{r};\rho) = \int d\mathbf{r}_0 \int\limits_0^\infty dj \; j\varphi_{\mathrm{Lag}}(t;\mathbf{r},\rho,j|\mathbf{r}_0).$$

Problem 6 *Derive Eq. (3.9), page 44 from the dynamic system (1.25), (1.26), page 20 in the Lagrangian description.*

Problem 7 *Derive the Hopf equation for the functional*

$$\varphi[\mathbf{z}(\mathbf{K}')] = \exp\left\{ i \int d^4\mathbf{K}'\hat{\mathbf{u}}(\mathbf{K}')\mathbf{z}(\mathbf{K}') \right\}, \tag{3.29}$$

which is the solution of integral equation (1.59), page 29

Solution.

$$(i\omega + \nu\mathbf{k}^2)\frac{\delta\varphi[\mathbf{z}(\mathbf{K}')]}{\delta z_i(\mathbf{K})} = i\hat{f}_i(\mathbf{K})\,\varphi[\mathbf{z}(\mathbf{K}')]$$

$$-\frac{1}{2}\int d^4\mathbf{K}_1 \int d^4\mathbf{K}_2\, \Lambda_i^{\alpha\beta}(\mathbf{K}_1,\mathbf{K}_2,\mathbf{K})\, \frac{\delta\varphi[\mathbf{z}(\mathbf{K}')]}{\delta z_\alpha(\mathbf{K}_1)\,\delta z_\beta(\mathbf{K}_2)}. \tag{3.30}$$

Part II

Statistical description of stochastic systems

Chapter 4

Random quantities, processes and fields

Prior to consider statistical descriptions of dynamical systems, we discuss basic concepts of the theory of random quantities, processes, and fields.

4.1 Random quantities and their characteristics

The probability for a random quantity ξ to fall in interval $-\infty < \xi < z$ is the monotonous function

$$F(z) = P(-\infty < \xi < z) = \langle \theta(z - \xi) \rangle_\xi, \quad F(\infty) = 1, \tag{4.1}$$

where

$$\theta(z) = \begin{cases} 1, & \text{if } z > 0, \\ 0, & \text{if } z < 0 \end{cases}$$

is the Heaviside step function and $\langle \ldots \rangle_\xi$ denotes averaging over an ensemble of realizations of random quantity ξ. This function is called the *probability distribution function* or the *integral distribution function*. Definition (4.1) reflects the real-world procedure of finding the probability according to the rule

$$P(-\infty < \xi < z) = \lim_{N \to \infty} \frac{n}{N},$$

where n is the integer equal to the number of realizations of event $\xi < z$ in N independent trials. Consequently, the probability for a random quantity ξ to fall into interval $z < \xi < z + dz$, where dz is the infinitesimal increment, can be written in the form

$$P(z < \xi < z + dz) = p(z)dz,$$

where function $p(z)$ called the *probability density* is represented by the formula

$$p(z) = \frac{d}{dz} P(-\infty < \xi < z) = \langle \delta(z - \xi) \rangle_\xi, \tag{4.2}$$

where $\delta(z)$ is the Dirac delta function. In terms of the probability density $p(z)$, the integral distribution function is expressed by the formula

$$F(z) = P(-\infty < \xi < z) = \int\limits_{-\infty}^{z} d\xi\, p(\xi), \tag{4.3}$$

so that

$$p(z) > 0, \quad \int\limits_{-\infty}^{\infty} dz\, p(z) = 1.$$

Multiplying Eq. (4.2) by arbitrary function $f(z)$ and integrating over the whole domain of variable z, we express the mean value of the arbitrary function of random quantity in the following form

$$\langle f(\xi) \rangle_\xi = \int\limits_{-\infty}^{\infty} dz\, p(z) f(z). \tag{4.4}$$

The *characteristic function* defined by the equality

$$\Phi(v) = \left\langle e^{iv\xi} \right\rangle_\xi = \int\limits_{-\infty}^{\infty} dz\, e^{ivz} p(z)$$

is a very important quantity that exhaustively describes all characteristics of random quantity ξ. The characteristic function being known, we can obtain the probability density (via the Fourier transform)

$$p(x) = \frac{1}{2\pi} \int\limits_{-\infty}^{\infty} dv\, \Phi(v) e^{-ivx},$$

moments

$$M_n = \langle \xi^n \rangle = \int\limits_{-\infty}^{\infty} dz\, p(z) z^n = \left(\frac{d}{idv} \right)^n \Phi(v) \Bigg|_{v=0},$$

cumulants (or semi-invariants)

$$K_n = \left(\frac{d}{idv} \right)^n \Theta(v) \Bigg|_{v=0},$$

where $\Theta(v) = \ln \Phi(v)$, and other statistical characteristics. In terms of moments and cumulants of random quantity ξ, functions $\Theta(v)$ and $\Phi(v)$ are the Taylor series

$$\Phi(v) = \sum_{n=0}^{\infty} \frac{i^n}{n!} M_n v^n, \quad \Theta(v) = \sum_{n=1}^{\infty} \frac{i^n}{n!} K_n v^n. \tag{4.5}$$

In the case of multidimensional random quantity $\boldsymbol{\xi} = \{z_1, ..., z_n\}$, the exhaustive statistical description assumes the multidimensional characteristic function

$$\Phi(\mathbf{v}) = \left\langle e^{i\mathbf{v}\boldsymbol{\xi}} \right\rangle_{\boldsymbol{\xi}}, \quad \mathbf{v} = \{v_1, ..., v_n\}. \tag{4.6}$$

The corresponding joint probability density for quantities $\xi_1, ..., \xi_n$ is the Fourier transform of characteristic function $\Phi(\mathbf{v})$, i.e.,

$$P(\mathbf{x}) = \frac{1}{(2\pi)^n} \int d\mathbf{v}\, \Phi(\mathbf{v}) e^{-i\mathbf{v}\mathbf{x}}, \quad \mathbf{x} = \{x_1, ..., x_n\}. \tag{4.7}$$

Substituting function $\Phi(\mathbf{v})$ defined by Eq. (4.6) in Eq. (4.7) and integrating the result over $\mathbf{v}$, we obtain the obvious equality

$$P(\mathbf{x}) = \langle \delta(\boldsymbol{\xi} - \mathbf{x}) \rangle_{\boldsymbol{\xi}} = \langle \delta(\xi_1 - x_1)...\delta(\xi_n - x_n) \rangle \tag{4.8}$$

that can serve the definition of the probability density of random vector quantity $\boldsymbol{\xi}$.

In this case, the moments and cumulants of random quantity $\boldsymbol{\xi}$ are defined by the expressions

$$M_{i_1,\dots,i_n} = \left.\frac{\partial^n}{i^n \partial v_{i_1}\dots\partial v_{i_n}}\Phi(\mathbf{v})\right|_{\mathbf{v}=0}, \quad K_{i_1,\dots,i_n} = \left.\frac{\partial^n}{i^n \partial v_{i_1}\dots\partial v_{i_n}}\Theta(\mathbf{v})\right|_{\mathbf{v}=0},$$

where $\Theta(\mathbf{v}) = \ln\Phi(\mathbf{v})$, and functions $\Theta(\mathbf{v})$ and $\Phi(\mathbf{v})$ are expressed in terms of moments $M_{i_1,\dots,i_n}$ and cumulants $K_{i_1,\dots,i_n}$ via the Taylor series

$$\Phi(\mathbf{v}) = \sum_{n=0}^{\infty}\frac{i^n}{n!}M_{i_1,\dots,i_n}v_{i_1}\dots v_{i_n}, \quad \Theta(\mathbf{v}) = \sum_{n=1}^{\infty}\frac{i^n}{n!}K_{i_1,\dots,i_n}v_{i_1}\dots v_{i_n}. \tag{4.9}$$

Note that, for quantities ξ assuming only discrete values ξ_i $(i = 1, 2, \dots)$ with probabilities p_i, formula (4.8) is replaced with its discrete analog

$$p_k = \left\langle \delta_{z,\xi_k} \right\rangle,$$

where $\delta_{i,k}$ is the Kronecker delta ($\delta_{i,k} = 1$ for $i = k$ and 0 otherwise).

Consider now statistical average $\langle \xi f(\xi) \rangle_\xi$, where $f(z)$ is arbitrary deterministic function such that the above average exists. We calculate this average using the procedure that will be widely used in what follows. Instead of $f(\xi)$, we consider function $f(\xi + \eta)$, where η is arbitrary deterministic quantity. Expand function $f(\xi + \eta)$ in the Taylor series in powers of ξ, i.e., represent it the form

$$f(\xi + \eta) = \sum_{n=0}^{\infty}\frac{1}{n!}f^{(n)}(\eta)\xi^n = e^{\xi\frac{d}{d\eta}}f(\eta),$$

where we introduced the shift operator with respect to η. Then we can write the equality

$$\langle \xi f(\xi + \eta) \rangle_\xi = \left\langle \xi e^{\xi\frac{d}{d\eta}} \right\rangle_\xi f(\eta)$$

$$= \frac{\left\langle \xi e^{\xi\frac{d}{d\eta}} \right\rangle_\xi}{\left\langle e^{\xi\frac{d}{d\eta}} \right\rangle_\xi}\left\langle e^{\xi\frac{d}{d\eta}} \right\rangle_\xi f(\eta) = \Omega\left(\frac{d}{id\eta}\right)\langle f(\xi + \eta) \rangle_\xi, \tag{4.10}$$

where function

$$\Omega(v) = \frac{\left\langle \xi e^{i\xi v} \right\rangle_\xi}{\left\langle e^{i\xi v} \right\rangle_\xi} = \frac{d}{idv}\ln\Phi(v) = \frac{d}{idv}\Theta(v),$$

and $\Phi(v)$ is the characteristic function of random quantity ξ. Using now the Taylor series (4.5) for function $\Theta(v)$, we obtain function $\Omega(v)$ in the form of the series

$$\Omega(v) = \sum_{n=0}^{\infty}\frac{i^n}{n!}K_{n+1}v^n. \tag{4.11}$$

Because variable η appears in the right-hand side of Eq. (4.10) only as a term of the sum $\xi + \eta$, we can replace differentiation with respect to η with differentiation with respect to

ξ (in so doing, operator $\Omega(d/id\xi)$ should be introduced into averaging brackets) and set $\eta = 0$. As a result, we obtain the equality

$$\langle \xi f(\xi) \rangle_\xi = \left\langle \Omega\left(\frac{d}{id\xi}\right) f(\xi) \right\rangle_\xi,$$

which can be rewritten, using expansion (4.11) for $\Omega(v)$, as the series in cumulants K_n

$$\langle \xi f(\xi) \rangle_\xi = \sum_{n=0}^{\infty} \frac{1}{n!} K_{n+1} \left\langle \frac{d^n f(\xi)}{d\xi^n} \right\rangle_\xi. \tag{4.12}$$

Note that, setting $f(\xi) = \xi^{n-1}$ in Eq. (4.12), we obtain the recurrence formula

$$M_n = \sum_{k=1}^{n} \frac{(n-1)!}{(k-1)!(n-k)!} K_k M_{n-k} \quad (M_0 = 1, \quad n = 1, 2, ...) \tag{4.13}$$

that relates moments and cumulants of random quantity ξ.

To illustrate practicability of the above formulas, we consider two types of random quantities ξ as examples.

1. Let ξ be the Gaussian random quantity with the probability density

$$p(z) = \frac{1}{\sqrt{2\pi}\sigma} \exp\left\{ -\frac{z^2}{2\sigma^2} \right\}.$$

Then, we have

$$\Phi(v) = \exp\left\{ -\frac{v^2\sigma^2}{2} \right\}, \quad \Theta(v) = -\frac{v^2\sigma^2}{2},$$

so that

$$M_1 = K_1 = \langle \xi \rangle = 0, \quad M_2 = K_2 = \sigma^2 = \langle \xi^2 \rangle, \quad K_{n>2} = 0.$$

In this case, the recurrence formula (4.13) assumes the form

$$M_n = (n-1)\sigma^2 M_{n-2}, \quad n = 2, ..., \tag{4.14}$$

from which follows that

$$M_{2n+1} = 0, \quad M_{2n} = (2n-1)!!\sigma^{2n}.$$

For average (4.12) we obtain in this case the expression

$$\langle \xi f(\xi) \rangle_\xi = \sigma^2 \left\langle \frac{df(\xi)}{d\xi} \right\rangle_\xi. \tag{4.15}$$

If we deal with the random Gaussian vector whose components are ξ_i ($\langle \xi_i \rangle_{\vec{\xi}} = 0$, $i = 1, ..., n$), then the characteristic function is given by the equality

$$\Phi(\mathbf{v}) = \exp\left\{ -\frac{1}{2} B_{ij} v_i v_j \right\}, \quad \Theta(\mathbf{v}) = -\frac{1}{2} B_{ij} v_i v_j,$$

where matrix $B_{ij} = \langle \xi_i \xi_j \rangle$ and repeated indexes assume summation. In this case, Eq. (4.15) is replaced with the equality

$$\langle \xi_i f(\boldsymbol{\xi}) \rangle_{\boldsymbol{\xi}} = B_{ij} \left\langle \frac{df(\boldsymbol{\xi})}{d\xi_j} \right\rangle_{\boldsymbol{\xi}}. \tag{4.16}$$

2. Let $\xi \equiv n$ be the integer random quantity governed by the Poisson distribution

$$p_n = \frac{\bar{n}^n}{n!} e^{-\bar{n}},$$

where $\bar{n}$ is the average value of quantity n. In this case, we have

$$\Phi(v) = \exp\left\{\bar{n}\left(e^{iv} - 1\right)\right\}, \quad \Theta(v) = \bar{n}\left(e^{iv} - 1\right),$$

so that all cumulants are identical, i.e.,

$$K_n = \bar{n}.$$

The recurrence formula (4.13) and Eq. (4.12) assume for this random quantity the forms

$$M_l = \bar{n} \sum_{k=0}^{l-1} \frac{(l-1)!}{k!(l-1-k)!} M_k \equiv \bar{n}\left\langle (n+1)^{l-1} \right\rangle, \quad \langle nf(n) \rangle = \bar{n}\langle f(n+1) \rangle. \tag{4.17}$$

4.2   Random processes, fields, and their characteristics

4.2.1   General remarks

If we deal with random function (random process) $z(t)$, then all function's statistical characteristics at any fixed instant t are exhaustively described in terms of the one-point (one-time) probability density

$$P(t; z) = \langle \delta\left(z(t) - z\right) \rangle \tag{4.18}$$

dependent parametrically on time t by the following relationship

$$\langle f\left(z(t)\right) \rangle = \int\limits_{-\infty}^{\infty} dz f(z) P(t; z).$$

The integral distribution function for this process, i.e. the probability of the event that process $z(t) < Z$ at instant t, is calculated by the formula

$$F(t, Z) = P(z(t) < Z) = \int\limits_{-\infty}^{Z} dz P(t; z)$$

from which follows that

$$F(t, Z) = \langle \theta(Z - z(t)) \rangle, \quad F(t, \infty) = 1, \tag{4.19}$$

where $\theta(z)$ is the Heaviside step function equal to zero for $z < 0$ and unity for $z > 0$.

Note that the singular Dirac delta function

$$\varphi(t; z) = \delta\left(z(t) - z\right)$$

appeared in Eq. (4.18) in angle brackets of averaging and parametrically dependent on time is called the *indicator function*.

Similar definitions hold for the two-point probability density

$$P(t_1, t_2; z_1, z_2) = \langle \varphi(t_1, t_2, ; z_1, z_2) \rangle$$

and for the general case of the n-point probability density

$$P(t_1, ..., t_n; z_1, ..., z_n) = \langle \varphi(t_1, ..., t_n; z_1, ..., z_n) \rangle,$$

where

$$\varphi(t_1, ..., t_n; z_1, ..., z_n) = \delta(z(t_1) - z_1)...\delta(z(t_n) - z_n)$$

is the n-point indicator function.

Process $z(t)$ is called *stationary* if all its statistical characteristics are invariant with respect to arbitrary temporal shift, i.e., if

$$P(t_1 + \tau, ..., t_n + \tau; z_1, ..., z_n) = P(t_1, ..., t_n; z_1, ..., z_n).$$

In particular, the one-point probability density of stationary process is at all independent of time, and the correlation function depends only on the difference of times,

$$B_z(t_1, t_2) = \langle z(t_1) z(t_2) \rangle = B_z(t_1 - t_2).$$

Temporal scale τ_0 characteristic of correlation function $B_z(t)$ is called the temporal correlation radius of process $z(t)$. We can determine this scale for example by the equality

$$\int\limits_0^\infty \langle z(t + \tau) z(t) \rangle \, d\tau = \tau_0 \left\langle z^2(t) \right\rangle. \tag{4.20}$$

Note that the Fourier transform of the stationary process correlation function

$$\Phi_z(\omega) = \int\limits_{-\infty}^\infty dt B_z(t) e^{i\omega t}$$

is called the *temporal spectral function* (or simply *temporal spectrum*).

For random field $f(\mathbf{x}, t)$, the one- and n-point probability densities parametrically dependent on space-time points are defined similarly

$$P(t, \mathbf{x}; f) = \langle \varphi(t, \mathbf{x}; f) \rangle, \tag{4.21}$$

$$P(t_1, ..., t_n, \mathbf{x}_1, ...\mathbf{x}_n; f_1, ...f_n) = \langle \varphi(t_1, ..., t_n, \mathbf{x}_1, ...\mathbf{x}_n; f_1, ...f_n) \rangle, \tag{4.22}$$

where indicator functions are defined as follows:

$$\begin{aligned} \varphi(t, \mathbf{x}; f) &= \delta(f(\mathbf{x}, t) - f), \\ \varphi(t_1, ..., t_n, \mathbf{x}_1, ...\mathbf{x}_n; f_1, ...f_n) &= \delta\left(f(\mathbf{x}_1, t_1) - f_1\right)...\delta\left(f(\mathbf{x}_n, t_n) - f_n\right). \end{aligned} \tag{4.23}$$

For clarity, we use here variables $\mathbf{x}$ and t as spatial and temporal coordinates; however, in many physical problems, some preferred spatial coordinate can play the role of the temporal coordinate.

Random field $f(\mathbf{x}, t)$ is called the spatially homogeneous field if all its statistical characteristics are invariant relative to spatial translations by arbitrary vector $\mathbf{a}$, i.e., if

$$P(t_1, ..., t_n, \mathbf{x}_1 + \mathbf{a}, ...\mathbf{x}_n + \mathbf{a}; f_1, ...f_n) = P(t_1, ..., t_n, \mathbf{x}_1, ...\mathbf{x}_n; f_1, ...f_n).$$

In this case, the one-point probability density is independent of $\mathbf{x}$, $P(t, \mathbf{x}; f) = P(t; f)$, and the spatial correlation function $B_f(\mathbf{x}_1, t_1; \mathbf{x}_2, t_2)$ depends on the difference $\mathbf{x}_1 - \mathbf{x}_2$

$$B_f(\mathbf{x}_1, t_1; \mathbf{x}_2, t_2) = \langle f(\mathbf{x}_1, t_1) f(\mathbf{x}_2, t_2) \rangle = B_f(\mathbf{x}_1 - \mathbf{x}_2; t_1, t_2).$$

If random field $f(\mathbf{x}, t)$ is additionally invariant with respect to rotation of all vectors $\mathbf{x}_i$ by arbitrary angle, i.e., with respect to rotations of the reference system, then field $f(\mathbf{x}, t)$ is called the homogeneous isotropic random field. In this case, the correlation function depends on the length $|\mathbf{x}_1 - \mathbf{x}_2|$:

$$B_f(\mathbf{x}_1, t_1; \mathbf{x}_2, t_2) = \langle f(\mathbf{x}_1, t_1) f(\mathbf{x}_2, t_2) \rangle = B_f(|\mathbf{x}_1 - \mathbf{x}_2|; t_1, t_2).$$

The Fourier transform of correlation function of the homogeneous random field with respect to spatial variables defines the spatial spectral function (called also the angular spectrum)

$$\Phi_f(\mathbf{k}, t) = \int d\mathbf{x} B_f(\mathbf{x}, t) e^{i\mathbf{k}\mathbf{x}},$$

and the Fourier transform of correlation function of the stationary and homogeneous random field $f(\mathbf{x}, t)$ with respect to both spatial and temporal variables defines the space-time spectrum

$$\Phi_f(\mathbf{k}, \omega) = \int d\mathbf{x} \int\limits_{-\infty}^{\infty} dt B_f(\mathbf{x}, t) e^{i(\mathbf{k}\mathbf{x}+\omega t)}.$$

In the case of isotropic random field $f(\mathbf{x}, t)$, the space-time spectrum appears isotropic in the $\mathbf{k}$-space:

$$\Phi_f(\mathbf{k}, \omega) = \Phi_f(k, \omega).$$

An exhaustive description of random function $z(t)$ can be given in terms of the characteristic functional

$$\Phi[v(\tau)] = \left\langle \exp\left\{ i \int\limits_{-\infty}^{\infty} d\tau v(\tau) z(\tau) \right\} \right\rangle,$$

where $v(t)$ is arbitrary (but sufficiently smooth) function. Functional $\Phi[v(\tau)]$ being known, one can determine such characteristics of random function $z(t)$ as *mean value* $\langle z(t) \rangle$, *correlation function* $\langle z(t_1) z(t_2) \rangle$, *n-point moment functions* $\langle z(t_1)...z(t_n) \rangle$, etc.

Indeed, expanding functional $\Phi[v(\tau)]$ in the functional Taylor series, we obtain the representation of characteristic functional in terms of the moment functions of process $z(t)$:

$$\Phi[v(\tau)] = \sum_{n=0}^{\infty} \frac{i^n}{n!} \int\limits_{-\infty}^{\infty} dt_1 ... \int\limits_{-\infty}^{\infty} dt_n M_n(t_1, ..., t_n) v(t_1)...v(t_n),$$

$$M_n(t_1, ..., t_n) = \langle z(t_1)...z(t_n) \rangle = \frac{1}{i^n} \frac{\delta^n}{\delta v(t_1)...\delta v(t_n)} \Phi[v(\tau)]\Big|_{v=0}.$$

Consequently, moment functions of random process $z(t)$ are expressed in terms of the variational derivatives of characteristic functional.

Represent now functional $\Phi[v(\tau)]$ in the form $\Phi[v(\tau)] = \exp\{\Theta[v(\tau)]\}$. Functional $\Theta[v(\tau)]$ also can be expanded in the functional Taylor series

$$\Theta[v(\tau)] = \sum_{n=1}^{\infty} \frac{i^n}{n!} \int\limits_{-\infty}^{\infty} dt_1 ... \int\limits_{-\infty}^{\infty} dt_n K_n(t_1, ..., t_n) v(t_1)...v(t_n), \tag{4.24}$$

where function

$$K_n(t_1,...,t_n) = \frac{1}{i^n} \frac{\delta^n}{\delta v(t_1)...\delta v(t_n)} \Theta[v(\tau)]\Big|_{v=0}$$

is called the n-th order *cumulant function* of random process $z(t)$.

The characteristic functional and the n-th order cumulant functions of scalar random field $f(\mathbf{x},t)$ are defined similarly

$$\Phi[v(\mathbf{x}',\tau)] = \left\langle \exp\left\{ i \int d\mathbf{x} \int\limits_{-\infty}^{\infty} dt v(\mathbf{x},t) f(\mathbf{x},t) \right\} \right\rangle = \exp\left\{ \Theta[v(\mathbf{x}',\tau)] \right\},$$

$$M_n(\mathbf{x}_1,t_1,...,\mathbf{x}_n,t_n) = \frac{1}{i^n} \frac{\delta^n}{\delta v(\mathbf{x}_1,t_1)...\delta v(\mathbf{x}_n,t_n)} \Phi[v(\mathbf{x}',\tau)]\Big|_{v=0},$$

$$K_n(\mathbf{x}_1,t_1,...,\mathbf{x}_n,t_n) = \frac{1}{i^n} \frac{\delta^n}{\delta v(\mathbf{x}_1,t_1)...\delta v(\mathbf{x}_n,t_n)} \Theta[v(\mathbf{x}',\tau)]\Big|_{v=0}.$$

In the case of vector random field $\mathbf{f}(\mathbf{x},t)$, we must assume that $\mathbf{v}(\mathbf{x},t)$ is the vector function.

As we noted earlier, the exhaustive description of random processes and fields requires the use of characteristic functionals. On the other hand, even simple one-point probability densities can provide an insight into the temporal behavior and spatial structure of random processes for arbitrary long temporal intervals. Ideas of statistical topography of random processes and fields can assist in obtaining this insight.

4.2.2 Statistical topography of random processes and fields

Random processes

We discuss first the concept of typical realization curve of random process $z(t)$. This concept concerns the fundamental features of the behavior of a separate process realization as a whole for temporal intervals of arbitrary duration.

Random process typical realization curve. We will call the typical realization curve of random process $z(t)$ the deterministic curve $z^*(t)$, which is the *median of the integral distribution function* (4.19) and is determined as the solution to the algebraic equation

$$F(t, z^*(t)) = 1/2. \tag{4.25}$$

The reason to this definition rests on the median property consisting in the fact that, for any temporal interval (t_1, t_2), random process $z(t)$ entwines about curve $z^*(t)$ in a way to force the identity of average times during which the inequalities $z(t) > z^*(t)$ and $z(t) < z^*(t)$ hold (Fig. 4.1):

$$\left\langle T_{z(t)>z^*(t)} \right\rangle = \left\langle T_{z(t)<z^*(t)} \right\rangle = \frac{1}{2}(t_2 - t_1). \tag{4.26}$$

Indeed, integrating Eq. (4.25) over temporal interval (t_1, t_2), we obtain

$$\int\limits_{t_1}^{t_2} dt F(t, z^*(t)) = \frac{1}{2}(t_2 - t_1). \tag{4.27}$$

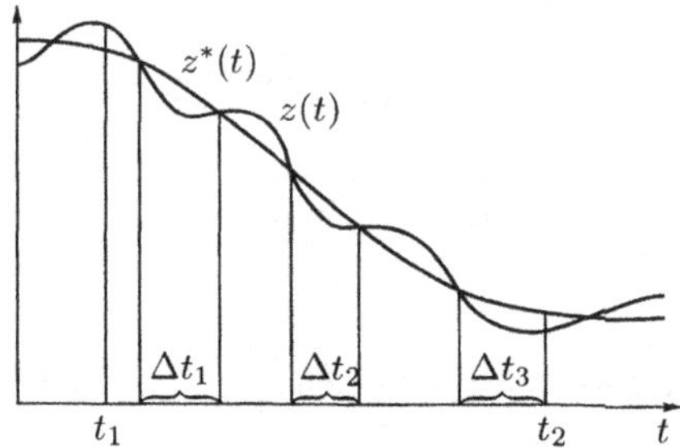

Figure 4.1: To definition of the typical realization curve of a random process.

On the other hand, in view of the definition of integral distribution function (4.19), the integral in the right-hand side of Eq. (4.27) can be represented as

$$\int\limits_{t_1}^{t_2} dt F\left(t, z^*(t)\right) = \langle T(t_1, t_2) \rangle, \tag{4.28}$$

where $T(t_1, t_2) = \sum\limits_{1}^{N} \Delta t_k$ is the combined time during which the realization of process $z(t)$ appears above curve $z^*(t)$ in interval (t_1, t_2). Combining Eqs. (4.27) and (4.28), we obtain Eq. (4.26).

Curve $z^*(t)$ can significantly differ from any particular realization of process $z(t)$ and cannot describe possible magnitudes of spikes. Nevertheless, the definitional domain of typical realization curve $z^*(t)$ derived from the one-point probability density of random process $z(t)$ is the whole of temporal axis $t \in (0, \infty)$.

Consideration of specific random processes allow obtaining an additional information concerning the realization's spikes relative to the typical realization curve.

Statistics of random process cross points with a line. The one-point probability density (4.18) of random process $z(t)$ is a result of averaging the singular indicator function over an ensemble of realizations of this process. This function is concentrated at points at which process $z(t)$ crosses line $z = \mathrm{const}$. Because the cross points are determined as roots of algebraic equation

$$z(t_n) = z \quad (n = 0, 1, ..., \infty),$$

we can rewrite the indicator function in the following form

$$\varphi(t; z) = \sum\limits_{k=1}^{n} \frac{1}{|p(t_k)|} \delta(t - t_k),$$

where $p(t) = \frac{d}{dt} z(t)$.

The number of cross points by itself is obviously a random quantity described by the formula

$$n(t, z) = \int\limits_{-\infty}^{t} d\tau |p(\tau)| \varphi(\tau; z). \tag{4.29}$$

As a consequence, the average number of points where process $z(t)$ crosses line $z = $ const can be described in terms of the correlation between the process derivative with respect to time and the process indicator function, or in terms of the joint one-point probability density of process $z(t)$ and its derivative with respect to time $\frac{d}{dt}z(t)$

$$\langle n(t,z)\rangle = \int\limits_{-\infty}^{t} d\tau \, \langle |p(\tau)|\varphi(\tau;z)\rangle \, .$$

In a similar way, we can determine certain elements of statistics related to some other specific points (such as points of maxima or minima) of random process $z(t)$.

Random fields

Similarly to common topography of mountain ranges, the statistical topography studies the systems of contours (level lines in the two-dimensional case and surfaces of constant values in the three-dimensional case) specified by the equality $f(\mathbf{r},t) = f = $ const.

For analyzing a system of contours (in this section, we will deal for simplicity with the two-dimensional case and assume $\mathbf{r} = \mathbf{R}$), we introduce the singular indicator function (4.23) concentrated on these contours.

The convenience of function (4.23) consists, in particular, in the fact that it allows simple expressions for quantities such as the total area of regions where $f(\mathbf{R},t) > f$ (i.e., inside level lines $f(\mathbf{R},t) = f$)

$$S(t;f) = \int \theta(f(\mathbf{R},t) - f)d\mathbf{R} = \int\limits_{f}^{\infty} df' \int d\mathbf{R}\varphi(t,\mathbf{R};f'),$$

and the total "mass" of the field within these regions

$$M(t;f) = \int f(\mathbf{R},t)\theta(f(\mathbf{R},t) - f)d\mathbf{R} = \int\limits_{f}^{\infty} f'df' \int d\mathbf{R}\varphi(t,\mathbf{R};f').$$

As we mentioned earlier, the mean value of indicator function (4.23) over an ensemble of realizations determines the one-time (in time) and one-point (in space) probability density

$$P(t,\mathbf{R};f) = \langle\varphi(t,\mathbf{R};f)\rangle = \langle\delta\left(f(\mathbf{R},t)-f\right)\rangle \, .$$

Consequently, this probability density immediately determines ensemble-averaged values of the above expressions.

If we include into consideration the spatial gradient $\mathbf{p}(\mathbf{R},t) = \nabla f(\mathbf{R},t)$, we can obtain additional information on structural details of field $f(\mathbf{R},t)$. For example, quantity

$$l(t;f) = \int d\mathbf{R}\,|\mathbf{p}(\mathbf{R},t)|\,\delta(f(\mathbf{R},t) - f) = \oint dl \tag{4.30}$$

is the total length of contours and extends formula (4.29) to random fields.

The integrand in Eq. (4.30) is described in terms of the extended indicator function

$$\varphi(t,\mathbf{R};f,\mathbf{p}) = \delta\left(f(\mathbf{R},t) - f\right)\delta\left(\mathbf{p}(\mathbf{R},t)-\mathbf{p}\right), \tag{4.31}$$

so that the average value of total length (4.30) is related to the joint one-time probability density of field $f(\mathbf{R}, t)$ and its gradient $\mathbf{p}(\mathbf{R}, t)$, which is defined as the ensemble average of indicator function (4.31), i.e., as the function

$$P(t, \mathbf{R}; f, \mathbf{p}) = \langle \delta\left(f(\mathbf{R}, t) - f\right) \delta\left(\mathbf{p}(\mathbf{R}, t) - \mathbf{p}\right) \rangle.$$

Inclusion of second-order spatial derivatives into consideration allows estimating the total number of contours $f(\mathbf{R}, t) = f = \mathrm{const}$ by the approximate formula (neglecting unclosed lines)

$$N(t; f) = N_{\mathrm{in}}(t; f) - N_{\mathrm{out}}(t; f) = \frac{1}{2\pi} \int d\mathbf{R}\kappa(t, \mathbf{R}; f) \, |\mathbf{p}(\mathbf{R}, t)| \, \delta\left(f(\mathbf{R}, t) - f\right),$$

where $N_{\mathrm{in}}(t; f)$ and $N_{\mathrm{out}}(t; f)$ are the numbers of contours for which vector $\mathbf{p}$ is directed along internal and external normals, respectively; and $\kappa(t, \mathbf{R}; f)$ is the curvature of the level line.

Recall that, in the case of the spatially homogeneous field $f(\mathbf{R}, t)$, the corresponding one-point probability densities $P(t, \mathbf{R}; f)$ and $P(t, \mathbf{R}; f, \mathbf{p})$ are independent of $\mathbf{R}$. In this case, if statistical averages of the above expressions (without integration over $\mathbf{R}$) exist, they will characterize the corresponding specific (per unit area) values of these quantities.

Consider now several examples of random processes.

4.2.3 Gaussian random process

We start the discussion with continuous processes; namely, we consider the Gaussian random process $z(t)$ with zero-valued mean ($\langle z(t) \rangle = 0$) and correlation function $B(t_1, t_2) = \langle z(t_1)z(t_2) \rangle$. The corresponding characteristic functional assumes the form

$$\Phi[v(\tau)] = \exp\left\{ -\frac{1}{2} \int\limits_{-\infty}^{\infty} \int\limits_{-\infty}^{\infty} dt_1 dt_2 B(\tau_1, \tau_2) v(\tau_1) v(\tau_2) \right\}.$$

Only one cumulant function (the correlation function $K_2(t_1, t_2) = B(t_1, t_2)$) is different from zero for this process, so that

$$\Theta[v(\tau)] = -\frac{1}{2} \int\limits_{-\infty}^{\infty} \int\limits_{-\infty}^{\infty} dt_1 dt_2 B(\tau_1, \tau_2) v(\tau_1) v(\tau_2). \tag{4.32}$$

Consider the n-th order variational derivative of functional $\Phi[v(\tau)]$. It satisfies the following line of equalities:

$$\frac{\delta^n}{\delta v(t_1)...\delta v(t_n)} \Phi[v(\tau)] = \frac{\delta^{n-1}}{\delta v(t_2)...\delta v(t_n)} \frac{\delta\Theta[v(\tau)]}{\delta v(t_1)} \Phi[v(\tau)]$$

$$= \frac{\delta^2\Theta[v(\tau)]}{\delta v(t_1)\delta v(t_2)} \frac{\delta^{n-2}}{\delta v(t_3)...\delta v(t_n)} \Phi[v(\tau)] + \frac{\delta^{n-2}}{\delta v(t_3)...\delta v(t_n)} \frac{\delta\Theta[v(\tau)]}{\delta v(t_1)} \frac{\delta\Phi[v(\tau)]}{\delta v(t_2)}.$$

Setting now $v = 0$, we obtain that moment functions of the Gaussian process $z(t)$ satisfy the recurrence formula

$$M_n(t_1, ..., t_n) = \sum_{k=2}^{n} B(t_1, t_2) M_{n-2}(t_2, ..., t_{k-1}, t_{k+1}, ..., t_n). \tag{4.33}$$

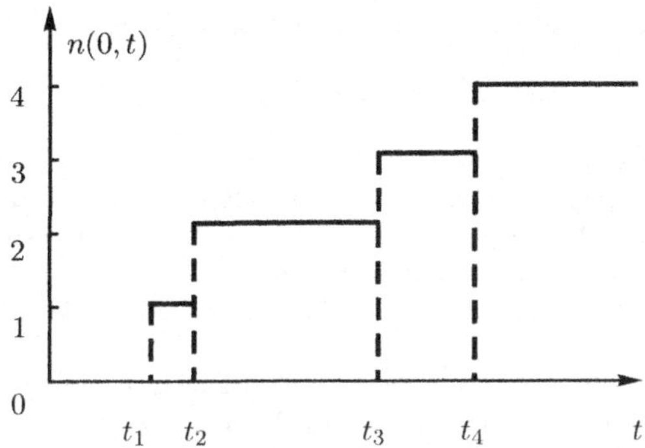

Figure 4.2: A possible realization of process $n(0, t)$.

From this formula follows that, for the Gaussian process with zero-valued mean, all moment functions of odd orders are identically equal to zero and the moment functions of even orders are represented as sums of terms which are the products of averages of all possible pairs $z(t_i)z(t_k)$.

If we assume that function $v(\tau)$ in Eq. (4.32) is different from zero only in interval $0 < \tau < t$, the characteristic function

$$\Phi[t; v(\tau)] = \left\langle \exp\left(i \int_0^t d\tau\, z(\tau) v(\tau) \right) \right\rangle = \exp\left\{ -\int_0^t d\tau_1 \int_0^{\tau_1} d\tau_2\, B(\tau_1, \tau_2) v(\tau_1) v(\tau_2) \right\}$$

(4.34)

becomes a function of time t and satisfies the ordinary differential equation

$$\frac{d}{dt}\Phi[t; v(\tau)] = -v(t) \int_0^t d\tau\, B(t, \tau) v(\tau) \Phi[t; v(\tau)], \quad \Phi[0; v(\tau)] = 1.$$

(4.35)

4.2.4 Discontinuous random processes

Consider now some examples of discontinuous processes. The discontinuous processes are the random functions that change their time-dependent behavior at discrete instants $t_1, t_2, \ldots$ given statistically. The description of discontinuous processes requires first of all either the knowledge of the statistics of these instants, or the knowledge of the statistics of number $n(0, t)$ of instants t_i falling in time interval $(0, t)$. In the latter case, we have the equality

$$n(0, t) = n(0, t') + n(t', t), \quad 0 \le t' \le t.$$

The quantity $n(0, t)$ by itself is a random process, and Fig. 4.2 shows its possible realization.

The set of points of discontinuity $t_1, t_2, \ldots$ of process $z(t)$ is called the *stream of points*. In what follows, we will consider the *Poisson stationary stream of points* in which the probability of falling n points in interval (t_1, t_2) is specified by the Poisson formula

$$P_{n(t_1, t_2) = n} = \frac{\left[\overline{n(t_1, t_2)} \right]^n}{n!} e^{-\overline{n(t_1, t_2)}}$$

(4.36)

with mean number of points in interval (t_1, t_2) given by the formula

$$\overline{n(t_1, t_2)} = \nu|t_1 - t_2|,$$

where ν is the mean number of points per unit time. It is assumed here that the numbers of points falling in nonoverlapping intervals are statistically independent and the instants at which points were fallen in interval (t_1, t_2) under the condition that their total number was n are also statistically independent and uniformly distributed over the interval (t_1, t_2). The length of the interval between adjacent points of discontinuity satisfies the exponential distribution.

The Poisson stream of points is an example of the Markovian process.

Consider now random processes whose points of discontinuity form Poisson streams of points. Currently, three types of such processes — the *Poisson process*, *telegrapher's process*, and *generalized telegrapher's process* — are mainly used in model problems of physics. Below, we focus our attention on these processes.

Poisson (impulse) random process

Poisson (impulse) random process $z(t)$ is the process described by the formula

$$z(t) = \sum_{i=1}^{n} \xi_i g(t - t_i), \tag{4.37}$$

where random quantities ξ_i are statistically independent and distributed with probability density $p(\xi)$; random points t_k are uniformly distributed on interval $(0, T)$, so that their number n obeys the Poisson law with parameter $\bar{n} = \nu T$; and function $g(t)$ is the deterministic function that describes the pulse envelope ($g(t) = 0$ for $t < 0$).

The characteristic functional of Poisson random process $z(t)$ assumes the form

$$\Phi[t; v(\tau)] = \exp\left\{\nu \int_0^t dt' \left[W\left(\int_{t'}^t d\tau v(\tau)g(t - t')\right) - 1\right]\right\},$$

where

$$W(v) = \int_{-\infty}^{\infty} d\xi p(\xi)e^{i\xi v}$$

is the characteristic function of random quantity ξ. Consequently, functional $\Theta[t; v(\tau)]$ and cumulant functions assume the forms

$$\Theta[t; v(\tau)] = \nu \int_0^t dt' \int_{-\infty}^{\infty} d\xi p(\xi) \left\{\exp\left[i\xi \int_{t'}^t d\tau v(\tau)g(t - t')\right] - 1\right\}, \tag{4.38}$$

$$K_n(t_1, ..., t_n) = \nu \langle \xi^n \rangle \int_0^{\min\{t_1,...,t_n\}} dt' g(t - t')...g(t_n - t').$$

We consider Poisson processes of two types important for applications.

1. Let

$$g(t) = \theta(t) = \begin{cases} 1, & t > 0, \\ 0, & t < 0, \end{cases}$$

i.e.,

$$z(t) = \sum_{i=1}^{n} \xi_i \theta(t - t_i).$$

In this case,

$$K_n(t_1, ..., t_n) = \nu \langle \xi^n \rangle \min\{t_1, ..., t_n\}.$$

If additionally $\xi = 1$, then process $z(t) \equiv n(0,t)$, and we have

$$K_n(t_1, ..., t_n) = \nu \min\{t_1, ..., t_n\}, \quad \Theta[t; v(\tau)] = \nu \int_0^t dt' \left\{ \exp\left[i \int_{t'}^t d\tau v(\tau) \right] - 1 \right\}. \tag{4.39}$$

2. Let now $g(t) = \delta(t)$. In this case, process

$$z(t) = \sum_{i=1}^{n} \xi_i \delta(t - t_i)$$

is usually called the *shot noise process*. This process is a special case of *delta-correlated processes*. For such a process, functional $\Theta[t; v(\tau)]$ and cumulant functions assume the forms

$$\Theta[t; v(\tau)] = \nu \int_0^t d\tau \int_{-\infty}^{\infty} d\xi p(\xi) \left\{ e^{i\xi v(\tau)} - 1 \right\}, \tag{4.40}$$

$$K_n(t_1, ..., t_n) = \nu \langle \xi^n \rangle \, \delta(t_1 - t_2)\delta(t_2 - t_3)...\delta(t_{n-1} - t_n).$$

Telegrapher's random process

Consider now statistical characteristics of telegrapher's (binary, or two-state) random process (Fig. 4.3) defined by the formula

$$z(t) = a(-1)^{n(0,t)} \quad (z(0) = a, \quad z^2(t) \equiv a^2), \tag{4.41}$$

where $n(t_1, t_2)$ is the random sequence of integers equal to the number of points of discontinuity in interval (t_1, t_2).

We consider two cases.

1. We will assume first that amplitude a is the deterministic quantity.

For the two firsts moment functions of process $z(t)$, we have the expressions

$$\langle z(t) \rangle = a \sum_{n(0,t)=0}^{\infty} (-1)^{n(0,t)} P_{n(0,t)} = ae^{-2\overline{n(0,t)}} = ae^{-2\nu t},$$

$$\langle z(t_1)z(t_2) \rangle = a^2 \left\langle (-1)^{n(0,t_1)+n(0,t_2)} \right\rangle = a^2 \left\langle (-1)^{n(t_2,t_1)} \right\rangle$$

$$= a^2 e^{-2\overline{n(t_2,t_1)}} = a^2 e^{-2\nu(t_1-t_2)} \quad (t_1 \geq t_2). \tag{4.42}$$

The higher moment functions for $t_1 \geqslant t_2 \geqslant ... \geqslant t_n$ satisfy the recurrence relationship

$$M_n(t_1, ..., t_n) = \langle z(t_1)...z(t_n) \rangle = a^2 \left\langle (-1)^{n(0,t_1)+n(0,t_2)+n(0,t_3)+...+n(0,t_n)} \right\rangle$$

$$= a^2 \left\langle (-1)^{n(t_2,t_1)} \right\rangle \left\langle (-1)^{n(0,t_3)+...+n(0,t_n)} \right\rangle = \langle z(t_1)z(t_2) \rangle M_{n-2}(t_3, ..., t_n). \tag{4.43}$$

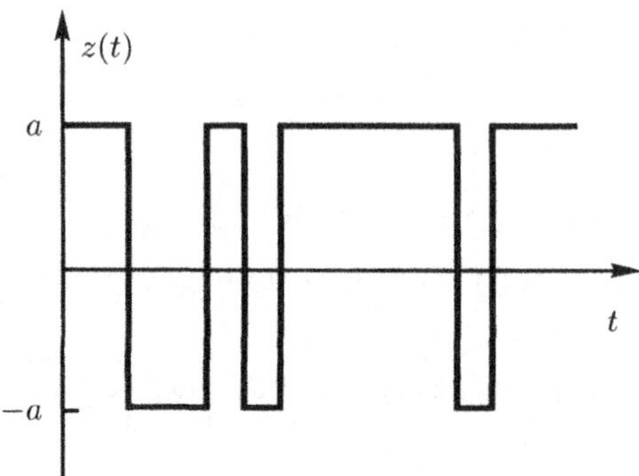

Figure 4.3: A possible realization of telegrapher's random process.

This relationship is very similar to Eq. (4.33) for the Gaussian process with correlation function $B(t_1, t_2)$. The only difference is that the right-hand side of Eq. (4.43) coincides with only one term of the sum in Eq. (4.33), namely, with the term that corresponds to the above order of times.

Consider now the characteristic functional of this process

$$\Phi_a[t; v(\tau)] = \left\langle \exp\left\{ i \int_0^t d\tau\, z(\tau) v(\tau) \right\} \right\rangle,$$

where index a means that amplitude a is the deterministic quantity. Expanding the characteristic functional in the functional Taylor series and using recurrence formula (4.43), we obtain the integral equation

$$\Phi_a[t; v(\tau)] = 1 + ia \int_0^t dt_1\, e^{-2\nu t_1} v(t_1)$$

$$- a^2 \int_0^t dt_1 \int_0^{t_1} dt_2\, e^{-2\nu(t_1 - t_2)} v(t_1) v(t_2) \Phi_a[t_2; v(\tau)]. \tag{4.44}$$

Differentiating Eq. (4.44) with respect to t, we obtain the integro-differential equation

$$\frac{d}{dt}\Phi_a[t; v(\tau)] = iae^{-2\nu t} v(t) - a^2 v(t) \int_0^t dt_1\, e^{-2\nu(t - t_1)} v(t_1) \Phi_a[t_1; v(\tau)]. \tag{4.45}$$

2. Let now amplitude a be the random quantity with probability density $p(a)$. To obtain the characteristic functional of process $z(t)$ in this case, we should average Eq. (4.45) with respect to random amplitude a. In the general case, such averaging cannot be performed analytically. Analytical averaging of Eq. (4.45) appears possible only if probability density of random amplitude a has the form

$$p(a) = \frac{1}{2}\left[\delta\left(a - a_0\right) + \delta\left(a + a_0\right)\right]$$

with $\langle a \rangle = 0$ and $\langle a^2 \rangle = a_0^2$ (in fact, this very case is what is called usually telegrapher's process). As a result, we obtain the integro-differential equation

$$\frac{d}{dt}\Phi[t;v(\tau)] = -a_0^2 v(t) \int_0^t dt_1 e^{-2\nu(t-t_1)} v(t_1)\Phi[t_1;v(\tau)]. \qquad (4.46)$$

Now, we dwell on an important limiting theorem concerning telegrapher's random processes.

Consider the random process

$$\xi_N(t) = z_1(t) + ... + z_N(t),$$

where all $z_k(t)$ are statistically independent telegrapher's processes with zero-valued means and correlation functions

$$\langle z(t)z(t+\tau)\rangle = \frac{\sigma^2}{N}e^{-\alpha|\tau|}.$$

In the limit $N \to \infty$, we obtain that process $\xi(t) = \lim_{N\to\infty}\xi_N(t)$ is the Gaussian random process with the exponential correlation function

$$\langle \xi(t)\xi(t+\tau)\rangle = \sigma^2 e^{-\alpha|\tau|},$$

i.e., the Gaussian Markovian process. Thus, process $\xi_N(t)$ for finite N is the finite-number-of-states process approximating the Gaussian Markovian process.

Generalized telegrapher's random process

Consider now generalized telegrapher's process defined by the formula

$$z(t) = a_{n(0,t)}. \qquad (4.47)$$

Here, $n(0,t)$ is the sequence of integers described above and quantities a_k are assumed statistically independent with distribution function $p(a)$. Figure 4.4 shows a possible realization of such a process.

For process $z(t)$, we have

$$\langle z(t)\rangle = \sum_{k=0}^{\infty}\left\langle a_k \delta_{k,n(0,t)}\right\rangle = \langle a \rangle,$$

$$\langle z(t_1)z(t_2)\rangle = \sum_{k=0}^{\infty}\sum_{k=0}^{\infty}\langle a_k a_l\rangle\left\langle \delta_{k,n(0,t_1)}\delta_{l,n(0,t_2)}\right\rangle$$

$$= \left\langle a^2\right\rangle\left\{\sum_{k=0}^{\infty}\left\langle \delta_{k,n(0,t_1)}\right\rangle\left\langle \delta_{0,n(t_2,t_1)}\right\rangle + 1 - \sum_{k=0}^{\infty}\left\langle \delta_{k,n(0,t_2)}\right\rangle\left\langle \delta_{0,n(t_2,t_1)}\right\rangle\right\}$$

$$= \left\langle a^2\right\rangle e^{-\nu(t_1-t_2)} + \left\langle a^2\right\rangle\left(1 - e^{-\nu(t_1-t_2)}\right) \quad (t_1 \geqslant t_2),$$

and so on. In addition, the probability of absence of points of discontinuity in interval (t_2,t_1) is given by the formula

$$P_{n(t_2,t_1)=0} = \left\langle \delta_{0,n(t_2,t_1)}\right\rangle = e^{-\nu|t_1-t_2|}.$$

Earlier, we noted that the Poisson stream of points and processes based on these streams are the Markovian processes. Below, we consider this important class of random processes in detail.

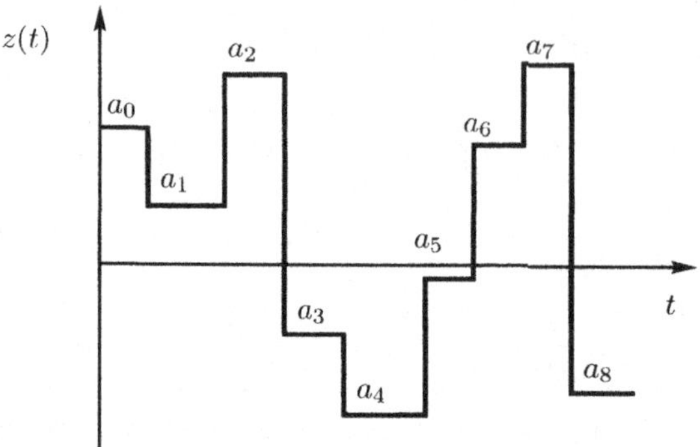

Figure 4.4: A possible realization of generalized telegrapher's random process.

4.3 Markovian processes

4.3.1 General properties

In the foregoing section, we considered the characteristic functional that describes all statistical characteristics of random process $z(t)$. Specification of the argument of the functional in the form

$$v(t) = \sum_{k=1}^{n} v_k \delta(t - t_k)$$

transforms the characteristic functional into the joint characteristic function of random quantities $z_k = z(t_k)$

$$\Phi_n(v_1, ..., v_n) = \left\langle \exp\left\{ i \sum_{k=1}^{n} v_k z(t_k) \right\} \right\rangle,$$

whose Fourier transform is the joint probability density of process $z(t)$ at discrete instants

$$P_n(z_1, t_1; ...; z_n, t_n) = \langle \delta(z(t_1) - z_1)...\delta(z(t_n) - z_n) \rangle. \tag{4.48}$$

Assume that the above instants are ordered according to the line of inequalities

$$t_1 \geq t_2 \geq ... \geq t_n.$$

Then, by the definition of conditional probability, we have

$$P_n(z_1, t_1; ...; z_n, t_n) = p_n(z_1, t_1 | z_2, t_2; ...; z_n, t_n) P_{n-1}(z_2, t_2; ...; z_n, t_n), \tag{4.49}$$

where p_n is the conditional probability density of value z_1 of process $z(t)$ at instant t_1 under the condition that function $z(t)$ was equal to z_k at instants t_k for $k = 2, ..., n$ ($z(t_k) = z_k$, $k = 2, ..., n$). If process $z(t)$ is such that the conditional probability density for all $t_1 > t_2$ is unambiguously determined by the value z_2 of the process at instant t_2 and is independent of the previous history, i.e., if

$$p_n(z_1, t_1 | z_2, t_2; ...; z_n, t_n) = p(z_1, t_1 | z_2, t_2), \tag{4.50}$$

then this process is called the Markovian process, or the memoryless process. In this case, function

$$p(z, t | z_0, t_0) = \langle \delta(z(t) - z) | z(t_0) = z_0 \rangle \quad (t > t_0) \tag{4.51}$$

is called the *transition probability density*. Setting $t = t_0$ in Eq. (4.51), we obtain the equality

$$p(z, t_0 | z_0, t_0) = \delta(z - z_0).$$

Substituting expression (4.49) in Eq. (4.48), we obtain the recurrence formula for the n-time probability density of process $z(t)$. Iterating this formula, we find the expression of probability density P_n in terms of the one-time probability density $(t_1 \geq t_2 \geq ... \geq t_n)$

$$P_n(z_1, t_1; ...; z_n, t_n) = p(z_1, t_1 | z_2, t_2)...p(z_{n-1}, t_{n-1} | z_n, t_n) P(t_n, z_n). \tag{4.52}$$

Thus, only two functions — transition probability density $p(z, t | z_0, t_0)$ and one-time probability density $P(t, z)$ — are sufficient to exhaustively describe all statistical characteristics of the Markovian process $z(t)$. This inference, clearly, holds also for multidimensional processes, i.e., for vector functions $\mathbf{z}(t)$.

Nevertheless, statistical analysis of stochastic equations often requires the knowledge of the characteristic functional of random process $z(t)$.

4.3.2 Characteristic functional of the Markovian process

For the Markovian process $z(t)$, no closed equation can be derived in the general case for the characteristic functional $\Phi[t; v(\tau)] = \langle \varphi[t; v(\tau)] \rangle$, where

$$\varphi[t; v(\tau)] = \exp\left\{ i \int\limits_0^t d\tau z(\tau) v(\tau) \right\}.$$

Instead, we can derive the closed equation for the functional

$$\Psi[z, t; v(\tau)] = \langle \delta(z(t) - z) \varphi[t; v(\tau)] \rangle \tag{4.53}$$

describing correlations of process $z(t)$ with its prehistory. The characteristic functional $\Phi[t; v(\tau)]$ can be obtained from functional $\Psi[z, t; v(\tau)]$ by the formula

$$\Phi[t; v(\tau)] = \int\limits_{-\infty}^{\infty} dz \Psi[z, t; v(\tau)]. \tag{4.54}$$

To derive the equation for functional $\Psi[z, t; v(\tau)]$, we note that the following equality

$$\varphi[t; v(\tau)] = 1 + i \int\limits_0^t dt_1 z(t_1) v(t_1) \varphi[t_1; v(\tau)] \tag{4.55}$$

holds. Substituting Eq. (4.55) in Eq. (4.53), we obtain the expression

$$\Psi[t, z; v(\tau)] = P(t, z) + i \int\limits_0^t dt_1 v(t_1) \langle \delta(z(t) - z) z(t_1) \varphi[t_1; v(\tau)] \rangle, \tag{4.56}$$

where $P(t, z) = \langle \delta(z(t) - z) \rangle$ is the one-point probability density of random quantity $z(t)$.

We rewrite Eq. (4.56) in the form

$$\Psi[t, z; v(\tau)] = P(t, z)$$
$$+ i \int_0^t dt_1 v(t_1) \int_{-\infty}^{\infty} dz_1 z_1 \langle \delta(z(t) - z) \delta(z(t_1) - z_1) \varphi[t_1; v(\tau)] \rangle . \qquad (4.57)$$

Taking into account the fact that process $z(t)$ is the Markovian process, we can perform averaging in (4.57) to obtain the closed integral equation

$$\Psi[t, z; v(\tau)] = P(t, z) + i \int_0^t dt_1 v(t_1) \int_{-\infty}^{\infty} dz_1 z_1 p(z, t | z_1, t_1) \Psi[t_1, z_1; v(\tau)], \qquad (4.58)$$

where $p(z, t; z_0, t_0)$ is the transition probability density.

Integrating Eq. (4.58) with respect to z, we obtain an additional relationship between the characteristic functional $\Phi[t; v(\tau)]$ and functional $\Psi[z, t; v(\tau)]$. This relationship has the form

$$\frac{1}{iv(t)} \frac{d}{dt} \Phi[t; v(\tau)] = \int_{-\infty}^{\infty} dz_1 z_1 \Psi[t, z_1; v(\tau)] = \Psi[t; v(\tau)]. \qquad (4.59)$$

Multiplying Eq. (4.58) by z and integrating the result over z, we obtain the relationship between functionals $\Psi[t; v(\tau)]$ and $\Psi[t, z; v(\tau)]$

$$\Psi[t; v(\tau)] = \langle z(t) \rangle + i \int_0^t dt_1 v(t_1) \int_{-\infty}^{\infty} dz_1 \langle z(t) | z_1, t_1 \rangle \Psi[t_1, z_1; v(\tau)]. \qquad (4.60)$$

Equation (4.58) is in the general case a complicated integral equation whose explicit form depends on functions $P(t, z)$ and $p(z, t; z_0, t_0)$, i.e., on parameters of the Markovian process.

In some specific cases, the situation is simplified. For example, for telegrapher's process, we have from Eq. (4.42)

$$\langle z(t) | z_1, t_1 \rangle = z_1 e^{-2\nu(t - t_1)}, \quad \langle z(t) \rangle = 0,$$

and we obtain Eq. (4.16).

Problems

Problem 8 *Show that random process*

$$\xi_N(t) = z_1(t) + \ldots + z_N(t),$$

where all $z_k(t)$ are statistically independent telegrapher's processes with zero-valued means and correlation functions

$$\langle z(t) z(t + \tau) \rangle = \frac{\sigma^2}{N} e^{-\alpha |\tau|}$$

in the limit $N \to \infty$ is the Gaussian Markovian random process $\xi(t) = \lim_{N \to \infty} \xi_N(t)$ with the exponential correlation function

$$\langle \xi(t) \xi(t + \tau) \rangle = \sigma^2 e^{-\alpha |\tau|}.$$

Remark. Consider the differential equation for the characteristic functional of random process $\xi_N(t)$

$$\Phi_N[t; v(\tau)] = \left\langle \exp\left\{ i \int_0^t d\tau\, \xi_N(\tau) v(\tau) \right\} \right\rangle.$$

Chapter 5

Correlation splitting

5.1 General remarks

For simplicity, we content ourselves here with the one-dimensional random processes (extensions to multidimensional cases are obvious). We need the ability of calculating correlators of the form $\langle z(t)R[z(\tau)]\rangle$, where $R[z(\tau)]$ is the functional that can depend on process $z(t)$ both explicitly and implicitly.

To calculate this average, we consider auxiliary functional $R[z(\tau) + \eta(\tau)]$, where $\eta(t)$ is arbitrary deterministic function, and calculate the correlator

$$\langle z(t)R[z(\tau) + \eta(\tau)]\rangle.$$

The correlator of interest will be obtained by setting $\eta(\tau) = 0$ in the final result.

We can expand the above auxiliary functional $R[z(\tau) + \eta(\tau)]$ in the functional Taylor series with respect to $z(\tau)$. The result can be represented in the form

$$R\left[z(\tau) + \eta(\tau)\right] = e^{\int\limits_{-\infty}^{\infty} d\tau z(\tau)\frac{\delta}{\delta\eta_2(\tau)}} R\left[\eta(\tau)\right],$$

where we introduced the functional shift operator. With this representation, we can obtain the following expression for the correlation

$$\langle z(t)R[z(\tau) + \eta(\tau)]\rangle = \Omega\left[t; \frac{\delta}{i\delta\eta(\tau)}\right] \langle R[z(\tau) + \eta(\tau)]\rangle, \tag{5.1}$$

where functional

$$
\begin{aligned}
\Omega[t; v(\tau)] &= \frac{\left\langle z(t)\exp\left\{i\int\limits_{-\infty}^{\infty} d\tau z(\tau)v(\tau)\right\}\right\rangle}{\left\langle \exp\left\{i\int\limits_{-\infty}^{\infty} d\tau z(\tau)v(\tau)\right\}\right\rangle} \\
&= \frac{1}{\Phi[v(\tau)]}\frac{\delta}{i\delta v(t)}\Phi[v(\tau)] = \frac{\delta}{i\delta v(t)}\Theta[v(\tau)].
\end{aligned} \tag{5.2}
$$

Here $\Theta[v(\tau)] = \ln\Phi[v(\tau)]$ and $\Phi[v(\tau)]$ is the characteristic functional of random process $z(t)$.

Replacing now differentiation with respect to $\eta(\tau)$ by differentiation with respect to $z(\tau)$ and setting $\eta(\tau) = 0$, we obtain the expression

$$\langle z(t) R[z(\tau)] \rangle = \left\langle \Omega \left[t; \frac{\delta}{i\delta z(\tau)} \right] R[z(\tau)] \right\rangle . \tag{5.3}$$

If we expand functional $\Theta[v(\tau)]$ in the functional Taylor series (4.24)

$$\Omega[t; v(\tau)] = \sum_{n=0}^{\infty} \frac{i^n}{n!} \int\limits_{-\infty}^{\infty} dt_1 ... \int\limits_{-\infty}^{\infty} dt_n K_{n+1}(t, t_1, ..., t_n) v(t_1)...v(t_n),$$

then expression (5.3) will assume the form

$$\langle z(t) R[z(\tau)] \rangle = \sum_{n=0}^{\infty} \frac{1}{n!} \int\limits_{-\infty}^{\infty} dt_1 ... \int\limits_{-\infty}^{\infty} dt_n K_{n+1}(t, t_1, ..., t_n) \left\langle \frac{\delta^n R[z(\tau)]}{\delta z(t_1)...\delta z(t_n)} \right\rangle . \tag{5.4}$$

In physical problems obeying the dynamic causality condition in time t, statistical characteristics of the solution at instant t depend on statistical characteristics of process $z(\tau)$ for $0 \leq \tau \leq t$, which are completely described by the characteristic functional

$$\Phi[t; v(\tau)] = \exp\{\Theta[t; v(\tau)]\} = \left\langle \exp\left\{ i \int\limits_0^t d\tau z(\tau)v(\tau) \right\} \right\rangle .$$

In this case, the obtained formulas hold also for calculating statistical averages $\langle z(t') R[t; z(\tau)] \rangle$ for $t' < t$, $\tau \leq t$, i.e., we have the equality

$$\langle z(t') R[t; z(\tau)] \rangle = \left\langle \Omega \left[t', t; \frac{\delta}{i\delta z(\tau)} \right] R[t; z(\tau)] \right\rangle \quad (0 < t' < t), \tag{5.5}$$

where

$$\Omega[t', t; v(\tau)] = \frac{\delta}{i\delta v(t')} \Theta[t; v(\tau)]$$

$$+ \sum_{n=0}^{\infty} \frac{i^n}{n!} \int\limits_0^t dt_1 ... \int\limits_0^t dt_n K_{n+1}(t', t_1, ..., t_n) v(t_1)...v(t_n). \tag{5.6}$$

For $t' = t - 0$, formula (5.5) holds as before, i.e.

$$\langle z(t) R[t; z(\tau)] \rangle = \left\langle \Omega \left[t, t; \frac{\delta}{i\delta z(\tau)} \right] R[t; z(\tau)] \right\rangle . \tag{5.7}$$

However, expansion (5.6) not always gives the correct result in the limit $t' \to t - 0$ (which means that the limiting process and the procedure of expansion in the functional Fourier series can appear non-commutable). In this case,

$$
\Omega[t, t; v(\tau)] = \frac{\left\langle z(t) \exp\left\{ i \int\limits_0^t d\tau z(\tau)v(\tau) \right\} \right\rangle}{\left\langle \exp\left\{ i \int\limits_0^t d\tau z(\tau)v(\tau) \right\} \right\rangle}
$$

$$
= \frac{1}{\Phi[t; v(\tau)]} \frac{d}{iv(t)dt} \Phi[t; v(\tau)] = \frac{d}{iv(t)dt} \Theta[t; v(\tau)], \tag{5.8}
$$

and statistical averages in Eqs. (5.5) and (5.7) can be discontinuous at $t' = t - 0$.

Consider several examples of random processes.

5.2　Gaussian process

In the case of the Gaussian random process $z(t)$, all formulas obtained in the previous section become significantly simpler. In this case, the logarithm of characteristic functional $\Phi[v(\tau)]$ is given by Eq. (4.32) (we assume that the mean value of process $z(t)$ is zero), and functional $\Theta[t, v(\tau)]$ assumes the form

$$
\Theta[t, v(\tau)] = -\frac{1}{2} \int\limits_{-\infty}^{\infty} \int\limits_{-\infty}^{\infty} d\tau_1 d\tau_2 B(\tau_1, \tau_2) v(\tau_1) v(\tau_2).
$$

As a consequence, functional $\Omega[t; v(\tau)]$ (5.2) is the linear functional

$$
\Omega[t; v(\tau)] = i \int\limits_{-\infty}^{\infty} d\tau_1 B(t, \tau_1) v(\tau_1),
\tag{5.9}
$$

and Eq. (5.1) assumes the form

$$
\langle z(t) R[z(\tau) + \eta(\tau)] \rangle = \int\limits_{-\infty}^{\infty} d\tau_1 B(t, \tau_1) \frac{\delta}{\delta\eta(\tau_1)} \langle R[z(\tau) + \eta(\tau)] \rangle.
\tag{5.10}
$$

Replacing differentiation with respect to $\eta(\tau)$ by differentiation with respect to $z(\tau)$ and setting $\eta(\tau) = 0$, we obtain the equality

$$
\langle z(t) R[z(\tau)] \rangle = \int\limits_{-\infty}^{\infty} d\tau_1 B(t, \tau_1) \left\langle \frac{\delta}{\delta z(\tau_1)} R[z(\tau)] \right\rangle
\tag{5.11}
$$

commonly known in physics as the *Furutsu–Novikov formula* [3, 33].

One can easily obtain the multi-dimensional extension of Eq. (5.11); it can be written in the form

$$
\langle z_{i_1,\dots,i_n}(\mathbf{r}) R[\mathbf{z}] \rangle = \int d\mathbf{r}' \, \langle z_{i_1,\dots,i_n}(\mathbf{r}) z_{j_1,\dots,j_n}(\mathbf{r}') \rangle \left\langle \frac{\delta R[\mathbf{z}]}{\delta z_{j_1,\dots,j_n}(\mathbf{r}')} \right\rangle,
\tag{5.12}
$$

where $\mathbf{r}$ stands for all continuous arguments of random vector field $\mathbf{z}(\mathbf{r})$ and $i_1, \dots, i_n$ are the discrete (index) arguments. Repeated index arguments in the right-hand side of Eq. (5.12) assume summation.

Formulas (5.11) and (5.12) extend formulas (4.15), (4.16) to the Gaussian random processes.

If random process $z(\tau)$ is defined only on time interval $[0, t]$, then functional $\Theta[t, v(\tau)]$ assumes the form

$$
\Theta[t, v(\tau)] = -\frac{1}{2} \int\limits_{0}^{t} \int\limits_{0}^{t} d\tau_1 d\tau_2 B(\tau_1, \tau_2) v(\tau_1) v(\tau_2),
\tag{5.13}
$$

and functionals $\Omega[t', t; v(\tau)]$ and $\Omega[t, t; v(\tau)]$ are the linear functionals

$$
\Omega[t', t; v(\tau)] = \frac{\delta}{i\delta v(t')} \Theta[t, v(\tau)] = i \int\limits_{0}^{t} d\tau B(t', \tau) v(\tau),
$$

$$
\Omega[t, t; v(\tau)] = \frac{d}{iv(t)dt} \Theta[t, v(\tau)] = i \int\limits_{0}^{t} d\tau B(t, \tau) v(\tau).
\tag{5.14}
$$

As a consequence, Eqs. (5.5), (5.7) assume the form

$$\langle z(t') R[t, z(\tau)] \rangle = \int_0^t d\tau\, B(t', \tau) \left\langle \frac{\delta R[z(\tau)]}{\delta z(\tau)} \right\rangle \quad (t' \leqslant t) \tag{5.15}$$

that coincides with Eq. (5.11) if the condition

$$\frac{\delta R[t; z(\tau)]}{\delta z(\tau)} = 0 \quad \text{for} \quad \tau < 0, \quad \tau > t \tag{5.16}$$

holds.

5.3   Poisson process

The Poisson process is defined by Eq. (4.37), and its characteristic functional logarithm is given by Eq. (4.38). In this case, formulas (5.6) and (5.8) for functionals $\Omega[t', t; v(\tau)]$ and $\Omega[t, t; v(\tau)]$ assume the forms

$$\Omega[t', t; v(\tau)] = \frac{\delta}{i\delta v(t')} \Theta[t, v(\tau)] = -i \int_0^{t'} d\tau\, g(t' - \tau) \dot{W}\left(\int_\tau^t d\tau_1 v(\tau_1) g(\tau_1 - \tau)\right),$$

$$\Omega[t, t; v(\tau)] = \frac{d}{iv(t)dt} \Theta[t, v(\tau)] = -i \int_0^t d\tau\, g(t - \tau) \dot{W}\left(\int_\tau^t d\tau_1 v(\tau_1) g(\tau_1 - \tau)\right),$$

$$\tag{5.17}$$

where $\dot{W}(v) = \frac{dW(v)}{dv} = i \int_{-\infty}^{\infty} d\xi\, \xi p(\xi) e^{i\xi v}$.

Changing the integration order, we can rewrite equalities (5.17) in the form

$$\Omega[t', t; v(\tau)] = i \int_{-\infty}^{\infty} d\xi\, \xi p(\xi) \int_0^{t'} d\tau\, g(t' - \tau) \exp\left\{ i\xi \int_\tau^t d\tau_1 v(\tau_1) g(\tau_1 - \tau) \right\} \quad (t' \leqslant t). \tag{5.18}$$

As a result, we obtain that correlations of Poisson random process $z(t)$ with functionals of this process are described by the expression

$$\langle z(t') R[t; z(\tau)] \rangle = \nu \int_{-\infty}^{\infty} d\xi\, \xi p(\xi) \int_0^{t'} d\tau'\, g(t' - \tau') \langle R[t; z(\tau) + \xi g(\tau - \tau')] \rangle \quad (t' \leqslant t). \tag{5.19}$$

As we mentioned earlier, random process $n(0, t)$ describing the number of jumps during temporal interval $(0, t)$ is the special case of the Poisson process. In this case, $p(\xi) = \delta(\xi - 1)$ and $g(t) = \theta(t)$, so that Eq. (5.19) assumes the extra-simple form

$$\langle n(0, t) R[t; n(0, \tau)] \rangle = \nu \int_0^{t'} d\tau\, \langle R[t; n(0, \tau) + \theta(t' - \tau)] \rangle \quad (t' \leqslant t). \tag{5.20}$$

Equality (5.20) extends formula (4.17) for Poisson random quantities to Poisson random processes.

5.4 Telegrapher's random process

Now, we dwell on telegrapher's random process defined by formula (4.41)

$$z(t) = a(-1)^{n(0,t)}, \tag{5.21}$$

where a is the deterministic quantity.

As we mentioned earlier, calculation of correlator $\langle z(t)R[t; z(\tau)]\rangle$ for $\tau \leq t$ assumes the knowledge of the characteristic functional of process $z(t)$; unfortunately, the characteristic functional is unavailable in this case. We know only Eqs. (4.45) and (4.46) that yield the relationship between the functional

$$\Psi[t; v(\tau)] = \frac{1}{iv(t)} \frac{d}{dt} \Phi[t; v(\tau)] = \left\langle z(t) \exp\left\{ i \int_0^t d\tau\, z(\tau) v(\tau) \right\} \right\rangle$$

and the characteristic functional $\Phi[t; v(\tau)]$ in the form

$$\left\langle z(t) \exp\left\{ i \int_0^t d\tau\, z(\tau) v(\tau) \right\} \right\rangle$$

$$= a e^{-2\nu t} + i a^2 \int_0^t dt_1\, e^{-2\nu(t-t_1)} v(t_1) \left\langle \exp\left\{ i \int_0^{t_1} d\tau\, z(\tau) v(\tau) \right\} \right\rangle.$$

This relationship is sufficient to split correlator $\langle z(t)R[t; z(\tau)]\rangle$, i.e., to express it immediately in terms of the average $\langle R[t; z(\tau)]\rangle$.

Indeed, the equality

$$\langle z(t)R[t; z(\tau) + \eta(\tau)]\rangle = \left\langle z(t) \exp\left\{ \int_0^t d\tau\, z(\tau) \frac{\delta}{\delta\eta(\tau)} \right\} \right\rangle R[t; \eta(\tau)]$$

$$= \left\{ a e^{-2\nu t} + a^2 \int_0^t dt_1\, e^{-2\nu(t-t_1)} \frac{\delta}{\delta\eta(t_1)} \left\langle \exp\left\{ \int_0^{t_1} d\tau\, z(\tau) \frac{\delta}{\delta\eta(\tau)} \right\} \right\rangle \right\} R[t; \eta(\tau)]$$

$$= a R[t; \eta(\tau)] e^{-2\nu t} + a^2 \int_0^t dt_1\, e^{-2\nu(t-t_1)} \frac{\delta}{\delta\eta(t_1)} \langle R[t; z(\tau)\theta(t_1 - \tau) + \eta(\tau)]\rangle, \tag{5.22}$$

where $\eta(t)$ is arbitrary deterministic function, holds for any functional $R[t; z(\tau)]$ for $\tau \leq t$. The final result is obtained by the limit process $\eta \to 0$:

$$\langle z(t)R[t; z(\tau)]\rangle = a R[t; 0] e^{-2\nu t} + a^2 \int_0^t dt_1\, e^{-2\nu(t-t_1)} \left\langle \frac{\delta}{\delta z(t_1)} \tilde{R}[t, t_1; z(\tau)] \right\rangle, \tag{5.23}$$

where functional $\tilde{R}[t, t_1; z(\tau)]$ is defined by the formula

$$\tilde{R}[t, t_1; z(\tau)] = R[t; z(\tau)\theta(t_1 - \tau + 0)]. \tag{5.24}$$

Let now quantity a be the random quantity with probability distribution density

$$p(a) = \frac{1}{2} \left[\delta(a - a_0) + \delta(a + a_0) \right]. \tag{5.25}$$

Then, an additional averaging of Eq. (5.23) over a yields the equality

$$\langle z(t)R[t;z(\tau)]\rangle = a_0^2 \int_0^t dt_1 e^{-2\nu(t-t_1)} \left\langle \frac{\delta}{\delta z(t_1)} \tilde{R}[t,t_1;z(\tau)] \right\rangle. \tag{5.26}$$

Formula (5.26) is very similar to the formula for splitting the correlator of the Gaussian process $z(t)$ characterized by the exponential correlation function (i.e., the Gaussian Markovian process) with functional $R[t;z(\tau)]$. The difference consists in the fact that the right-hand side of Eq. (5.26) depends on the functional $\tilde{R}[t,t_1;z(\tau)]$, which is cut by the process $z(\tau)$, rather than on functional $R[t;z(\tau)]$ itself.

Let us differentiate Eq. (5.23) with respect to time t. Taking into account the fact that there is no need in differentiating with respect to the upper limit of the integral in the right-hand side of Eq. (5.23), we obtain the expression [34, 35]

$$\left(\frac{d}{dt} + 2\nu \right) \langle z(t)R[t;z(\tau)]\rangle = \left\langle z(t) \frac{d}{dt} R[t;z(\tau)] \right\rangle \tag{5.27}$$

called usually the *differentiation formula*.

One essential point should be noticed. Functional $R[t;z(\tau)]$ in differentiation formula (5.27) is an arbitrary functional and can simply coincide with process $z(t-0)$. In the general case, the realization of telegrapher's process is the generalized function. The derivative of this process is also a generalized function (the sequence of delta-functions), so that

$$z(t) \frac{d}{dt} z(t) \neq \frac{1}{2} \frac{d}{dt} z^2(t) \equiv 0.$$

in the general case. These generalized functions, as any generalized functions, are defined only in terms of functionals constructed on them. In the case of our interest, such functionals are the average quantities denoted by angle brackets $\langle ... \rangle$, and the above differentiation formula describes the differential constraint between different functionals related to random process $z(t)$ and its one-sided derivatives for $t \to t - 0$, such as dz/dt, d^2z/dt^2. For example, formula (5.27) allows derivation of equalities, such as

$$\left\langle z(t) \frac{d}{d\tau} z(\tau) \right\rangle \bigg|_{\tau=t-0} = 2\nu \left\langle z^2 \right\rangle, \quad \left\langle z(t) \frac{d^2}{d\tau^2} z(\tau) \right\rangle \bigg|_{\tau=t-0} = 4\nu^2 \left\langle z^2 \right\rangle.$$

It is clear that these formulas can be obtained immediately by differentiating the correlation function $\langle z(t)z(t')\rangle$ with respect to t' $(t' < t)$ for $t' \to t - 0$.

Now, we dwell on some extensions of the above formulas for the vector Markovian process $\mathbf{z}(t) = \{z_1(t), ..., z_N(t)\}$. For example, all above Markovian processes having the exponential correlation function

$$\langle z(t)z(t+\tau)\rangle = \left\langle z^2 \right\rangle e^{-\alpha|\tau|} \tag{5.28}$$

satisfy the equality

$$\left(\frac{\partial}{\partial t} + \alpha k \right) \langle z_1(t)....z_k(t)R[t;\mathbf{z}(\tau)]\rangle = \left\langle z_1(t)....z_k(t) \frac{\partial}{\partial t} R[t;\mathbf{z}(\tau)] \right\rangle. \tag{5.29}$$

Formula (5.29) defines the rule of factoring the operation of differentiating with respect to time out of angle brackets of averaging; in particular, we have

$$\left\langle z_1(t)....z_k(t) \frac{\partial^n}{\partial t^n} R[t;\mathbf{z}(\tau)] \right\rangle = \left(\frac{\partial}{\partial t} + \alpha k \right)^n \langle z_1(t)....z_k(t)R[t;\mathbf{z}(\tau)]\rangle, \tag{5.30}$$

where $k = 1, ..., N$.

5.5   Delta-correlated random processes

Random processes $z(t)$ that can be treated as delta-correlated processes are of special interest in physics. The importance of this approximation follows first of all from the fact that it is physically obvious in the context of many physical problems and the corresponding dynamic systems allow obtaining closed equations for probability densities of the solutions to these systems.

For the Gaussian delta-correlated (in time) process, correlation function has the form

$$B(t_1, t_2) = \langle z(t_1)z(t_2) \rangle = B(t_1)\delta(t_1 - t_2), \quad (\langle z(t) \rangle = 0).$$

In this case, functionals $\Theta[t; v(\tau)]$, $\Omega[t', t; v(\tau)]$ and $\Omega[t, t; v(\tau)]$ (5.13), (5.14) introduced above are

$$\Theta[t; v(\tau)] = -\frac{1}{2}\int_0^t d\tau B(\tau)v^2(\tau),$$

$$\Omega[t', t; v(\tau)] = iB(t')v(t'), \quad \Omega[t, t; v(\tau)] = \frac{i}{2}B(t)v(t),$$

and Eqs. (5.15) appear significantly simpler and assume the forms

$$\langle z(t')R[t; v(\tau)] \rangle = B(t')\left\langle \frac{\delta}{\delta z(t')}R[t; v(\tau)] \right\rangle \quad (0 < t' < t),$$

$$\langle z(t)R[t; v(\tau)] \rangle = \frac{1}{2}B(t)\left\langle \frac{\delta}{\delta z(t)}R[t; v(\tau)] \right\rangle. \tag{5.31}$$

Formulas (5.31) show that statistical averages of the Gaussian delta-correlated process considered here are discontinuous at $t' = t$. This discontinuity is dictated entirely by the fact that this process is delta-correlated; if the process at hand is not delta-correlated, no discontinuity occurs (see Eq. (5.15)).

The Poisson delta-correlated random process corresponds to the limit process

$$g(t) \to \delta(t).$$

In this case, the logarithm of the characteristic functional has the simple form (see (4.40)); as a consequence, Eqs. (5.18) for functionals $\Omega[t', t; v(\tau)]$ and $\Omega[t, t; v(\tau)]$ assume the forms

$$\Omega[t', t; v(\tau)] = i\int_{-\infty}^{\infty} d\xi \xi p(\xi)e^{i\xi v(t')} \quad (t' < t),$$

$$\Omega[t, t; v(\tau)] = \frac{\nu}{iv(t)}\int_{-\infty}^{\infty} d\xi \xi p(\xi)\left[e^{i\xi v(t)} - 1\right] = \nu\int_{-\infty}^{\infty} d\xi p(\xi)\int_0^{\xi} d\eta e^{i\eta v(t)},$$

and we obtain the following expression for the correlation of Poisson random process $z(t)$ with a functional of this process

$$\langle z(t')R[t; z(\tau)] \rangle = \nu\int_{-\infty}^{\infty} d\xi \xi p(\xi)\langle R[t; z(\tau) + \xi\delta(\tau - \tau')] \rangle \quad (t' \leqslant t),$$

$$\langle z(t)R[t; z(\tau)] \rangle = \nu\int_{-\infty}^{\infty} d\xi p(\xi)\int_0^{\xi} d\eta \langle R[t; z(\tau) + \eta\delta(t - \tau)] \rangle. \tag{5.32}$$

These expressions also show that statistical averages are discontinuous at $t' = t$. As in the case of the Gaussian process, this discontinuity is dictated entirely by the fact that this process is delta-correlated.

In the general case of delta-correlated process $z(t)$, we can expand the logarithm of characteristic functional in the functional Taylor series

$$\Theta[t; v(\tau)] = \sum_{n=1}^{\infty} \frac{i^n}{n!} \int_0^t d\tau \, K_n(\tau) v^n(\tau), \qquad (5.33)$$

where cumulant functions assume the form

$$K_n(t_1, ..., t_n) = K_n(t_1)\delta(t_1 - t_2)...\delta(t_{n-1} - t_n).$$

As can be seen from Eq. (5.33), a characteristic feature of these processes consists in the validity of the equality

$$\dot{\Theta}[t; v(\tau)] = \dot{\Theta}[t; v(t)] \quad \left(\dot{\Theta}[t; v(\tau)] = \frac{d}{dt}\Theta[t; v(\tau)] \right), \qquad (5.34)$$

which is of fundamental significance. This equality shows that, in the case of arbitrary delta-correlated process, quantity $\Theta[t; v(\tau)]$ appears not a functional, but simply a function of time t. In this case, functionals $\Omega[t', t; v(\tau)]$ and $\Omega[t, t; v(\tau)]$ are

$$\Omega[t', t; v(\tau)] = \sum_{n=0}^{\infty} \frac{i^n}{n!} K_{n+1}(t') v^n(t') \quad (t' < t),$$

$$\Omega[t, t; v(\tau)] = \sum_{n=0}^{\infty} \frac{i^n}{(n+1)!} K_{n+1}(t) v^n(t),$$

and formulas for splitting correlations assume the forms

$$\langle z(t') R[t; z(\tau)] \rangle = \sum_{n=0}^{\infty} \frac{i^n}{n!} K_{n+1}(t') \left\langle \frac{\delta^n R[t; z(\tau)]}{\delta z^n(t')} \right\rangle \quad (t' < t),$$

$$\langle z(t') R[t; z(\tau)] \rangle = \sum_{n=0}^{\infty} \frac{i^n}{(n+1)!} K_{n+1}(t) \left\langle \frac{\delta^n R[t; z(\tau)]}{\delta z^n(t)} \right\rangle. \qquad (5.35)$$

These formulas describe the discontinuity of statistical averages at $t' = t$ in the general case of delta-correlated processes.

Note that, for $t' > t$, delta-correlated processes satisfy the obvious equality

$$\langle z(t') R[t; z(\tau)] \rangle = \langle z(t') \rangle \, \langle R[t; z(\tau)] \rangle. \qquad (5.36)$$

Now, we dwell on the concept of random delta-correlated (in time) fields.

We will deal with vector field $\mathbf{f}(\mathbf{x}, t)$, where $\mathbf{x}$ describes the spatial coordinates and t is the temporal coordinate. In this case, the logarithm of characteristic functional can be expanded in the Taylor series with coefficients expressed in terms of cumulant functions of random field $\mathbf{f}(\mathbf{x}, t)$ (see Sect. 4.2). In the special case of cumulant functions

$$K_n^{i_1, ..., i_m}(\mathbf{x}_1, t_1; ..., \mathbf{x}_n, t_n) = K_n^{i_1, ..., i_m}(\mathbf{x}_1, ..., \mathbf{x}_n; t_1)\delta(t_1 - t_2)...\delta(t_{n-1} - t_n), \qquad (5.37)$$

we will call field $\mathbf{f}(\mathbf{x}, t)$ the random field delta-correlated in time t. In this case, functional $\Theta[t; \psi(\mathbf{x}', \tau)]$ assumes the form

$$\Theta[t; \psi(\mathbf{x}', \tau)]$$
$$= \sum_{n=1}^{\infty} \frac{i^n}{n!} \int\limits_0^t d\tau \int \ldots \int d\mathbf{x}_1 \ldots d\mathbf{x}_n K_n^{i_1, \ldots, i_m}(\mathbf{x}_1, \ldots, \mathbf{x}_n; \tau) \psi_{i_1}(\mathbf{x}_1, \tau) \ldots \psi_{i_m}(\mathbf{x}_n, \tau).$$

$$(5.38)$$

An important feature of this functional is the fact that it satisfies the equality similar to Eq. (5.34):

$$\dot\Theta[t; \psi(\mathbf{x}', \tau)] = \dot\Theta[t; \psi(\mathbf{x}', t)].$$

$$(5.39)$$

5.5.1 Asymptotic meaning of delta-correlated processes and fields

The nature knows no delta-correlated processes. All actual processes and fields are characterized by a finite temporal correlation radius, and delta-correlated processes and fields appear as a result of asymptotic expansions in temporal correlation radii.

We illustrate the appearance of delta-correlated processes using the stationary Gaussian process with correlation radius τ_0 as an example. In this case, the logarithm of the characteristic functional is described by the expression

$$\Theta[t; v(\tau)] = - \int\limits_0^t d\tau_1 v(\tau_1) \int\limits_0^{\tau_1} d\tau_2 B\left(\frac{\tau_1 - \tau_2}{\tau_0}\right) v(\tau_2).$$

$$(5.40)$$

Setting $\tau_1 - \tau_2 = \xi\tau_0$, we transform Eq. (5.40) to the form

$$\Theta[t; v(\tau)] = -\tau_0 \int\limits_0^t d\tau_1 v(\tau_1) \int\limits_0^{\tau_1/\tau_0} d\xi B(\xi) v(\tau_1 - \xi\tau_0).$$

Assume now that $\tau_0 \to 0$. In this case, the leading term of an asymptotic expansion in parameter τ_0 is given by the formula

$$\Theta[t; v(\tau)] = -\tau_0 \int\limits_0^{\infty} d\xi B(\xi) \int\limits_0^t d\tau_1 v^2(\tau_1)$$

that can be represented in the form

$$\Theta[t; v(\tau)] = -B^{\text{eff}} \int\limits_0^t d\tau_1 v^2(\tau_1),$$

$$(5.41)$$

where

$$B^{\text{eff}} = \int\limits_0^{\infty} d\tau B\left(\frac{\tau}{\tau_0}\right) = \frac{1}{2} \int\limits_{-\infty}^{\infty} d\tau B\left(\frac{\tau}{\tau_0}\right).$$

$$(5.42)$$

Certainly, asymptotic expression (5.41) holds only for functions $v(t)$ that slightly vary during times about τ_0, rather than for arbitrary functions $v(t)$. Indeed, if we specify this function as $v(t) = v\delta(t - t_0)$, asymptotic expression (5.41) appears invalid; in this case, we must replace Eq. (5.40) with the expression

$$\Theta[t; v(\tau)] = -\frac{1}{2}B(0)v^2 \quad (t > t_0)$$

corresponding to the characteristic function of process $z(t)$ at a fixed time $t = t_0$.

Consider now correlation $\langle z(t)R[t; z(\tau)]\rangle$. According to the Furutsu–Novikov formula (5.15), it is given by the expression

$$\langle z(t)R[t; z(\tau)]\rangle = \int_0^t dt_1 B\left(\frac{t - t_1}{\tau_0}\right)\left\langle \frac{\delta}{\delta z(t_1)}R[t; z(\tau)]\right\rangle.$$

The change of integration variable $t - t_1 \to \xi\tau_0$ transforms this expression to the form

$$\langle z(t)R[t; z(\tau)]\rangle = \tau_0 \int_0^{t/\tau_0} d\xi B(\xi)\left\langle \frac{\delta}{\delta z(t - \xi\tau_0)}R[t; z(\tau)]\right\rangle, \tag{5.43}$$

which grades for $\tau_0 \to 0$ into the equality obtained earlier for the Gaussian delta-correlated process

$$\langle z(t)R[t; z(\tau)]\rangle = B^{\text{eff}}\left\langle \frac{\delta}{\delta z(t)}R[t; z(\tau)]\right\rangle$$

provided that the variational derivative in Eq. (5.43) varies only slightly during times about τ_0.

Thus, the approximation of process $z(t)$ by the delta-correlated one is conditioned by small variations of functionals of this process during times of about its temporal correlation radius.

Consider now telegrapher's and generalized telegrapher's processes. In the case of telegrapher's process, the characteristic functional satisfies Eq. (4.46). The correlation radius of this process is $\tau_0 = 1/2\nu$, and, for $\nu \to \infty$ ($\tau_0 \to 0$), this equation grades for sufficiently smooth functions $v(t)$ into the equation

$$\frac{d}{dt}\Phi[t; v(\tau)] = -\frac{a_0^2}{2\nu}v^2(t)\Phi[t; v(\tau)], \tag{5.44}$$

which corresponds to the Gaussian delta-correlated process. If we additionally assume that $a_0^2 \to \infty$ and

$$\lim_{\nu \to \infty} a_0^2/2\nu = \sigma_0^2,$$

then Eq. (5.44) appears independent of parameter ν. This fact does not mean of cause that telegrapher's process looses its telegrapher's properties for $\nu \to \infty$. Indeed, for $\nu \to \infty$, the one-point probability distribution of process $z(t)$ will as before correspond to telegrapher's process, i.e., to the process with two possible states. As regards the correlation function and higher-order moment functions, they will possess for $\nu \to \infty$ all properties of delta-functions in view of the fact that

$$\lim_{\nu \to \infty}\left\{2\nu e^{-2\nu|\tau|}\right\} = \begin{cases} 0, & \text{if } \tau \neq 0, \\ \infty, & \text{if } \tau = 0. \end{cases}$$

Such functions should be considered the generalized functions; their delta-functional behavior will manifest itself in the integrals of them. As Eq. (5.44) shows, the limit process

$\nu \to \infty$ is equivalent for these quantities to the replacement of process $z(t)$ by the Gaussian delta-correlated process. This situation is completely similar to the approximation of the Gaussian random process with a finite correlation radius τ_0 by the delta-correlated process for $\tau_0 \to 0$.

Problems

Problem 9 *Calculate statistical average* $\langle g(\xi)f(\xi)\rangle_\xi$, *where* ξ *is the random quantity.*

Solution.

$$\langle g(\xi + \eta_1)f(\xi + \eta_2)\rangle_\xi$$
$$= \exp\left\{\Theta\left(\frac{d}{id\eta_1} + \frac{d}{id\eta_2}\right) - \Theta\left(\frac{d}{id\eta_1}\right) - \Theta\left(\frac{d}{id\eta_2}\right)\right\} \langle g(\xi + \eta_1)\rangle \langle f(\xi + \eta_2)\rangle.$$

Problem 10 *Calculate statistical average* $\left\langle e^{\omega\xi}f(\xi)\right\rangle_\xi$, *where parameter* ω *can assume complex values.*

Solution.

$$\left\langle e^{\omega\xi}f(\xi + \eta)\right\rangle_\xi = \exp\left\{\Theta\left(\frac{1}{i}\left(\omega + \frac{d}{d\eta}\right)\right) - \Theta\left(\frac{d}{id\eta}\right)\right\} \langle f(\xi + \eta)\rangle_\xi.$$

Problem 11 *Calculate statistical averages* $\left\langle e^{\omega\xi}\right\rangle_\xi$ *and* $\left\langle \xi e^{\omega\xi}\right\rangle_\xi$ *assuming that* ξ *is the Gaussian random quantity with zero-valued mean.*

Solution.

$$\left\langle e^{\omega\xi}\right\rangle_\xi = \exp\left\{\frac{\omega^2\sigma^2}{2}\right\}, \quad \left\langle \xi e^{\omega\xi}\right\rangle_\xi = \omega\sigma^2 \exp\left\{\frac{\omega^2\sigma^2}{2}\right\}.$$

Problem 12 *Calculate statistical average* $\left\langle e^{\omega\xi}f(\xi)\right\rangle_\xi$ *assuming that* ξ *is the Gaussian random quantity with zero-valued mean.*

Solution.

$$\left\langle e^{\omega\xi}f(\xi)\right\rangle_\xi = \exp\left\{\frac{\omega^2\sigma^2}{2}\right\} \left\langle f(z + \omega\sigma^2)\right\rangle.$$

Problem 13 *Split the correlator* $\langle F[z(\tau)]R[z(\tau)]\rangle$.

Solution.

$$\langle F[z(\tau) + \eta_1(\tau)]R[z(\tau) + \eta_2(\tau)]\rangle$$
$$= \left\langle \exp\left\{\Theta\left[\frac{1}{i}\left(\frac{\delta}{\delta\eta_1(\tau)} + \frac{\delta}{\delta\eta_2(\tau)}\right)\right] - \Theta\left[\frac{1}{i}\frac{\delta}{\delta\eta_1(\tau)}\right] - \Theta\left[\frac{1}{i}\frac{\delta}{\delta\eta_2(\tau)}\right]\right\}\right\rangle$$
$$\times \langle F[z(\tau) + \eta_1(\tau)]\rangle \langle R[z(\tau) + \eta_2(\tau)]\rangle,$$

where $\eta_1(\tau)$ and $\eta_2(\tau)$ are arbitrary deterministic functions.

Problem 14 *Split the correlator $\langle F[z(\tau)]R[z(\tau)]\rangle$ assuming that $z(t)$ is the Gaussian random process.*

Solution.

$$\langle F[z(\tau) + \eta_1(\tau)]R[z(\tau) + \eta_2(\tau)]\rangle$$

$$= \exp\left\{\int\limits_{-\infty}^{\infty}\int\limits_{-\infty}^{\infty} d\tau_1 d\tau_2 B(\tau_1,\tau_2)\frac{\delta}{\delta\eta_1(\tau_1)}\frac{\delta}{\delta\eta_2(\tau_2)}\right\}$$

$$\times \langle F[z(\tau) + \eta_1(\tau)]\rangle\,\langle R[z(\tau) + \eta_2(\tau)]\rangle,$$

or

$$\langle F[z(\tau)]R[z(\tau) + \eta(\tau)]\rangle$$

$$= F\left[z(\tau) + \int\limits_{-\infty}^{\infty} d\tau_1 B(\tau,\tau_1)\frac{\delta}{\delta\eta(\tau_1)}\right]\langle R[z(\tau) + \eta(\tau)]\rangle.$$

Problem 15 *Split the correlator $\left\langle e^{i\int\limits_{-\infty}^{\infty} d\tau z(\tau)v(\tau)} R[z(\tau)]\right\rangle$ assuming that $z(t)$ is the Gaussian random process.*

Solution.

$$\left\langle e^{i\int\limits_{-\infty}^{\infty} d\tau z(\tau)v(\tau)} R[z(\tau)]\right\rangle$$

$$= \Phi[v(\tau)]\left\langle R\left[z(\tau) + i\int\limits_{-\infty}^{\infty} d\tau_1 B(\tau,\tau_1)v(\tau_1)\right]\right\rangle.$$

Problem 16 *Split the correlator $\left\langle e^{i\int\limits_{0}^{t} d\tau z(\tau)v(\tau)} R[z(\tau)]\right\rangle$ assuming that $z(t)$ is the Gaussian random process.*

Solution.

$$\left\langle e^{i\int\limits_{0}^{t} d\tau z(\tau)v(\tau)} R[z(\tau)]\right\rangle$$

$$= \Phi[t; v(\tau)]\left\langle R\left[t; z(\tau) + i\int\limits_{0}^{t} d\tau_1 B(\tau,\tau_1)v(\tau_1)\right]\right\rangle,$$

where $\Phi[t; v(\tau)]$ is the characteristic functional of the Gaussian random process $z(t)$.

Problem 17 *Calculate the characteristic functional of process $z(t) = \xi^2(t)$, where $\xi(t)$ is the Gaussian random process with the parameters*

$$\langle\xi(t)\rangle = 0, \quad \langle\xi(t_1)\xi(t_2)\rangle = B(t_1 - t_2).$$

Instruction. The characteristic functional

$$\Phi[t; v(\tau)] = \langle \varphi[t; \xi(\tau)] \rangle, \quad \text{where} \quad \varphi[t; \xi(\tau)] = \exp\left\{ i \int_0^t d\tau v(\tau)\xi^2(\tau) \right\}$$

satisfies the stochastic equation

$$\frac{d}{dt}\Phi[t; v(\tau)] = iv(t) \left\langle \xi^2(t)\varphi[t; \xi(\tau)] \right\rangle.$$

One needs to obtain the integral equation for quantity $\Psi(t_1, t) = \langle \xi(t_1)\xi(t)\varphi[t; \xi(\tau)] \rangle$,

$$\Psi(t_1, t) = B(t_1 - t)\Phi[t; v(\tau)] + 2i \int_0^t d\tau B(t_1 - \tau)v(\tau)\Psi(\tau, t),$$

and represent function $\Psi(t_1, t)$ in the form

$$\Psi(t_1, t) = S(t_1, t)\Phi[t; v(\tau)].$$

Solution. The characteristic functional $\Phi[t; v(\tau)]$ has the form

$$\Phi[t; v(\tau)] = \exp\left\{ i \int_0^t d\tau v(\tau)S(\tau, \tau) \right\},$$

where

$$S(t, t) = \sum_{n=0}^{\infty} S^{(n)}(t, t), \qquad S^{(n)}(t, t)$$

$$= (2i)^n \int_0^t \dots \int_0^t d\tau_1 \dots d\tau_n v(\tau_1) \dots v(\tau_n) B(t - \tau_1)B(\tau_1 - \tau_2) \dots B(\tau_n - t).$$

Consequently, the n-th cumulant of process $z(t) = \xi^2(t)$ is expressed in terms of quantity $S^{(n-1)}(t, t)$.

Problem 18 *Calculate the characteristic functional of process* $z(t) = \xi^2(t)$, *where* $\xi(t)$ *is the Gaussian Markovian process with the parameters*

$$\langle \xi(t) \rangle = 0, \quad \langle \xi(t_1)\xi(t_2) \rangle = B(t_1 - t_2) = \sigma^2 \exp\{-\alpha|t_1 - t_2|\}.$$

Solution. The characteristic functional $\Phi[t; v(\tau)]$ has the form

$$\Phi[t; v(\tau)] = \exp\left\{ i \int_0^t d\tau v(\tau)S(\tau, \tau) \right\},$$

where function $S(t, t)$ can be described in terms of the initial-value problem

$$\frac{d}{dt}S(t, t) = -2\alpha \left[S(t, t) - \sigma^2 \right] + 2iv(t)S^2(t, t), \quad S(t, t)|_{t=0} = \sigma^2.$$

Problem 19 *Show that the expression*

$$\langle z(t_1)z(t_2)R[z(\tau)]\rangle = \langle z(t_1)z(t_2)\rangle \langle R[z(\tau)]\rangle,$$

holds for arbitrary functional $R[z(\tau)]$ and $\tau \le t_2 \le t_1$ under the condition that $z(t)$ is telegrapher's process [36].

Instruction. Expand functional $R[z(\tau)]$ in the Taylor series in $z(\tau)$ and use formulas (4.43).

Problem 20 *Show that the expression [36]*

$$\langle F[z(\tau_1)]z(t_1)z(t_2)R[z(\tau_2)]\rangle$$
$$= \langle F[z(\tau_1)]\rangle \langle z(t_1)z(t_2)\rangle \langle R[z(\tau_2)]\rangle + \langle F[z(\tau_1)]z(t_1)\rangle \langle z(t_2)R[z(\tau_2)]\rangle,$$

holds for any $\tau_1 \le t_1 \le t_2 \le \tau_2$ and arbitrary functionals $F[z(\tau_1)]$ and $R[z(\tau_2)]$ under the condition that $z(t)$ is telegrapher's process with random parameter a distributed according to the probability density

$$p(a) = \frac{1}{2}\left[\delta(a - a_0) + \delta(a + a_0)\right].$$

Instruction. In terms of the Taylor expansion in $z(\tau)$, functional $R[z(\tau_2)]$ can be represented in the form

$$R[z(\tau_2)] = \sum_{2k} + \sum_{2k+1},$$

where the first sum consists of terms with the even number of co-factors $z(\tau)$ and the second sum consists of terms with the odd number of such co-factors. The desired result immediately follows from the fact that

$$\langle R[z(\tau_2)]\rangle = \left\langle \sum_{2k}\right\rangle, \quad \langle z(t_2)R[z(\tau_2)]\rangle = \left\langle z(t_2)\sum_{2k+1}\right\rangle.$$

Problem 21 *Show that the equality*

$$\langle z(t')R[t; z(\tau)]\rangle = e^{-2\nu(t'-t)}\langle z(t)R[t; z(\tau)]\rangle \quad (t' \ge t)$$

holds for telegrapher's process $z(t)$.

Instruction. Rewrite the correlator $\langle z(t')R[t; z(\tau)]\rangle$ $(t' \ge t, \tau < t)$ in the form

$$\langle z(t)R[t; z(\tau)]\rangle = \frac{1}{a_0^2}\langle z(t')z(t)z(t)R[t; z(\tau)]\rangle.$$

Chapter 6

General approaches to analyzing stochastic dynamic systems

In this chapter, we will consider basic methods of determining statistical characteristics of solutions to the stochastic equations.

Consider for example a linear (differential, integro-differential, or integral) stochastic equation. Averaging of such an equation with fluctuating parameters over an ensemble of realizations will not result generally in a closed equation for the corresponding average value. To obtain the closed equation, we must deal with an additional extended space whose dimension appears infinite in most cases. This approach will allow us to obtain for the average quantity of interest the linear equation containing variational derivatives.

Consider some special types of dynamic systems.

6.1 Ordinary differential equations

Let dynamics of vector function $\mathbf{x}(t)$ is described by the ordinary differential equation

$$\frac{d}{dt}\mathbf{x}(t) = \mathbf{v}(\mathbf{x}, t) + \mathbf{f}(\mathbf{x}, t), \quad \mathbf{x}(t_0) = \mathbf{x}_0. \tag{6.1}$$

Here, function $\mathbf{v}(\mathbf{x}, t)$ is the deterministic function and $\mathbf{f}(\mathbf{x}, t)$ is the random function.

The solution to Eq. (6.1) is a functional of $\mathbf{f}(\mathbf{y}, \tau) + \mathbf{v}(\mathbf{y}, \tau)$ with $\tau \in (t_0, t)$, i.e.,

$$\mathbf{x}(t) = \mathbf{x}[t; \mathbf{f}(\mathbf{y}, \tau) + \mathbf{v}(\mathbf{y}, \tau)].$$

From this fact follows the equality

$$\frac{\delta}{\delta f_j(\mathbf{y}, \tau)} F(\mathbf{x}(t)) = \frac{\delta}{\delta v_j(\mathbf{y}, \tau)} F(\mathbf{x}(t)) = \frac{\partial F(\mathbf{x}(t))}{\partial x_i} \frac{\delta x_i(t)}{\delta f_j(\mathbf{y}, \tau)},$$

valid for arbitrary function $F(\mathbf{x})$. In addition, we have

$$\frac{\delta}{\delta f_j(\mathbf{y}, t - 0)} x_i(t) = \frac{\delta}{\delta v_j(\mathbf{y}, t - 0)} x_i(t) = \delta_{ij} \delta\left(\mathbf{x}(t) - \mathbf{y}\right).$$

The corresponding Liouville equation for the indicator function $\varphi(\mathbf{x}, t) = \delta(\mathbf{x}(t) - \mathbf{x})$ follows from Eq. (6.1) and has the form

$$\frac{\partial}{\partial t}\varphi(\mathbf{x}, t) = -\frac{\partial}{\partial \mathbf{x}}\left\{[\mathbf{v}(\mathbf{x}, t) + \mathbf{f}(\mathbf{x}, t)]\varphi(\mathbf{x}, t)\right\}, \quad \varphi(\mathbf{x}, t_0) = \delta(\mathbf{x} - \mathbf{x}_0), \tag{6.2}$$

from which follows the equality

$$\frac{\delta}{\delta \mathbf{f}(\mathbf{y}, t - 0)} \varphi(\mathbf{x}, t) = \frac{\delta}{\delta \mathbf{v}(\mathbf{y}, t - 0)} \varphi(\mathbf{x}, t) = -\frac{\partial}{\partial \mathbf{x}} \left\{ \delta(\mathbf{x} - \mathbf{y}) \varphi(\mathbf{x}, t) \right\}. \tag{6.3}$$

Using this equality, we can rewrite Eq. (6.2) in the form, which may look at first sight more complicated

$$\left(\frac{\partial}{\partial t} + \frac{\partial}{\partial \mathbf{x}} \mathbf{v}(\mathbf{x}, t) \right) \varphi(\mathbf{x}, t) = \int d\mathbf{y} \mathbf{f}(\mathbf{y}, t) \frac{\delta}{\delta \mathbf{v}(\mathbf{y}, t)} \varphi(\mathbf{x}, t). \tag{6.4}$$

Consider now the one-time probability density for solution $\mathbf{x}(t)$ of Eq. (6.1)

$$P(\mathbf{x}, t) = \langle \varphi(\mathbf{x}, t) \rangle = \langle \delta(\mathbf{x}(t) - \mathbf{x}) \rangle.$$

Here, $\mathbf{x}(t)$ is the solution of Eq. (6.1) corresponding to the particular realization of random field $\mathbf{f}(\mathbf{x}, t)$, and angle brackets $\langle ... \rangle$ denote averaging over an ensemble of realizations of field $\mathbf{f}(\mathbf{x}, t)$.

Averaging Eq. (6.4) over an ensemble of realizations of field $\mathbf{f}(\mathbf{x}, t)$, we obtain the expression

$$\left(\frac{\partial}{\partial t} + \frac{\partial}{\partial \mathbf{x}} \mathbf{v}(\mathbf{x}, t) \right) P(\mathbf{x}, t) = \int d\mathbf{y} \frac{\delta}{\delta \mathbf{v}(\mathbf{y}, t)} \langle \mathbf{f}(\mathbf{y}, t) \varphi(\mathbf{x}, t) \rangle. \tag{6.5}$$

Quantity $\langle \mathbf{f}(\mathbf{y}, t) \varphi(\mathbf{x}, t) \rangle$ in the right-hand side of Eq. (6.5) is the correlation between random field $\mathbf{f}(\mathbf{y}, t)$ and function $\varphi(\mathbf{x}, t)$, which is a functional of random field $\mathbf{f}(\mathbf{y}, \tau)$ and is described either by Eq. (6.2), or by Eq. (6.4).

The characteristic functional

$$\Phi[t, t_0; \mathbf{u}(\mathbf{y}, \tau)] = \left\langle \exp \left\{ i \int_{t_0}^{t} d\tau \int d\mathbf{y} \mathbf{f}(\mathbf{y}, \tau) \mathbf{u}(\mathbf{y}, \tau) \right\} \right\rangle = \exp \left\{ \Theta[t, t_0; \mathbf{u}(\mathbf{y}, \tau)] \right\}$$

exhaustively describes all statistical characteristics of random field $\mathbf{f}(\mathbf{y}, \tau)$ for $\tau \in (t_0, t)$.

We split correlator $\langle \mathbf{f}(\mathbf{y}, t) \varphi(\mathbf{x}, t) \rangle$ using the technique of functionals. Introducing functional shift operator with respect to field $\mathbf{v}(\mathbf{y}, \tau)$, we represent functional $\varphi[t, \mathbf{x}; \mathbf{f}(\mathbf{y}, \tau) + \mathbf{v}(\mathbf{y}, \tau)]$ in the operator form

$$\varphi[t, \mathbf{x}; \mathbf{f}(\mathbf{y}, \tau) + \mathbf{v}(\mathbf{y}, \tau)] = \exp \left[\int_{t_0}^{t} d\tau \int d\mathbf{y} \mathbf{f}(\mathbf{y}, \tau) \frac{\delta}{\delta \mathbf{v}(\mathbf{y}, \tau)} \right] \varphi[t, \mathbf{x}; \mathbf{v}(\mathbf{y}, \tau)].$$

With this representation, the term in the right-hand side of Eq. (6.5) assumes the form

$$\int d\mathbf{y} \frac{\delta}{\delta v_j(\mathbf{y}, t)} \frac{\left\langle f_j(\mathbf{y}, t) \exp \left\{ \int_{t_0}^{t} d\tau \int d\mathbf{y}' \mathbf{f}(\mathbf{y}', \tau) \frac{\delta}{\delta \mathbf{v}(\mathbf{y}', \tau)} \right\} \right\rangle}{\left\langle \exp \left\{ \int_{t_0}^{t} d\tau \int d\mathbf{y}' \mathbf{f}(\mathbf{y}', \tau) \frac{\delta}{\delta \mathbf{v}(\mathbf{y}', \tau)} \right\} \right\rangle} P(\mathbf{x}, t)$$

$$= \dot{\Theta}_t \left[t, t_0; \frac{\delta}{i\delta \mathbf{v}(\mathbf{y}, \tau)} \right] P(\mathbf{x}, t),$$

where we introduced the functional

$$\dot{\Theta}_t[t, t_0; \mathbf{u}(\mathbf{y}, \tau)] = \frac{d}{dt} \ln \Phi[t, t_0; \mathbf{u}(\mathbf{y}, \tau)].$$

Consequently, we can rewrite Eq. (6.5) in the form

$$\left(\frac{\partial}{\partial t} + \frac{\partial}{\partial \mathbf{x}}\mathbf{v}(\mathbf{x},t)\right) P(\mathbf{x},t) = \dot{\Theta}_t\left[t,t_0; \frac{\delta}{i\delta\mathbf{v}(\mathbf{y},\tau)}\right] P(\mathbf{x},t). \tag{6.6}$$

Equation (6.6) is the closed equation with variational derivatives in the functional space of all possible functions $\{\mathbf{v}(\mathbf{y},\tau)\}$. However, for a fixed function $\mathbf{v}(\mathbf{x},t)$, we arrive at unclosed equation

$$\begin{aligned}
\left(\frac{\partial}{\partial t} + \frac{\partial}{\partial \mathbf{x}}\mathbf{v}(\mathbf{x},t)\right) P(\mathbf{x},t) &= \left\langle \dot{\Theta}_t\left[t,t_0; \frac{\delta}{i\delta\mathbf{f}(\mathbf{y},t)}\right] \varphi[t,\mathbf{x};\mathbf{f}(\mathbf{y},\tau)]\right\rangle, \\
P(\mathbf{x},t_0) &= \delta(\mathbf{x}-\mathbf{x}_0).
\end{aligned} \tag{6.7}$$

Equation (6.7) is the exact consequence of the initial dynamic equation (6.1). Statistical characteristics of random field $\mathbf{f}(\mathbf{x},t)$ appear in this equation only through functional $\Theta_t[t,t_0;\mathbf{u}(\mathbf{y},\tau)]$ whose expansion in the functional Taylor series in powers of $\mathbf{u}(\mathbf{x},\tau)$ depends on all space-time cumulant functions of random field $\mathbf{f}(\mathbf{x},t)$.

As we mentioned earlier, Eq. (6.7) is not closed with respect to function $P(\mathbf{x},t)$ in the general case, because quantity

$$\dot{\Theta}_t\left[t,t_0; \frac{\delta}{i\delta\mathbf{f}(\mathbf{y},\tau)}\right] \delta(\mathbf{x}(t)-\mathbf{x})$$

appearing in averaging brackets depends on the solution $\mathbf{x}(t)$ (which is a functional of random field $\mathbf{f}(\mathbf{y},\tau)$) for all times $t_0 < \tau < t$. However, in some cases, the variational derivative in Eq. (6.7) can be expressed in terms of ordinary differential operators. In such conditions, equations like Eq. (6.7) will be the closed equations for the corresponding probability densities. The corresponding examples will be given below.

Proceeding in a similar way, one can obtain the equation similar to Eq. (6.7) for the m-time probability density that refers to m different instants $t_1 < t_2 < ... < t_m$

$$P_m(\mathbf{x}_1,t_1;...;\mathbf{x}_m,t_m) = \langle\varphi_m(\mathbf{x}_1,t_1;...;\mathbf{x}_m,t_m)\rangle, \tag{6.8}$$

where the indicator function is defined by the equality

$$\varphi_m(\mathbf{x}_1,t_1;...;\mathbf{x}_m,t_m) = \delta(\mathbf{x}(t_1)-\mathbf{x}_1)...\delta(\mathbf{x}(t_m)-\mathbf{x}_m).$$

Differentiating Eq. (6.8) with respect to time t_m and using then dynamic equation (6.1), one can obtain the equation

$$\begin{aligned}
&\left(\frac{\partial}{\partial t_m} + \frac{\partial}{\partial \mathbf{x}_m}\mathbf{v}(\mathbf{x}_m,t_m)\right) P_m(\mathbf{x}_1,t_1;...;\mathbf{x}_m,t_m) \\
&= \left\langle \dot{\Theta}_{t_m}\left[t_m,t_0; \frac{\delta}{i\delta\mathbf{f}(\mathbf{y},\tau)}\right] \varphi_m(\mathbf{x}_1,t_1;...;\mathbf{x}_m,t_m)\right\rangle.
\end{aligned} \tag{6.9}$$

No summation over index m is performed here. The initial condition to Eq. (6.9) can be derived from Eq. (6.8). Setting $t_m = t_{m-1}$ in Eq. (6.8), we obtain the equality

$$P_m(\mathbf{x}_1,t_1;...;\mathbf{x}_m,t_{m-1}) = \delta(\mathbf{x}_m-\mathbf{x}_{m-1})P_{m-1}(\mathbf{x}_1,t_1;...;\mathbf{x}_{m-1},t_{m-1}),$$

which just determines the initial condition for Eq. (6.9).

6.2 Completely solvable stochastic dynamic systems

Consider now several dynamic systems that allow adequate statistical analysis for arbitrary random parameters.

6.2.1 Ordinary differential equations

Multiplicative action

As the first example, we consider the vector stochastic equation with initial condition

$$\frac{d}{dt}\mathbf{x}(t) = z(t)g(t)\mathbf{F}(\mathbf{x}), \quad \mathbf{x}(0) = \mathbf{x}_0, \tag{6.10}$$

where $g(t)$ and $F_i(\mathbf{x})$, $i = 1, ..., N$, are the deterministic functions and $z(t)$ is the random process whose statistical characteristics are described by the characteristic functional

$$\Phi[t; v(\tau)] = \left\langle \exp\left\{ i \int_0^t d\tau z(\tau)v(\tau) \right\} \right\rangle = e^{\Theta[t;v(\tau)]}.$$

Equation (6.10) has a feature that offers a possibility of determining statistical characteristics of its solutions in the general case of arbitrarily distributed process $z(t)$. The point is that introduction of new 'random' time

$$T = \int_0^t d\tau z(\tau)g(\tau)$$

reduces Eq. (6.10) to the equation formally looking deterministic

$$\frac{d}{dT}\mathbf{x}(T) = \mathbf{F}(\mathbf{x}), \quad \mathbf{x}(0) = \mathbf{x}_0,$$

so that the solution to Eq. (6.10) has the following structure

$$\mathbf{x}(t) = \mathbf{x}(T) = \mathbf{x}\left(\int_0^t d\tau z(\tau)g(\tau) \right). \tag{6.11}$$

Varying Eq. (6.11) with respect to $z(\tau)$ and using Eq. (6.10), we obtain the equality

$$\frac{\delta}{\delta z(\tau)}\mathbf{x}(t) = g(\tau)\frac{d}{dT}\mathbf{x}(T) = g(\tau)\mathbf{F}(\mathbf{x}(t)). \tag{6.12}$$

Thus, variational derivatives of solution $\mathbf{x}(t)$ is expressed in terms of the same solution at the same time. This fact makes it possible to immediately write the closed equations for statistical characteristics of problem (6.10).

Let us derive the equation for the one-point probability density $P(\mathbf{x}, t) = \langle \delta(\mathbf{x}(t) - \mathbf{x}) \rangle$. It has the form

$$\frac{\partial}{\partial t}P(\mathbf{x}, t) = \left\langle \dot{\Theta}_t \left[t; \frac{\delta}{i\delta z(\tau)} \right] \delta\left(\mathbf{x}(t) - \mathbf{x}\right) \right\rangle. \tag{6.13}$$

Consider now the result of operator $\delta/\delta z(\tau)$ applied to the indicator function $\varphi(\mathbf{x}, t) = \delta(\mathbf{x}(t) - \mathbf{x})$. Using formula (6.12), we obtain the expression

$$\frac{\delta}{\delta z(\tau)}\delta\left(\mathbf{x}(t) - \mathbf{x}\right) = -g(\tau)\frac{\partial}{\partial \mathbf{x}}\left\{\mathbf{F}(\mathbf{x})\varphi(\mathbf{x}, t)\right\}.$$

Consequently, we can rewrite Eq. (6.13) in the form of the closed operator equation

$$\frac{\partial}{\partial t}P(\mathbf{x}, t) = \dot{\Theta}_t\left[t; ig(\tau)\frac{\partial}{\partial \mathbf{x}}\mathbf{F}(\mathbf{x})\right]P(\mathbf{x}, t), \quad P(\mathbf{x}, 0) = \delta(\mathbf{x} - \mathbf{x}_0), \tag{6.14}$$

whose particular form depends on the behavior of process $z(t)$.

For the two-time probability density

$$P(\mathbf{x}, t; \mathbf{x}_1, t_1) = \langle\delta(\mathbf{x}(t) - \mathbf{x})\delta(\mathbf{x}(t_1) - \mathbf{x}_1)\rangle$$

we obtain similarly the equation (for $t > t_1$)

$$\frac{\partial}{\partial t}P(\mathbf{x}, t; \mathbf{x}_1, t_1) = \dot{\Theta}_t\left[t; ig(\tau)\left\{\frac{\partial}{\partial \mathbf{x}}\mathbf{F}(\mathbf{x}) + \theta(t_1 - \tau)\frac{\partial}{\partial \mathbf{x}_1}\mathbf{F}(\mathbf{x}_1)\right\}\right]P(\mathbf{x}, t; \mathbf{x}_1, t_1) \tag{6.15}$$

with the initial condition

$$P(\mathbf{x}, t_1; \mathbf{x}_1, t_1) = \delta(\mathbf{x} - \mathbf{x}_1)P(\mathbf{x}_1, t_1),$$

where function $P(\mathbf{x}_1, t_1)$ satisfies Eq. (6.14).

One can see from Eq. (6.15) that multidimensional probability density cannot be factorized in terms of the transition probability (see Sect. 4.3), so that process $\mathbf{x}(t)$ is not the Markovian process. The particular forms of Eqs. (6.14) and (6.15) is governed by the statistics of process $z(t)$.

If $z(t)$ is the Gaussian process whose mean value and correlation function are

$$\langle z(t)\rangle = 0, \quad B(t, t') = \langle z(t)z(t')\rangle,$$

then functional $\Theta[t; v(\tau)]$ has the form

$$\Theta[t; v(\tau)] = -\frac{1}{2}\int\limits_0^t dt_1\int\limits_0^t dt_2 B(t_1, t_2)v(t_1)v(t_2),$$

and Eq. (6.14) assumes the following form

$$\frac{\partial}{\partial t}P(\mathbf{x}, t) = g(t)\int\limits_0^t d\tau\, B(t, \tau)g(\tau)\frac{\partial}{\partial x_j}F_j(\mathbf{x})\frac{\partial}{\partial x_k}F_k(\mathbf{x})P(\mathbf{x}, t) \tag{6.16}$$

and can be considered as the *extended Fokker–Planck equation.*

The class of problems formulated in terms of the system of equations

$$\frac{d}{dt}\mathbf{x}(t) = z(t)\mathbf{F}(\mathbf{x}) - \lambda\mathbf{x}(t), \quad \mathbf{x}(0) = \mathbf{x}_0, \tag{6.17}$$

where $\mathbf{F}(\mathbf{x})$ are the homogeneous polynomials of power k, can be reduced to problem (6.10). Indeed, introducing new functions

$$\mathbf{x}(t) = \tilde{\mathbf{x}}(t)e^{-\lambda t},$$

we arrive at problem (6.10) with function $g(t) = e^{-\lambda(k-1)t}$. In the important special case with $k = 2$ and functions $\mathbf{F}(\mathbf{x})$ such that $\mathbf{x}\mathbf{F}(\mathbf{x}) = 0$, the system of equations (6.10) describes hydrodynamic systems with the linear friction. In this case, the interaction between the components appears random.

If $\lambda = 0$, energy conservation holds in hydrodynamic systems for any realization of process $z(t)$. For $t \to \infty$, there is the steady-state probability distribution $P(\mathbf{x})$, which is, under the assumption that no additional integrals of motion exist, the uniform distribution over sphere $x_i^2 = E_0$. If additional integrals of motion exist (as it is the case for finite-dimensional approximation of the two-dimensional motion of liquid), the domain of the steady-state probability distribution will coincide with the phase space region allowed by the integrals of motion.

Note that in the special case of the Gaussian process $z(t)$ satisfying the one-dimensional linear equation of type Eq. (6.17)

$$\frac{d}{dt}x(t) = -\lambda x(t) + z(t)x(t), \quad x(0) = 1,$$

which determines the simplest logarithmic-normal random process, we obtain, instead of Eq. (6.16), the extended Fokker–Planck equation

$$\left(\frac{\partial}{\partial t} + \lambda\frac{\partial}{\partial x}x\right)P(x,t) = \int\limits_0^t d\tau\, B(t,\tau)\frac{\partial}{\partial x}x\frac{\partial}{\partial x}xP(x,t), \quad P(x,0) = \delta(x-1). \tag{6.18}$$

Additive action

Consider now the class of linear equations

$$\frac{d}{dt}\mathbf{x}(t) = A(t)\mathbf{x}(t) + \mathbf{f}(t), \quad \mathbf{x}(0) = \mathbf{x}_0, \tag{6.19}$$

where $A(t)$ is the deterministic matrix and $\mathbf{f}(t)$ is the random vector function whose characteristic functional $\Phi[t; \mathbf{v}(\tau)]$ is known.

For the probability density of the solution to Eq. (6.19), we have

$$\frac{\partial}{\partial t}P(\mathbf{x},t) = -\frac{\partial}{\partial x_i}(A_{ik}(t)x_kP(\mathbf{x},t)) + \left\langle\dot\Theta_t\left[t; \frac{\delta}{i\delta\mathbf{f}(\tau)}\right]\delta(\mathbf{x}(t)-\mathbf{x})\right\rangle. \tag{6.20}$$

In the problem under consideration, the variational derivative $\delta\mathbf{x}(t)/\delta\mathbf{f}(\tau)$ also satisfies (for $\tau < t$) the linear equation with the initial condition

$$\frac{d}{dt}\frac{\delta}{\delta\mathbf{f}(\tau)}x_i(t) = A_{ik}(t)\frac{\delta}{\delta\mathbf{f}(\tau)}x_k(t), \quad \left.\frac{\delta}{\delta f_l(\tau)}x_i(t)\right|_{t=\tau} = \delta_{il}. \tag{6.21}$$

Equation (6.21) has no randomness and governs Green's function $G_{il}(t,\tau)$ of homogeneous system (6.19), which means that

$$\frac{\delta}{\delta f_l(\tau)}x_i(t) = G_{il}(t,\tau).$$

As a consequence, we have

$$\frac{\delta}{\delta f_l(\tau)}\delta(\mathbf{x}(t) - \mathbf{x}) = -\frac{\partial}{\partial x_k}G_{kl}(t,\tau)\delta(\mathbf{x}(t) - \mathbf{x}),$$

and Eq. (6.20) appears converted into the closed equation

$$\frac{\partial}{\partial t}P(\mathbf{x},t) = -\frac{\partial}{\partial x_i}\left(A_{ik}(t)x_k P(\mathbf{x},t)\right) + \dot{\Theta}_t\left[t; iG_{kl}(t,\tau)\frac{\partial}{\partial x_k}\right]P(\mathbf{x},t). \tag{6.22}$$

From Eq. (6.22) follows that any moment of quantity $\mathbf{x}(t)$ will satisfy a closed linear equation that will include only a finite number of cumulant functions whose orders will not exceed the order of the moment of interest.

For the two-time probability density

$$P(\mathbf{x},t;\mathbf{x}_1,t_1) = \langle\delta(\mathbf{x}(t)-\mathbf{x})\delta(\mathbf{x}(t_1)-\mathbf{x}_1)\rangle,$$

we quite similarly obtain the equation

$$\frac{\partial}{\partial t}P(\mathbf{x},t;\mathbf{x}_1,t_1) = -\frac{\partial}{\partial x_i}\left(A_{ik}(t)x_k P(\mathbf{x},t;\mathbf{x}_1,t_1)\right)$$
$$+\dot{\Theta}_t\left[t; i\left\{G_{kl}(t,\tau)+G_{kl}(t_1,\tau)\right\}\frac{\partial}{\partial x_l}\right]P(\mathbf{x},t;\mathbf{x}_1,t_1) \quad (t>t_1) \tag{6.23}$$

with the initial condition

$$P(\mathbf{x},t_1;\mathbf{x}_1,t_1) = \delta(\mathbf{x}-\mathbf{x}_1)P(\mathbf{x}_1,t_1),$$

where $P(\mathbf{x}_1,t_1)$ is the one-point probability density satisfying Eq. (6.22). From Eq. (6.23) follows that $\mathbf{x}(t)$ is not the Markovian process. The particular form of Eqs. (6.22) and (6.23) depends on the structure of functional $\Phi[t;v(\tau)]$, i.e., on the random behavior of function $\mathbf{f}(t)$.

For the Gaussian vector process $\mathbf{f}(t)$ whose mean value and correlation function are as follows

$$\langle\mathbf{f}(t)\rangle = 0, \quad B_{ij}(t,t') = \langle f_i(t)f_j(t')\rangle,$$

Eq. (6.22) assumes the form of the extended Fokker–Planck equation

$$\frac{\partial}{\partial t}P(\mathbf{x},t) = -\frac{\partial}{\partial x_i}\left(A_{ik}(t)x_k P(\mathbf{x},t)\right)$$
$$+\int_0^t d\tau\, B_{jl}(t,\tau)G_{kj}(t,t)G_{ml}(t,\tau)\frac{\partial^2}{\partial x_k \partial x_m}P(\mathbf{x},t). \tag{6.24}$$

Consider the dynamics of a particle under random forces in the presence of friction [37] as an example of such a problem.

Inertial particle under random forces. The simplest example of particle diffusion under the action of random external force $\mathbf{f}(t)$ and linear friction is described by the linear system of equations (1.43)

$$\frac{d}{dt}\mathbf{r}(t) = \mathbf{v}(t), \quad \frac{d}{dt}\mathbf{v}(t) = -\lambda\left[\mathbf{v}(t)-\mathbf{f}(t)\right],$$
$$\mathbf{r}(0) = 0, \quad \mathbf{v}(0) = 0. \tag{6.25}$$

The stochastic solution to Eqs. (6.25) has the form

$$\mathbf{v}(t) = \lambda\int_0^t d\tau\, e^{-\lambda(t-\tau)}\mathbf{f}(\tau), \quad \mathbf{r}(t) = \int_0^t d\tau\left[1-e^{-\lambda(t-\tau)}\right]\mathbf{f}(\tau). \tag{6.26}$$

In the case of stationary random process $\mathbf{f}(t)$ with the correlation tensor $\langle f_i(t)f_j(t')\rangle = B_{ij}(t - t')$ and temporal correlation radius τ_0 determined from the relationship

$$\int\limits_0^\infty d\tau\, B_{ii}(\tau) = \tau_0 B_{ii}(0),$$

Eq. (6.25) allows obtaining the analytical expressions for correlators between particle velocity components and coordinates

$$\langle v_i(t)v_j(t)\rangle = \lambda \int\limits_0^t d\tau\, B_{ij}(\tau)\left[e^{-\lambda\tau} - e^{-\lambda(2t-\tau)}\right],$$

$$\frac{1}{2}\frac{d}{dt}\langle r_i(t)r_j(t)\rangle = \langle r_i(t)v_j(t)\rangle = \int\limits_0^t d\tau\, B_{ij}(\tau)\left[1 - e^{-\lambda t}\right]\left[1 - e^{-\lambda(t-\tau)}\right]. \quad (6.27)$$

In the steady-state regime when $\lambda t \gg 1$ and $t/\tau_0 \gg 1$, but parameter $\lambda\tau_0$ can be arbitrary, particle velocity is the stationary process with the correlation tensor

$$\langle v_i(t)v_j(t)\rangle = \lambda \int\limits_0^\infty d\tau\, B_{ij}(\tau)e^{-\lambda\tau}, \qquad (6.28)$$

and correlators $\langle r_i(t)v_j(t)\rangle$ and $\langle r_i(t)r_j(t)\rangle$ are as follows

$$\langle r_i(t)v_j(t)\rangle = \int\limits_0^\infty d\tau\, B_{ij}(\tau), \quad \langle r_i(t)r_j(t)\rangle = 2t\int\limits_0^\infty d\tau\, B_{ij}(\tau). \qquad (6.29)$$

If we additionally assume that $\lambda\tau_0 \gg 1$, the correlation tensor grades into

$$\langle v_i(t)v_j(t)\rangle = B_{ij}(0), \qquad (6.30)$$

which is consistent with (6.25), because $\mathbf{v}(t) \equiv \mathbf{f}(t)$ in this limit.

If the opposite condition $\lambda\tau_0 \ll 1$ holds, then

$$\langle v_i(t)v_j(t)\rangle = \lambda \int\limits_0^\infty d\tau\, B_{ij}(\tau).$$

This result corresponds to random process $\mathbf{f}(t)$ in the delta-correlated approximation.

Introduce now the indicator function of the solution to Eq. (6.25)

$$\varphi(\mathbf{r}, \mathbf{v}; t) = \delta\left(\mathbf{r}(t) - \mathbf{r}\right)\delta\left(\mathbf{v}(t) - \mathbf{v}\right),$$

which satisfies the Liouville equation

$$\left(\frac{\partial}{\partial t} + \mathbf{v}\frac{\partial}{\partial \mathbf{r}} - \lambda\frac{\partial}{\partial \mathbf{v}}\mathbf{v}\right)\varphi(\mathbf{r}, \mathbf{v}; t) = -\lambda\mathbf{f}(t)\frac{\partial}{\partial \mathbf{v}}\varphi(\mathbf{r}, \mathbf{v}; t),$$

$$\varphi(\mathbf{r}, \mathbf{v}; 0) = \delta\left(\mathbf{r}\right)\delta\left(\mathbf{v}\right). \qquad (6.31)$$

The mean value of the indicator function $\varphi(\mathbf{r}, \mathbf{v}; t)$ over an ensemble of realizations of random process $\mathbf{f}(t)$ is the joint one-time probability density of particle position and velocity

$$P(\mathbf{r}, \mathbf{v}; t) = \langle\varphi(\mathbf{r}, \mathbf{v}; t)\rangle = \langle\delta\left(\mathbf{r}(t) - \mathbf{r}\right)\delta\left(\mathbf{v}(t) - \mathbf{v}\right)\rangle_{\mathbf{f}}.$$

Averaging Eq. (6.31) over an ensemble of realizations of random process $\mathbf{f}(t)$, we obtain the unclosed equation

$$\left(\frac{\partial}{\partial t} + \mathbf{v}\frac{\partial}{\partial \mathbf{r}} - \lambda\frac{\partial}{\partial \mathbf{v}}\mathbf{v}\right) P(\mathbf{r}, \mathbf{v}; t) = -\lambda\frac{\partial}{\partial \mathbf{v}}\left\langle \mathbf{f}(t)\varphi(\mathbf{r}, \mathbf{v}; t)\right\rangle,$$
$$P(\mathbf{r}, \mathbf{v}; 0) = \delta(\mathbf{r})\delta(\mathbf{v}). \tag{6.32}$$

This equation contains correlation $\langle \mathbf{f}(t)\varphi(\mathbf{r}, \mathbf{v}; t)\rangle$ and is equivalent to the equality

$$\left(\frac{\partial}{\partial t} + \mathbf{v}\frac{\partial}{\partial \mathbf{r}} - \lambda\frac{\partial}{\partial \mathbf{v}}\mathbf{v}\right) P(\mathbf{r}, \mathbf{v}; t) = \left\langle \dot{\Theta}\left[t; \frac{\delta}{i\delta\mathbf{f}(\tau)}\right] \varphi(\mathbf{r}, \mathbf{v}; t)\right\rangle,$$
$$P(\mathbf{r}, \mathbf{v}; 0) = \delta(\mathbf{r})\delta(\mathbf{v}), \tag{6.33}$$

where functional $\Theta[t; \mathbf{v}(\tau)]$ is related to the characteristic functional of random process $\mathbf{f}(t)$

$$\Phi[t; \boldsymbol{\psi}(\tau)] = \left\langle \exp\left\{i\int_0^t d\tau\, \boldsymbol{\psi}(\tau)\mathbf{f}(\tau)\right\}\right\rangle = e^{\Theta[t; \boldsymbol{\psi}(\tau)]}$$

by the formula

$$\dot{\Theta}[t; \boldsymbol{\psi}(\tau)] = \frac{d}{dt}\ln\Phi[t; \boldsymbol{\psi}(\tau)] = \frac{d}{dt}\Theta[t; \boldsymbol{\psi}(\tau)].$$

Functional $\Theta[t; \boldsymbol{\psi}(\tau)]$ can be expanded in the functional power series

$$\Theta[t; \boldsymbol{\psi}(\tau)] = \sum_{n=1}^{\infty} \frac{i^n}{n!}\int_0^t dt_1 ... \int_0^t dt_n K_{i_1,...,i_n}^{(n)}(t_1, ..., t_n)\psi_{i_1}(t_1)...\psi_{i_n}(t_n),$$

where functions

$$K_{i_1,...,i_n}^{(n)}(t_1, ..., t_n) = \frac{1}{i^n}\frac{\delta^n}{\delta\psi_{i_1}(t_1)...\delta\psi_{i_n}(t_n)}\Theta[t; \boldsymbol{\psi}(\tau)]\bigg|_{\boldsymbol{\psi}=0}$$

are the n-th order cumulant functions of random process $\mathbf{f}(t)$.

Consider the variational derivative

$$\frac{\delta}{\delta f_j(t')}\varphi(\mathbf{r}, \mathbf{v}; t) = -\left[\frac{\partial}{\partial r_k}\frac{\delta r_k(t)}{\delta f_j(t')} + \frac{\partial}{\partial v_k}\frac{\delta v_k(t)}{\delta f_j(t')}\right]\varphi(\mathbf{r}, \mathbf{v}; t). \tag{6.34}$$

In the context of dynamic problem (6.25), the variational derivatives of functions $\mathbf{r}(t)$ and $\mathbf{v}(t)$ in Eq. (6.34) can be calculated from Eqs. (6.26) and have the forms

$$\frac{\delta v_k(t)}{\delta f_j(t')} = \lambda\delta_{kj}e^{-\lambda(t-t')}, \quad \frac{\delta r_k(t)}{\delta f_j(t')} = \delta_{kj}\left[1 - e^{-\lambda(t-t')}\right]. \tag{6.35}$$

Using Eq. (6.35), we can now rewrite Eq. (6.34) in the form

$$\frac{\delta}{\delta\mathbf{f}(t')}\varphi(\mathbf{r}, \mathbf{v}; t) = -\left\{\left[1 - e^{-\lambda(t-t')}\right]\frac{\partial}{\partial \mathbf{r}} + \lambda e^{-\lambda(t-t')}\frac{\partial}{\partial \mathbf{v}}\right\}\varphi(\mathbf{r}, \mathbf{v}; t),$$

after which Eq. (6.33) assumes the closed form

$$\left(\frac{\partial}{\partial t} + \mathbf{v}\frac{\partial}{\partial \mathbf{r}} - \lambda\frac{\partial}{\partial \mathbf{v}}\mathbf{v}\right) P(\mathbf{r}, \mathbf{v}; t)$$
$$= \dot{\Theta}\left[t; i\left\{\left[1 - e^{-\lambda(t-t')}\right]\frac{\partial}{\partial \mathbf{r}} + \lambda e^{-\lambda(t-t')}\frac{\partial}{\partial \mathbf{v}}\right\}\right] P(\mathbf{r}, \mathbf{v}; t),$$
$$P(\mathbf{r}, \mathbf{v}; 0) = \delta(\mathbf{r})\delta(\mathbf{v}). \tag{6.36}$$

Note that from Eq. (6.36) follows that equations for n-th order moment functions include cumulant functions of order not higher than n.

Assume now that $\mathbf{f}(t)$ is the Gaussian stationary process with the zero mean value and correlation tensor

$$B_{ij}(t - t') = \langle f_i(t)f_j(t') \rangle.$$

In this case, the characteristic functional of process $\mathbf{f}(t)$ is

$$\Phi\left[t; \boldsymbol{\psi}(\tau)\right] = \exp\left\{ -\frac{1}{2} \int\limits_0^t \int\limits_0^t dt_1 dt_2 B_{ij}(t_1 - t_2)\psi_i(t_1)\psi_j(t_2) \right\},$$

functional $\dot{\Theta}[t; \boldsymbol{\psi}(\tau)]$ is given by the formula

$$\dot{\Theta}\left[t; \boldsymbol{\psi}(\tau)\right] = -\psi_i(t) \int\limits_0^t dt' B_{ij}(t - t')\psi_j(t'),$$

and Eq. (6.36) appears an extension of the Fokker–Planck equation

$$\left(\frac{\partial}{\partial t} + \mathbf{v}\frac{\partial}{\partial \mathbf{r}} - \lambda\frac{\partial}{\partial \mathbf{v}}\mathbf{v}\right) P(\mathbf{r}, \mathbf{v}; t) = \lambda^2 \int\limits_0^t d\tau\, B_{ij}(\tau) e^{-\lambda\tau} \frac{\partial^2}{\partial v_i \partial v_j} P(\mathbf{r}, \mathbf{v}; t)$$

$$+ \lambda \int\limits_0^t d\tau\, B_{ij}(\tau) \left[1 - e^{-\lambda\tau}\right] \frac{\partial^2}{\partial v_i \partial r_j} P(\mathbf{r}, \mathbf{v}; t),$$

$$P(\mathbf{r}, \mathbf{v}; 0) = \delta(\mathbf{r})\,\delta(\mathbf{v}). \tag{6.37}$$

Equation (6.37) is the exact equation and remains valid for arbitrary times t. From this equation follows that $\mathbf{r}(t)$ and $\mathbf{v}(t)$ are the Gaussian functions. For moment functions of processes $\mathbf{r}(t)$ and $\mathbf{v}(t)$, we obtain in the ordinary way the system of equations

$$\frac{d}{dt}\langle r_i(t)r_j(t)\rangle = 2\langle r_i(t)v_j(t)\rangle,$$

$$\left(\frac{d}{dt} + \lambda\right)\langle r_i(t)v_j(t)\rangle = \langle v_i(t)v_j(t)\rangle + \lambda \int\limits_0^t d\tau\, B_{ij}(\tau)\left[1 - e^{-\lambda\tau}\right],$$

$$\left(\frac{d}{dt} + 2\lambda\right)\langle v_i(t)v_j(t)\rangle = 2\lambda^2 \int\limits_0^t d\tau\, B_{ij}(\tau)e^{-\lambda\tau}. \tag{6.38}$$

From system (6.38) follows that steady-state values of all one-time correlations for $\lambda t \gg 1$ and $t/\tau_0 \gg 1$ are given by the expressions

$$\langle v_i(t)v_j(t)\rangle = \lambda \int\limits_0^\infty d\tau\, B_{ij}(\tau)e^{-\lambda\tau}, \quad \langle r_i(t)v_j(t)\rangle = D_{ij},$$

$$\langle r_i(t)r_j(t)\rangle = 2t D_{ij}, \tag{6.39}$$

where

$$D_{ij} = \int\limits_0^\infty d\tau\, B_{ij}(\tau) \tag{6.40}$$

is the spatial diffusion tensor, which agrees with expressions (6.28) and (6.29).

Remark. Temporal correlation tensor and temporal correlation radius of process $\mathbf{v}(t)$.

We can additionally calculate the temporal correlation radius of velocity $\mathbf{v}(t)$, i.e., correlation $\langle v_i(t)v_j(t_1)\rangle$. Using equalities (6.35), we obtain for $t_1 < t$ the equation

$$\left(\frac{d}{dt} + \lambda\right)\langle v_i(t)v_j(t_1)\rangle = \lambda^2 \int\limits_0^{t_1} dt'\, B_{ij}(t - t')e^{-\lambda(t_1 - t')}$$

$$= \lambda^2 e^{\lambda(t - t_1)} \int\limits_{t - t_1}^{t} d\tau\, B_{ij}(\tau)e^{-\lambda\tau}, \tag{6.41}$$

with the initial condition

$$\langle v_i(t)v_j(t_1)\rangle\,|_{t = t_1} = \langle v_i(t_1)v_j(t_1)\rangle. \tag{6.42}$$

In the steady-state regime, i.e., for $\lambda t \gg 1$ and $\lambda t_1 \gg 1$, but at fixed difference $(t - t_1)$, we obtain the equation with initial condition $(\tau = t - t_1)$

$$\left(\frac{d}{d\tau} + \lambda\right)\langle v_i(t + \tau)v_j(t)\rangle = \lambda^2 e^{\lambda\tau} \int\limits_{\tau}^{\infty} d\tau_1\, B_{ij}(\tau_1)e^{-\lambda\tau_1},$$

$$\langle v_i(t + \tau)v_j(t)\rangle_{\tau = 0} = \langle v_i(t)v_j(t)\rangle. \tag{6.43}$$

One can easily write the solution to Eq. (6.43); however, our interest here concerns only the temporal correlation radius $\tau_{\mathbf{v}}$ of random process $\mathbf{v}(t)$. To obtain this quantity, we integrate Eq. (6.43) with respect to parameter τ over the interval $(0, \infty)$. The result is

$$\lambda \int\limits_0^{\infty} d\tau\, \langle v_i(t + \tau)v_j(t)\rangle = \langle v_i(t)v_j(t)\rangle + \lambda \int\limits_0^{\infty} d\tau_1\, B_{ij}(\tau_1)\left[1 - e^{-\lambda\tau_1}\right],$$

and we, using Eq. (6.39), arrive at the expression

$$\tau_{\mathbf{v}}\left\langle \mathbf{v}^2(t)\right\rangle = D_{ii} = \tau_0 B_{ii}(0), \tag{6.44}$$

i.e.,

$$\tau_{\mathbf{v}} = \frac{\tau_0 B_{ii}(0)}{\langle \mathbf{v}^2(t)\rangle} = \frac{\tau_0 B_{ii}(0)}{\lambda \int\limits_0^{\infty} d\tau\, B_{ii}(\tau)e^{-\lambda\tau}} = \begin{cases} \tau_0, & \text{for } \lambda\tau_0 \gg 1, \\[2ex] 1/\lambda, & \text{for } \lambda\tau_0 \ll 1. \end{cases} \quad \blacklozenge \tag{6.45}$$

Integrating Eq.(6.37) over $\mathbf{r}$, we obtain the closed equation for the probability density of particle velocity

$$\left(\frac{\partial}{\partial t} - \lambda\frac{\partial}{\partial\mathbf{v}}\mathbf{v}\right)P(\mathbf{v}; t) = \lambda^2 \int\limits_0^{t} d\tau\, B_{ij}(\tau)e^{-\lambda\tau}\frac{\partial^2}{\partial v_i \partial v_j}P(\mathbf{r}, \mathbf{v}; t),$$

$$P(\mathbf{r}, \mathbf{v}; 0) = \delta(\mathbf{v}).$$

The solution to this equation corresponds to the Gaussian process $\mathbf{v}(t)$ with correlation tensor (6.27), which follows from the fact that the second equation of system (6.25) is closed. It can be shown that, if the steady-state probability density exists under the condition $\lambda t \gg 1$, then this probability density satisfies the equation

$$-\frac{\partial}{\partial \mathbf{v}} \mathbf{v} P(\mathbf{v};t) = \lambda \int\limits_0^\infty d\tau\, B_{ij}(\tau) e^{-\lambda \tau} \frac{\partial^2}{\partial v_i \partial v_j} P(\mathbf{v};t),$$

and the rate of establishing this distribution depends on parameter λ.

The equation for the probability density of particle coordinate $P(\mathbf{r};t)$ cannot be derived immediately from Eq. (6.37). Indeed, integrating Eq. (6.37) over $\mathbf{v}$, we obtain the equality

$$\frac{\partial}{\partial t} P(\mathbf{r},t) = -\frac{\partial}{\partial \mathbf{r}} \int \mathbf{v} P(\mathbf{r},\mathbf{v};t) d\mathbf{v}, \quad P(\mathbf{r},0) = \delta(\mathbf{r}). \tag{6.46}$$

For function $\int v_k P(\mathbf{r},\mathbf{v};t) d\mathbf{v}$, we have the equality

$$\left(\frac{\partial}{\partial t} + \lambda\right) \int v_k P(\mathbf{r},\mathbf{v};t) d\mathbf{v} = -\frac{\partial}{\partial \mathbf{r}} \int v_k \mathbf{v} P(\mathbf{r},\mathbf{v};t) d\mathbf{v}$$

$$-\lambda \int\limits_0^t d\tau\, B_{kj}(\tau) \left[1 - e^{-\lambda \tau}\right] \frac{\partial}{\partial r_j} P(\mathbf{r},t), \tag{6.47}$$

and so on, i.e., this approach results in an infinite system of equations.

Random function $\mathbf{r}(t)$ satisfies the first equation of system (6.25) and, if we would know the complete statistics of function $\mathbf{v}(t)$ (i.e., the multi-time statistics), we could calculate all statistical characteristics of function $\mathbf{r}(t)$. Unfortunately, Eq. (6.37) describes only one-time statistical quantities, and only the infinite system of equations similar to Eqs. (6.46), (6.47), and so on appears equivalent to the multi-time statistics of function $\mathbf{v}(t)$. Indeed, function $\mathbf{r}(t)$ can be represented in the form

$$\mathbf{r}(t) = \int\limits_0^t dt_1 \mathbf{v}(t_1),$$

so that the spatial diffusion coefficient in the steady-state regime assumes, in view of Eq. (6.44), the form

$$\frac{1}{2}\frac{d}{dt}\left\langle \mathbf{r}^2(t) \right\rangle = \int\limits_0^\infty d\tau\, \langle \mathbf{v}(t+\tau)\mathbf{v}(t)\rangle$$

$$= \tau_\mathbf{v} \left\langle \mathbf{v}^2(t) \right\rangle = D_{ii} = \tau_0 B_{ii}(0), \tag{6.48}$$

from which follows that it depends on the temporal correlation radius $\tau_\mathbf{v}$ and the variance of random function $\mathbf{v}(t)$.

However, in the case of this simplest problem, we know immediately variances and all correlations of functions $\mathbf{v}(t)$ and $\mathbf{r}(t)$ (see Eqs. (6.27)) and, consequently, can draw the equation for the probability density of particle coordinate $P(\mathbf{r};t)$. This equation is the diffusion equation

$$\frac{\partial}{\partial t} P(\mathbf{r};t) = D_{ij}(t) \frac{\partial^2}{\partial r_i \partial r_j} P(\mathbf{r},t), \quad P(\mathbf{r},0) = \delta(\mathbf{r}),$$

where

$$
\begin{aligned}
D_{ij}(t) &= \frac{1}{2}\frac{d}{dt}\langle r_i(t)r_j(t)\rangle = \frac{1}{2}\{\langle r_i(t)v_j(t)\rangle + \langle r_j(t)v_i(t)\rangle\} \\
&= \int_0^t d\tau\, B_{ij}(\tau)\left[1 - e^{-\lambda t}\right]\left[1 - e^{-\lambda(t-\tau)}\right]
\end{aligned}
$$

is the diffusion tensor (6.27). Under the condition $\lambda t \gg 1$, we obtain the equation

$$
\frac{\partial}{\partial t}P(\mathbf{r};t) = D_{ij}\frac{\partial^2}{\partial r_i \partial r_j}P(\mathbf{r},t), \quad P(\mathbf{r},0) = \delta(\mathbf{r}) \tag{6.49}
$$

with the diffusion tensor

$$
D_{ij} = \int_0^\infty d\tau\, B_{ij}(\tau). \tag{6.50}
$$

Note that conversion from Eq. (6.37) to the equation for the probability density of particle coordinate (6.49) with the diffusion coefficient (6.48) corresponds to the so-called *Kramers problem*.

Under the assumption that $\lambda\tau_0 \ll 1$, where τ_0 is the temporal correlation radius of process $\mathbf{f}(t)$, Eq. (6.37) becomes simpler

$$
\left(\frac{\partial}{\partial t} + \mathbf{v}\frac{\partial}{\partial \mathbf{r}} - \lambda\frac{\partial}{\partial \mathbf{v}}\mathbf{v}\right)P(\mathbf{r},\mathbf{v};t) = \lambda^2 \int_0^t d\tau\, B_{ij}(\tau)\frac{\partial^2}{\partial v_i \partial v_j}P(\mathbf{r},\mathbf{v};t),
$$
$$
P(\mathbf{r},\mathbf{v};0) = \delta(\mathbf{r})\delta(\mathbf{v}),
$$

and corresponds to the approximation of random function $\mathbf{f}(t)$ by the delta-correlated process. If $t \gg \tau_0$, we can replace the upper limit of the integral with the infinity and proceed to the standard diffusion Fokker–Planck equation

$$
\left(\frac{\partial}{\partial t} + \mathbf{v}\frac{\partial}{\partial \mathbf{r}} - \lambda\frac{\partial}{\partial \mathbf{v}}\mathbf{v}\right)P(\mathbf{r},\mathbf{v};t) = \lambda^2 D_{ij}\frac{\partial^2}{\partial v_i \partial v_j}P(\mathbf{r},\mathbf{v};t),
$$
$$
P(\mathbf{r},\mathbf{v};0) = \delta(\mathbf{r})\delta(\mathbf{v}), \tag{6.51}
$$

with diffusion tensor (6.48). In this approximation, the combined random process $\{\mathbf{r}(t), \mathbf{v}(t)\}$ is the Markovian process.

Under the condition $\lambda t \gg 1$, there are the steady-state equation for the probability density of particle velocity

$$
-\lambda\frac{\partial}{\partial \mathbf{v}}\mathbf{v}P(\mathbf{v}) = \lambda^2 D_{ij}\frac{\partial^2}{\partial v_i \partial v_j}P(\mathbf{v}),
$$

and the nonstationary equation for the probability density of particle coordinate

$$
\frac{\partial}{\partial t}P(\mathbf{r};t) = D_{ij}\frac{\partial^2}{\partial r_i \partial r_j}P(\mathbf{r},t), \quad P(\mathbf{r},0) = \delta(\mathbf{r}).
$$

Consider now the limit $\lambda\tau_0 \gg 1$. In this case, we can rewrite Eq. (6.37) in the form

$$\left(\frac{\partial}{\partial t} + \mathbf{v}\frac{\partial}{\partial \mathbf{r}} - \lambda\frac{\partial}{\partial \mathbf{v}}\mathbf{v}\right) P(\mathbf{r},\mathbf{v};t)$$

$$= \lambda B_{ij}(0)\left[1 - e^{-\lambda t}\right]\frac{\partial^2}{\partial v_i \partial v_j}P(\mathbf{r},\mathbf{v};t)$$

$$- B_{ij}(0)\left[1 - e^{-\lambda t}\right]\frac{\partial^2}{\partial v_i \partial r_j}P(\mathbf{r},\mathbf{v};t) + \lambda\int\limits_0^t d\tau\, B_{ij}(\tau)\frac{\partial^2}{\partial v_i \partial r_j}P(\mathbf{r},\mathbf{v};t),$$

$$P(\mathbf{r},\mathbf{v};0) = \delta\left(\mathbf{r}\right)\delta\left(\mathbf{v}\right).$$

Integrating this equation over $\mathbf{r}$, we obtain the equation for the probability density of particle velocity

$$\left(\frac{\partial}{\partial t} - \lambda\frac{\partial}{\partial \mathbf{v}}\mathbf{v}\right) P(\mathbf{v};t) = \lambda B_{ij}(0)\left[1 - e^{-\lambda t}\right]\frac{\partial^2}{\partial v_i \partial v_j}P(\mathbf{v};t),$$

$$P(\mathbf{v};0) = \delta\left(\mathbf{v}\right),$$

and, under the condition $\lambda t \gg 1$, we arrive at the steady-state Gaussian probability density with variance

$$\langle v_i(t)v_j(t)\rangle = B_{ij}(0).$$

As regards the probability density of particle position, it satisfies under the condition $\lambda t \gg 1$ the equation

$$\frac{\partial}{\partial t}P(\mathbf{r};t) = D_{ij}\frac{\partial^2}{\partial r_i \partial r_j}P(\mathbf{r},t), \quad P(\mathbf{r},0) = \delta\left(\mathbf{r}\right) \tag{6.52}$$

with the same diffusion coefficient as previously. This is a consequence of the fact that Eq. (6.44) is independent of parameter λ. Note that this equation corresponds to the limit process $\lambda \to \infty$ in Eq. (6.25)

$$\frac{d}{dt}\mathbf{r}(t) = \mathbf{v}(t), \quad \mathbf{v}(t) = \mathbf{f}(t), \quad \mathbf{r}(0) = 0.$$

In the limit $\lambda \to \infty$ (or $\lambda\tau_0 \gg 1$), we have the equality

$$\mathbf{v}(t) \approx \mathbf{f}(t), \tag{6.53}$$

and all multi-time statistics of random functions $\mathbf{v}(t)$ and $\mathbf{r}(t)$ will be described in terms of statistical characteristics of process $\mathbf{f}(t)$. In particular, the one-time probability density of particle velocity $\mathbf{v}(t)$ is the Gaussian probability density with variance $\langle v_i(t)v_j(t)\rangle = B_{ij}(0)$, and the spatial diffusion coefficient is

$$D = \frac{1}{2}\frac{d}{dt}\left\langle \mathbf{r}^2(t)\right\rangle = \int\limits_0^\infty d\tau\, B_{ii}(\tau) = \tau_0 B_{ii}(0).$$

As we have seen earlier, in the case of process $\mathbf{f}(t)$ such that it can be correctly described in the delta-correlated approximation (i.e., if $\lambda\tau_0 \ll 1$), the approximate equality (6.53) appears inappropriate to determine statistical characteristics of process $\mathbf{v}(t)$. Nevertheless, Eq. (6.52) with the same diffusion tensor remains as before valid for the one-time statistical characteristics of process $\mathbf{r}(t)$, which follows from the fact that Eq. (6.48) is valid for any parameter λ and arbitrary probability density of random process $\mathbf{f}(t)$.

Above, we considered several types of stochastic ordinary differential equations that allow obtaining closed statistical description in the general form. It is clear that similar situations can appear in dynamic systems formulated in terms of partial differential equations.

6.2.2 Partial differential equations

First of all, we note that the first-order partial differential equation

$$\left(\frac{\partial}{\partial t} + z(t)g(t)\frac{\partial}{\partial \mathbf{x}}\mathbf{F}(\mathbf{x})\right)\rho(\mathbf{r},t) = 0$$

is equivalent to the system of ordinary differential equations (6.10) and, consequently, also allows the complete statistical description of any random process $z(t)$.

Consider now the class of nonlinear partial differential equations whose parameters are independent of spatial variable $\mathbf{x}$,

$$\frac{\partial}{\partial t}q(t,\mathbf{x}) + \mathbf{z}(t)\frac{\partial}{\partial \mathbf{x}}q(t,\mathbf{x}) = F\left(t, q, \frac{\partial q}{\partial \mathbf{x}}, \frac{\partial}{\partial \mathbf{x}} \otimes \frac{\partial}{\partial \mathbf{x}}q, ...\right),$$

where $\mathbf{z}(t)$ is the vector random process and F is the deterministic function. Solution to this equation is representable in the form

$$q(t,\mathbf{x}) = Q\left(t, \mathbf{x} - \int\limits_0^t d\tau \mathbf{z}(\tau)\right),$$

where function $Q(t,\mathbf{x})$ satisfies the deterministic equation

$$\frac{\partial}{\partial t}Q(t,\mathbf{x}) = F\left(t, Q, \frac{\partial Q}{\partial \mathbf{x}}, \frac{\partial}{\partial \mathbf{x}} \otimes \frac{\partial}{\partial \mathbf{x}}Q, ...\right),$$

and, consequently,

$$\frac{\delta}{\delta z_i(\tau)}q(t,\mathbf{x}) = -\theta(t-\tau)\frac{\partial}{\partial z_i(\tau)}q(t,\mathbf{x}).$$

In the case of such problems, statistical characteristics of the solution can be determined immediately by averaging the corresponding expressions constructed from the solution to the last equation.

Unfortunately, there is only limited number of equations that allow sufficiently complete analysis. In the general case, the analysis of dynamic systems appears possible only on the basis of various asymptotic and approximate techniques. In physics, techniques based on approximating actual random processes and fields by the fields delta-correlated in time are often and successfully used.

6.3 Delta-correlated fields and processes

In the case of the delta-correlated (in time) random field $\mathbf{f}(\mathbf{x},t)$, the following equality holds (see Chapter 5)

$$\dot{\Theta}_t[t,t_0;\mathbf{v}(\mathbf{y},\tau)] \equiv \dot{\Theta}_t[t,t_0;\mathbf{v}(\mathbf{y},t)],$$

and situation becomes significantly simpler. The fact that field $\mathbf{f}(\mathbf{x}, t)$ is delta-correlated means that

$$\Theta[t, t_0; \mathbf{v}(\mathbf{y}, \tau)]$$
$$= \sum_{i=1}^{\infty} \frac{i^n}{n!} \int_{t_0}^{t} d\tau \int d\mathbf{y}_1 ... \int d\mathbf{y}_n K_n^{i_1,...,i_n}(\mathbf{y}_1, ..., \mathbf{y}_n; \tau) v_{i_1}(\mathbf{y}_1, \tau)...v_{i_n}(\mathbf{y}_n, \tau),$$

which, in turn, means that field $\mathbf{f}(\mathbf{x}, t)$ is characterized by cumulant functions of the form

$$K_n^{i_1,...,i_n}(\mathbf{y}_1, t_1; ...; \mathbf{y}_n, t_n) = K_n^{i_1,...,i_n}(\mathbf{y}_1, ..., \mathbf{y}_n; t_1)\delta(t_1 - t_2)...\delta(t_{n-1} - t_n).$$

In this case, Eqs. (6.7), and (6.9) appear, in view of Eq. (6.3), the closed operator equations in functions $P(\mathbf{x}, t)$, $p(\mathbf{x}, t|\mathbf{x}_0, t_0)$, and $P_m(\mathbf{x}_1, t_1; ...; \mathbf{x}_m, t_m)$. Indeed, Eq. (6.7) is reduced to the equation

$$\left(\frac{\partial}{\partial t} + \frac{\partial}{\partial \mathbf{x}}\mathbf{v}(\mathbf{x}, t)\right) P(\mathbf{x}, t) = \dot{\Theta}_t\left[t, t_0; i\frac{\partial}{\partial \mathbf{x}}\delta(\mathbf{y} - \mathbf{x})\right] P(\mathbf{x}, t), \quad P(\mathbf{x}, 0) = \delta(\mathbf{x} - \mathbf{x}_0),$$
$$(6.54)$$

whose concrete form is governed by functional $\Theta[t, t_0; \mathbf{v}(\mathbf{y}, \tau)]$, i.e., by statistical behavior of random field $\mathbf{f}(\mathbf{x}, t)$. Correspondingly, Eq. (6.9) for the m-time probability density is reduced to the operator equation $(t_1 \leq t_2 \leq ... \leq t_m)$

$$\left(\frac{\partial}{\partial t} + \frac{\partial}{\partial \mathbf{x}_m}\mathbf{v}(\mathbf{x}_m, t_m)\right) P_m(\mathbf{x}_1, t_1; ...; \mathbf{x}_m, t_m)$$
$$= \dot{\Theta}_{t_m}\left[t_m, t_0; i\frac{\partial}{\partial \mathbf{x}_m}\delta(\mathbf{y} - \mathbf{x}_m)\right] P_m(\mathbf{x}_1, t_1; ...; \mathbf{x}_m, t_m),$$
$$P_m(\mathbf{x}_1, t_1; ...; \mathbf{x}_m, t_{m-1}) = \delta(\mathbf{x}_m - \mathbf{x}_{m-1})P_{m-1}(\mathbf{x}_1, t_1; ...; \mathbf{x}_{m-1}, t_{m-1}). \quad (6.55)$$

We can seek the solution to Eq. (6.55) in the form

$$P_m(\mathbf{x}_1, t_1; ...; \mathbf{x}_m, t_m) = p(\mathbf{x}_m, t_m|\mathbf{x}_{m-1}, t_{m-1})P_{m-1}(\mathbf{x}_1, t_1; ...; \mathbf{x}_{m-1}, t_{m-1}). \quad (6.56)$$

Because all differential operations in Eq. (6.55) concern only t_m and $\mathbf{x}_m$, we can substitute Eq. (6.56) in Eq. (6.55) to obtain the following equation for the transition probability density

$$\left(\frac{\partial}{\partial t} + \frac{\partial}{\partial \mathbf{x}}\mathbf{v}(\mathbf{x}, t)\right) p(\mathbf{x}, t|\mathbf{x}_0, t_0) = \dot{\Theta}_t\left[t, t_0; i\frac{\partial}{\partial \mathbf{x}}\delta(\mathbf{y} - \mathbf{x})\right] p(\mathbf{x}, t|\mathbf{x}_0, t_0),$$
$$p(\mathbf{x}, t|\mathbf{x}_0, t_0)|_{t \to t_0} = \delta(\mathbf{x} - \mathbf{x}_0). \quad (6.57)$$

Here, we denoted variables $\mathbf{x}_m$ and t_m as $\mathbf{x}$ and t and variables $\mathbf{x}_{m-1}$ and t_{m-1} as $\mathbf{x}_0$ and t_0.

Using formula (6.56) $(m - 1)$ times, we obtain the relationship

$$P_m(\mathbf{x}_1, t_1; ...; \mathbf{x}_m, t_m) = p(\mathbf{x}_m, t_m|\mathbf{x}_{m-1}, t_{m-1})...p(\mathbf{x}_2, t_2|\mathbf{x}_1, t_1)P(\mathbf{x}_1, t_1), \quad (6.58)$$

where $P(\mathbf{x}_1, t_1)$ is the one-time probability density governed by Eq. (6.54). Equality (6.58) expresses the many-time probability density in terms of the product of transition probability densities, which means that random process $\mathbf{x}(t)$ is the Markovian process. The transition probability density is defined in this case as follows:

$$p(\mathbf{x}, t|\mathbf{x}_0, t_0) = \langle \delta(\mathbf{x}(t) - \mathbf{x})|\mathbf{x}_0, t_0 \rangle.$$

Special models of parameter fluctuations can significantly simplify the obtained equations.

For example, in the case of the Gaussian delta-correlated field $\mathbf{f}(\mathbf{x},t)$, the correlation tensor has the form ($\langle \mathbf{f}(\mathbf{x},t)\rangle = 0$)

$$B_{ij}(\mathbf{x},t;\mathbf{x}',t') = 2\delta(t - t')F_{ij}(\mathbf{x},\mathbf{x}';t). \tag{6.59}$$

Then, functional $\Theta[t, t_0; \mathbf{v}(\mathbf{y}, \tau)]$ assumes the form

$$\Theta[t, t_0; \mathbf{v}(\mathbf{y},\tau)] = -\int\limits_{t_0}^{t} d\tau \int d\mathbf{y}_1 \int d\mathbf{y}_2 F_{ij}(\mathbf{y}_1, \mathbf{y}_2; \tau)v_i(\mathbf{y}_1, \tau)v_j(\mathbf{y}_2, \tau),$$

and Eq. (6.54) reduces to the Fokker–Planck equation

$$\left(\frac{\partial}{\partial t} + \frac{\partial}{\partial x_k}\left[v_k(\mathbf{x},t) + A_k(\mathbf{x},t)\right]\right) P(\mathbf{x}, t) = \frac{\partial^2}{\partial x_k \partial x_l}\left[F_{kl}(\mathbf{x},\mathbf{x},t)P(\mathbf{x},t)\right], \tag{6.60}$$

where

$$A_k(\mathbf{x}, t) = \left.\frac{\partial}{\partial x_l'} F_{kl}(\mathbf{x}, \mathbf{x}';t)\right|_{\mathbf{x}'=\mathbf{x}}.$$

In view of the special role that the Gaussian delta-correlated field $\mathbf{f}(\mathbf{x},t)$ plays in physics, we give an alternative and more detailed discussion of this approximation commonly called the *approximation of the Gaussian delta-correlated field* in Chapter 8.

We illustrate the above general theory using several equations as examples.

6.3.1   One-dimensional nonlinear differential equation

Consider the one-dimensional stochastic equation

$$\frac{d}{dt}x(t) = f(x,t) + z(t)g(x,t), \quad x(0) = x_0, \tag{6.61}$$

where $f(x,t)$ and $g(x,t)$ are the deterministic functions and $z(t)$ is the random function of time. For indicator function $\varphi(x,t) = \delta(x(t) - x)$, we have the Liouville equation

$$\left(\frac{\partial}{\partial t} + \frac{\partial}{\partial x}f(x,t)\right)\varphi(x,t) = -z(t)\frac{\partial}{\partial x}\left\{g(x,t)\varphi(x,t)\right\},$$

so that the equation for one-time probability density $P(x,t)$ has the form

$$\left(\frac{\partial}{\partial t} + \frac{\partial}{\partial x}f(x,t)\right) P(x,t) = \left\langle \dot{\Theta}_t\left[t, \frac{\delta}{i\delta z(\tau)}\right]\varphi(x,t)\right\rangle.$$

In the case of delta-correlated random process $z(t)$, the equality

$$\dot{\Theta}_t[t, v(\tau)] = \dot{\Theta}_t[t, v(t)]$$

holds. Taking into account the equality

$$\frac{\delta}{\delta z(t - 0)}\varphi(x,t) = -\frac{\partial}{\partial x}\left\{g(x,t)\varphi(x,t)\right\},$$

we obtain the closed operator equation

$$\left(\frac{\partial}{\partial t} + \frac{\partial}{\partial x} f(x,t)\right) P(x,t) = \dot{\Theta}_t \left[t, i\frac{\partial}{\partial x} g(x,t)\right] P(x,t). \tag{6.62}$$

For the Gaussian delta-correlated process, we have

$$\Theta[t, v(\tau)] = -\frac{1}{2} \int\limits_0^t d\tau B(\tau) v^2(\tau), \tag{6.63}$$

and Eq. (6.62) assumes the form of the Fokker–Planck equation

$$\left(\frac{\partial}{\partial t} + \frac{\partial}{\partial x} f(x,t)\right) P(x,t) = \frac{1}{2} B(t) \frac{\partial}{\partial x} g(x,t) \frac{\partial}{\partial x} g(x,t) P(x,t). \tag{6.64}$$

For the Poisson delta-correlated process $z(t)$, we have

$$\Theta[t, v(\tau)] = \nu \int\limits_0^t d\tau \left\{ \int\limits_{-\infty}^{\infty} d\xi p(\xi) e^{i\xi v(\tau)} - 1 \right\}, \tag{6.65}$$

and Eq. (6.62) reduces to the form

$$\left(\frac{\partial}{\partial t} + \frac{\partial}{\partial x} f(x,t)\right) P(x,t) = \nu \left\{ \int\limits_{-\infty}^{\infty} d\xi p(\xi) e^{-\xi \frac{\partial}{\partial x} g(x,t)} - 1 \right\} P(x,t). \tag{6.66}$$

If we set $g(x,t) = 1$, Eq. (6.61) assumes the form

$$\frac{d}{dt} x(t) = f(x,t) + z(t), \quad x(0) = x_0.$$

In this case, the operator in the right-hand side of Eq. (6.66) is the shift operator, and Eq. (6.66) assumes the form of the Kolmogorov–Feller equation

$$\left(\frac{\partial}{\partial t} + \frac{\partial}{\partial x} f(x,t)\right) P(x,t) = \nu \int\limits_{-\infty}^{\infty} d\xi p(\xi) P(x - \xi, t) - \nu P(x,t).$$

Define now $g(x,t) = x$, so that Eq. (6.61) has the form

$$\frac{d}{dt} x(t) = f(x,t) + z(t)x(t), \quad x(0) = x_0.$$

In this case, Eq. (6.66) assumes the form

$$\left(\frac{\partial}{\partial t} + \frac{\partial}{\partial x} f(x,t)\right) P(x,t) = \nu \left\{ \int\limits_{-\infty}^{\infty} d\xi p(\xi) e^{-\xi \frac{\partial}{\partial x} x} - 1 \right\} P(x,t), \tag{6.67}$$

and we can represent Eq. (6.67) in the final form of the integro-differential equation similar to the Kolmogorov–Feller equation

$$\left(\frac{\partial}{\partial t} + \frac{\partial}{\partial x} f(x,t)\right) P(x,t) = \nu \int\limits_{-\infty}^{\infty} d\xi p(\xi) e^{-\xi} P(xe^{-\xi}, t) - \nu P(x,t).$$

6.3.2   Linear operator equation

Consider now the linear operator equation

$$\frac{d}{dt}\mathbf{x}(t) = \hat{A}(t)\mathbf{x}(t) + z(t)\hat{B}(t)\mathbf{x}(t), \quad \mathbf{x}(0) = \mathbf{x}_0, \tag{6.68}$$

where $\hat{A}(t)$ and $\hat{B}(t)$ are the deterministic operators (e.g., differential operators with respect to auxiliary variable, or regular matrixes). We will assume that function $z(t)$ is the random delta-correlated function.

Averaging system (6.68), we obtain according to general formulas

$$\frac{d}{dt}\left\langle \mathbf{x}(t) \right\rangle = \hat{A}(t)\left\langle \mathbf{x}(t) \right\rangle + \left\langle \dot{\Theta}_t\left[t, \frac{\delta}{i\delta z(t)}\right]\mathbf{x}(t) \right\rangle. \tag{6.69}$$

Then, taking into account the equality

$$\frac{\delta}{\delta z(t-0)}\mathbf{x}(t) = \hat{B}(t)\mathbf{x}(t)$$

that follows immediately from Eq. (6.68), we can rewrite Eq. (6.69) in the form

$$\frac{d}{dt}\left\langle \mathbf{x}(t) \right\rangle = \hat{A}(t)\left\langle \mathbf{x}(t) \right\rangle + \dot{\Theta}_t\left[t, -i\hat{B}\right]\left\langle \mathbf{x}(t) \right\rangle. \tag{6.70}$$

Thus, in the case of linear system (6.68), equations for average values are also the linear equations.

We can expand the characteristic functional logarithm $\Theta[t; v(\tau)]$ of delta-correlated processes in the functional Fourier series

$$\Theta[t; v(\tau)] = \sum_{n=1}^{\infty} \frac{i^n}{n!} \int_0^t d\tau\, K_n(\tau) v^n(\tau), \tag{6.71}$$

where $K_n(t)$ determine the cumulant functions of process $z(t)$. Substituting Eq. (6.71) in Eq. (6.70), we obtain the equation

$$\frac{d}{dt}\left\langle \mathbf{x}(t) \right\rangle = \hat{A}(t)\left\langle \mathbf{x}(t) \right\rangle + \sum_{n=1}^{\infty} \frac{1}{n!} K_n(t)\left[\hat{B}(t)\right]^n \left\langle \mathbf{x}(t) \right\rangle. \tag{6.72}$$

If there exists power l such that $\hat{B}^l(t) = 0$, then Eq. (6.72) assumes the form

$$\frac{d}{dt}\left\langle \mathbf{x}(t) \right\rangle = \hat{A}(t)\left\langle \mathbf{x}(t) \right\rangle + \sum_{n=1}^{l-1} \frac{1}{n!} K_n(t)\left[\hat{B}(t)\right]^n \left\langle \mathbf{x}(t) \right\rangle. \tag{6.73}$$

In this case, the equation for the average value depends only on a finite number of cumulants of process $z(t)$. This means that there is no necessity in knowledge of probability distribution of function $z(t)$ in the context of the equation for average value; sufficient information consists only of certain cumulants of process and knowledge of the fact that process $z(t)$ can be considered as delta-correlated random process. Statistical description of an oscillator with fluctuating frequency is a good example of such system in physics.

Problems

Problem 22 *Starting from the backward Liouville equation (3.4), page 43 describing the behavior of dynamic system (6.1), page 84 as a function of initial values t_0 and $\mathbf{x}_0$, derive the backward equation for probability density $P(\mathbf{x}, t | \mathbf{x}_0, t_0) = \langle \delta(\mathbf{x}, t | \mathbf{x}_0, t_0) \rangle$.*

Solution.

$$\left(\frac{\partial}{\partial t_0} + \mathbf{v}(\mathbf{x}_0, t_0) \frac{\partial}{\partial \mathbf{x}_0} \right) P(\mathbf{x}, t | \mathbf{x}_0, t_0) = \left\langle \dot{\Theta}_{t_0} \left[t, t_0; \frac{\delta}{i \delta \mathbf{f}(\mathbf{y}, \tau)} \right] \delta \left(\mathbf{x}(t | \mathbf{x}_0, t_0) - \mathbf{x} \right) \right\rangle,$$

where

$$\dot{\Theta}_{t_0}[t, t_0; \mathbf{u}(\mathbf{y}, \tau)] = \frac{d}{dt_0} \ln \Phi[t, t_0; \mathbf{u}(\mathbf{y}, \tau)]$$

and $\Phi[t, t_0; \mathbf{u}(\mathbf{y}, \tau)]$ is the characteristic functional of random field $\mathbf{f}(\mathbf{x}_0, t_0)$.

Problem 23 *Derive the equation for probability density of the solution to the one-dimensional problem on passive tracer diffusion*

$$\frac{\partial}{\partial t} \rho(t, x) + v(t) \frac{\partial}{\partial x} f(x) \rho(t, x) = 0,$$

where $v(t)$ is the Gaussian stationary random process with the parameters

$$\langle v(t) \rangle = 0, \quad \langle v(t) v(t') \rangle = B(t - t') \quad \left(B_v(0) = \left\langle v^2(t) \right\rangle \right)$$

and $f(x)$ is the deterministic function.

Solution.

$$\frac{\partial}{\partial t} P(t, x; \rho) = \int\limits_0^t d\tau\, B(\tau) \left\{ \frac{\partial}{\partial x} f(x) - \frac{df(x)}{dx} \left(1 + \frac{\partial}{\partial \rho} \rho \right) \right\}$$

$$\times \left\{ \frac{\partial}{\partial x} f(x) - \frac{df(x)}{dx} \left(1 + \frac{\partial}{\partial \rho} \rho \right) \right\} P(t, x; \rho),$$
$$P(0, x; \rho) = \delta(\rho_0(x) - \rho).$$

Problem 24 *Derive the equation for the average of the field satisfying the Burgers equation with random shear*

$$\frac{\partial}{\partial t} q(t, x) + (q + z(t)) \frac{\partial}{\partial x} q(t, x) = \nu \frac{\partial^2}{\partial x^2} q(t, x),$$

where $z(t)$ is the stationary Gaussian process with correlation function $B(t - t') = \langle z(t) z(t') \rangle$.

Solution.

$$\frac{\partial}{\partial t} \langle q(t, x) \rangle + \frac{1}{2} \frac{\partial}{\partial x} \sum_{n=0}^{\infty} \frac{2^n}{n!} \left[\int\limits_0^t d\tau (t - \tau) B(\tau) \right]^n \left[\frac{\partial^n}{\partial x^n} \langle q(t, x) \rangle \right]^2$$

$$= \left(\nu + \int\limits_0^t d\tau\, B(\tau) \right) \frac{\partial^2}{\partial x^2} \langle q(t, x) \rangle.$$

Problem 25 *Under the assumption that random process $z(t)$ is delta-correlated, derive the equation for probability density of the solution to stochastic equation (1.14), page 16 describing the stochastic parametric resonance.*

Instruction. Rewrite Eq. (1.14) in the form of the system of equations

$$\frac{d}{dt}x(t) = y(t), \quad \frac{d}{dt}y(t) = -\omega_0^2[1 + z(t)]x(t),$$
$$x(0) = x_0, \quad y(0) = y_0. \tag{6.74}$$

Solution.

$$\frac{\partial}{\partial t}P(t;x,y) = \left(-y\frac{\partial}{\partial x} + \omega_0^2 x\frac{\partial}{\partial y}\right)P(t;x,y) + \dot{\Theta}_t\left[t; -i\omega_0^2 x\frac{\partial}{\partial y}\right]P(t;x,y),$$
$$P(0;x,y) = \delta(x - x_0)\,\delta(y - y_0), \tag{6.75}$$

where $\dot{\Theta}_t[t; v(\tau)] = \frac{d}{dt}\Theta[t; v(\tau)]$, $\Theta[t; v(\tau)] = \ln \Phi[t; v(\tau)]$, and $\Phi[t; v(\tau)]$ is the characteristic functional of random process $z(t)$.

Problem 26 *Starting from system of equations (6.74), derive the equation for the average of quantity*
$$A_k(t) = x^k(t)y^{N-k}(t) \quad (k = 0, \ldots, N)$$
under the assumption that $z(t)$ is the delta-correlated random process.

Solution.

$$\frac{d}{dt}\langle A_k(t)\rangle = k\langle A_{k-1}(t)\rangle - \omega_0^2(N - k)\langle A_{k+1}(t)\rangle$$
$$+ \sum_{n=1}^{N}\frac{1}{n!}K_n\left[B^n\right]_{kl}\langle A_l(t)\rangle \quad (k = 0, \ldots, N),$$

where K_n are the cumulants of random process $z(t)$ and matrix $B_{ij} = -\omega_0^2(N - i)\delta_{i,j-1}$. The square of this matrix has the form $\hat{B}_{ij}^2 = -\omega_0^4(N - i)(N - j + i)\delta_{i,j-2}$ and so forth for higher powers, so that $\hat{B}^{N+1} \equiv 0$.

Problem 27 *Derive the Fokker–Planck equation for system of equations (6.74) in the case of the Gaussian delta-correlated process $z(t)$ with the parameters*

$$\langle z(t)\rangle = 0, \quad \langle z(t)z(t')\rangle = 2\sigma^2\tau_0\delta(t - t').$$

Solution.

$$\frac{\partial}{\partial t}P(t;x,y) = \left(-y\frac{\partial}{\partial x} + \omega_0^2 x\frac{\partial}{\partial y}\right)P(t;x,y) + D\omega_0^2 x^2\frac{\partial^2}{\partial y^2}P(t;x,y),$$
$$P(0;x,y) = \delta(x - x_0)\,\delta(y - y_0), \tag{6.76}$$

where $D = \sigma^2\tau_0\omega_0^2$ is the coefficient of diffusion in space $\{x, y/\omega_0\}$.

Problem 28 *Obtain the solutions for the first- and second-order moments, which follow from Eq. (6.76) for $D/\omega_0 \ll 1$ and $x(0) = 0$, $y(0) = \omega_0$.*

Solution.

$$\langle x(t) \rangle = \sin \omega_0 t, \quad \langle y(t) \rangle = \omega_0 \cos \omega_0 t,$$

which coincides with the solution to system (6.74) for absent fluctuations, and

$$
\begin{aligned}
\left\langle x^2(t) \right\rangle &= \frac{1}{2} \left\{ e^{Dt} - e^{-Dt/2} \left[\cos(2\omega_0 t) + \frac{3D}{4\omega_0} \sin(2\omega_0 t) \right] \right\}, \\
\langle x(t)y(t) \rangle &= \frac{\omega_0}{4} \left\{ 2e^{-Dt/2} \sin(2\omega_0 t) + \frac{D}{\omega_0} \left[e^{Dt} - e^{-Dt/2} \cos(2\omega_0 t) \right] \right\}, \\
\left\langle y^2(t) \right\rangle &= \frac{\omega_0^2}{2} \left\{ e^{Dt} + e^{-Dt/2} \left[\cos(2\omega_0 t) - \frac{D}{4\omega_0} \sin(2\omega_0 t) \right] \right\}
\end{aligned}
\tag{6.77}
$$

Remark. Solution (6.77) includes terms increasing with time, which corresponds to statistical parametric build-up of oscillations in dynamic system (6.74) because of fluctuations of frequency. Solutions to statistical problem (6.74) have two characteristic temporal scales $t_1 \sim 1/\omega_0$ and $t_2 \sim 1/D$. The first temporal scale corresponds to the period of oscillations in system (6.74) without fluctuations (fast processes), and the second scale characterizes slow variations caused in statistical characteristics by fluctuations (slow processes). The ratio of these scales is small:

$$t_1/t_2 = D/\omega_0 \ll 1.$$

We can explicitly obtain slow variations of statistical characteristics of processes $x(t)$ and $y(t)$ by excluding fast motions by averaging the corresponding quantities over the period $T = 2\pi/\omega_0$. Denoting such averaging with the overbar, we have

$$\overline{\langle x^2(t) \rangle} = \frac{1}{2} e^{Dt}, \quad \overline{\langle x(t)y(t) \rangle} = 0, \quad \overline{\langle y^2(t) \rangle} = \frac{\omega_0^2}{2} e^{Dt}.$$

Problem 29 *Starting from Eq. (6.76), derive the third-order equation for average potential energy of an oscillator* $\langle U(t) \rangle = \langle x^2(t) \rangle$.

Solution.

$$\frac{d^3}{dt^3} \langle U(t) \rangle + 4\omega_0^2 \frac{d}{dt} \langle U(t) \rangle - 4D\omega_0^2 \langle U(t) \rangle = 0, \tag{6.78}$$

Problem 30 *Derive the Fokker–Planck equation for the stochastic oscillator with linear friction*

$$\frac{d}{dt} x(t) = y(t), \quad \frac{d}{dt} y(t) = -2\gamma y(t) - \omega_0^2 [1 + z(t)]x(t),$$
$$x(0) = x_0, \quad y(0) = y_0, \tag{6.79}$$

and the condition of stochastic excitation of second moments.

Solution.

$$\frac{\partial}{\partial t} P(t; x, y) = \left(2\gamma \frac{\partial}{\partial y} y - y \frac{\partial}{\partial x} + \omega_0^2 x \frac{\partial}{\partial y} \right) P(t; x, y)$$
$$+ D\omega_0^2 x^2 \frac{\partial^2}{\partial y^2} P(t; x, y), \quad P(0; x, y) = \delta(x - x_0)\, \delta(y - y_0),$$

where $D = \sigma^2 \tau_0 \omega_0^2$ is the diffusion coefficient. If we substitute in the system of equations for second moments the solutions proportional to $\exp\{\lambda t\}$, we obtain the characteristic equation in λ of the form

$$\lambda^3 + 6\gamma\lambda^2 + 4(\omega_0^2 + 2\gamma^2)\lambda + 4\omega_0^2(2\gamma - D) = 0.$$

Under the condition

$$2\gamma < D, \tag{6.80}$$

second moments are the functions exponentially increasing with time, which means the statistical parametric excitation of second moments.

Problem 31 *Obtain steady-state values of second moments formed under the action of random forces on the stochastic parametric oscillator with friction*

$$\frac{d}{dt}x(t) = y(t), \quad x(0) = x_0,$$

$$\frac{d}{dt}y(t) = -2\gamma y(t) - \omega_0^2[1 + z(t)]x(t) + f(t), \quad y(0) = y_0. \tag{6.81}$$

Here, $f(t)$ is the Gaussian delta-correlated process statistically independent of process $z(t)$ and having the parameters

$$\langle f(t)\rangle = 0, \quad \langle f(t)f(t')\rangle = 2\sigma_f^2\tau_f\delta(t - t'),$$

where σ_f^2 is the variance and τ_f is the temporal correlation radius.

Solution. In the case of stochastic system of equations (6.81), the one-time probability density satisfies the Fokker–Planck equation

$$\frac{\partial}{\partial t}P(t; x, y) = \left(2\gamma\frac{\partial}{\partial y}y - y\frac{\partial}{\partial x} + \omega_0^2 x\frac{\partial}{\partial y}\right)P(t; x, y)$$

$$+ D\omega_0^2 x^2\frac{\partial^2}{\partial y^2}P(t; x, y) + \sigma_f^2\tau_f\frac{\partial^2}{\partial y^2}P(t; x, y),$$

$$P(0; x, y) = \delta(x - x_0)\,\delta(y - y_0).$$

Consequently, the steady-state solution for second moments exists for $t \to \infty$ under the condition (6.80) and has the form

$$\langle x(t)\rangle \;=\; 0, \quad \langle y(t)\rangle = 0,$$

$$\langle x(t)y(t)\rangle \;=\; 0, \quad \left\langle x^2(t)\right\rangle = \frac{\sigma_f^2\tau_f}{\omega_0^2(D - 2\gamma)}, \quad \left\langle y^2(t)\right\rangle = \frac{\sigma_f^2\tau_f}{D - 2\gamma}.$$

Problem 32 *Starting from Eq. (6.75), derive the Kolmogorov–Feller equation in the case of the Poisson delta-correlated random process $z(t)$.*

Solution.

$$\frac{\partial}{\partial t}P(t; x, y) = \left(-y\frac{\partial}{\partial x} + \omega_0^2 x\frac{\partial}{\partial y}\right)P(t; x, y)$$

$$+ \nu\int\limits_{-\infty}^{\infty} d\xi p(\xi)P(t; x, y + \xi\omega_0^2 x) - \nu P(t; x, y). \tag{6.82}$$

Remark. For sufficiently small ξ, Eq. (6.82) grades into the Fokker–Planck equation (6.76) with the diffusion coefficient

$$D = \frac{1}{2}\nu \left\langle \xi^2 \right\rangle \omega_0^2 x.$$

Problem 33 *Derive the equation for the characteristic functional of the solution to linear parabolic equation (1.49), page 26.*

Solution. Characteristic functional of the problem solution

$$\Phi[x; v, v^*] = \langle \varphi[x; v, v^*] \rangle,$$

where

$$\varphi[x; v, v^*] = \exp\left\{ i \int d\mathbf{R}' \left[u(x, \mathbf{R}')v(\mathbf{R}') + u^*(x, \mathbf{R}')v^*(\mathbf{R}') \right] \right\},$$

satisfies the equation

$$\frac{\partial}{\partial x}\Phi[x; v, v^*] = \left\langle \dot{\Theta}_x \left[x; \frac{\delta}{i\delta\varepsilon(\xi, \mathbf{R}')} \right] \varphi[x; v, v^*] \right\rangle$$
$$+ \frac{i}{2k}\left\{ \int d\mathbf{R}' \left[v(\mathbf{R}')\Delta_{\mathbf{R}'}\frac{\delta}{\delta v(\mathbf{R}')} - v^*(\mathbf{R}')\Delta_{\mathbf{R}'}\frac{\delta}{\delta v^*(\mathbf{R}')} \right] \right\} \Phi[x; v, v^*], \qquad (6.83)$$

where

$$\dot{\Theta}_x\left[x; \psi(\xi, \mathbf{R}') \right] = \frac{d}{dx}\ln \left\langle \exp\left\{ i \int\limits_0^x d\xi \int d\mathbf{R}'\varepsilon(\xi, \mathbf{R}')\psi(\xi, \mathbf{R}') \right\} \right\rangle$$

is the derivative of the characteristic functional logarithm of field $\varepsilon(x, \mathbf{R})$.

Problem 34 *Derive the equation for the characteristic functional of the solution to linear parabolic equation (1.49), page 26 assuming that $\varepsilon(x, \mathbf{R})$ is the homogeneous random field delta-correlated in x.*

Solution. Characteristic functional $\Phi[x; v, v^*]$ of the problem solution satisfies the equation

$$\frac{\partial}{\partial x}\Phi[x; v, v^*] = \dot{\Theta}_x\left[x, \frac{k}{2}\hat{M}(\mathbf{R}') \right] \Phi[x; v, v^*]$$
$$+ \frac{i}{2k}\left\{ \int d\mathbf{R}' \left[v(\mathbf{R}')\Delta_{\mathbf{R}'}\frac{\delta}{\delta v(\mathbf{R}')} - v^*(\mathbf{R}')\Delta_{\mathbf{R}'}\frac{\delta}{\delta v^*(\mathbf{R}')} \right] \right\} \Phi[x; v, v^*] \qquad (6.84)$$

with the Hermitian operator

$$\hat{M}(\mathbf{R}') = v(\mathbf{R}')\frac{\delta}{\delta v(\mathbf{R}')} - v^*(\mathbf{R}')\frac{\delta}{\delta v^*(\mathbf{R}')}.$$

Here, functional

$$\dot{\Theta}_x\left[x; \psi(\xi, \mathbf{R}') \right] = \frac{d}{dx}\ln \left\langle \exp\left\{ i \int\limits_0^x d\xi \int d\mathbf{R}'\varepsilon(\xi, \mathbf{R}')\psi(\xi, \mathbf{R}') \right\} \right\rangle$$

is the derivative of the logarithm of the characteristic functional of filed $\varepsilon(x, \mathbf{R})$.

Remark. Equation (6.84) yields the equations for the moment functions of field $u(x, \mathbf{R})$,

$$M_{m,n}(x; \mathbf{R}_1, ..., \mathbf{R}_m; \mathbf{R}'_1, ..., \mathbf{R}'_n) = \langle u(x, \mathbf{R}_1)...u(x, \mathbf{R}_m)u^*(x, \mathbf{R}'_1)...u^*(x, \mathbf{R}'_n) \rangle,$$

(for $m = n$, these functions are usually called the *coherence functions* of order $2n$)

$$\frac{\partial}{\partial x} M_{m,n} = \frac{i}{2k} \left(\sum_{p=1}^{m} \Delta_{\mathbf{R}_p} - \sum_{q=1}^{n} \Delta_{\mathbf{R}'_q} \right) M_{m,n}$$

$$+ \dot{\Theta}_x \left[x, \frac{1}{k} \left(\sum_{p=1}^{m} \delta(\mathbf{R}' - \mathbf{R}_p) - \sum_{q=1}^{n} \delta(\mathbf{R}' - \mathbf{R}'_q) \right) \right] M_{m,n}. \tag{6.85}$$

Problem 35 *Assuming that $\varepsilon(x, \mathbf{R})$ is the homogeneous Gaussian random process delta-correlated in x with the correlation function*

$$B_\varepsilon(x, \mathbf{R}) = A(\mathbf{R})\delta(x), \quad A(\mathbf{R}) = \int\limits_{-\infty}^{\infty} dx B_\varepsilon(x, \mathbf{R}),$$

derive the equations for the characteristic functional of wave field $u(x, \mathbf{R})$, which is the solution to linear parabolic equation (1.49), page 26, and this field moment functions [38].

Solution.

$$\frac{\partial}{\partial x} \Phi[x; v, v^*] = -\frac{k^2}{8} \int d\mathbf{R}' \int d\mathbf{R}\, A(\mathbf{R}' - \mathbf{R}) \hat{M}(\mathbf{R}') \hat{M}(\mathbf{R}) \Phi[x; v, v^*]$$

$$+ \frac{i}{2k} \int d\mathbf{R}' \left[v(\mathbf{R}') \Delta_{\mathbf{R}'} \frac{\delta}{\delta v(\mathbf{R}')} - v^*(\mathbf{R}') \Delta_{\mathbf{R}'} \frac{\delta}{\delta v^*(\mathbf{R}')} \right] \Phi[x; v, v^*],$$

$$\frac{\partial}{\partial x} M_{m,n} = \frac{i}{2k} \left(\sum_{p=1}^{m} \Delta_{\mathbf{R}_p} - \sum_{q=1}^{n} \Delta_{\mathbf{R}'_q} \right) M_{m,n} - \frac{k^2}{8} Q(\mathbf{R}_1, ..., \mathbf{R}_m; \mathbf{R}'_1, ..., \mathbf{R}'_m) M_{m,n},$$

where

$$Q(\mathbf{R}_1, ..., \mathbf{R}_m; \mathbf{R}'_1, ..., \mathbf{R}'_m)$$

$$= \sum_{i=1}^{m} \sum_{j=1}^{m} A(\mathbf{R}_i - \mathbf{R}_j) - 2 \sum_{i=1}^{m} \sum_{j=1}^{n} A(\mathbf{R}_i - \mathbf{R}'_j) + \sum_{i=1}^{n} \sum_{j=1}^{n} A(\mathbf{R}'_i - \mathbf{R}'_j).$$

Problem 36 *Derive the equation for the one-time characteristic functional of the solution to stochastic nonlinear integro-differential equation (1.57), page 29.*

Solution. Characteristic functional of the velocity field

$$\Phi[t; \mathbf{z}(\mathbf{k}')] = \Phi[t; \mathbf{z}] = \langle \varphi[t; \mathbf{z}(\mathbf{k}')] \rangle,$$

where

$$\varphi[t; \mathbf{z}(\mathbf{k}')] = \exp\left\{ i \int d\mathbf{k}' \hat{\mathbf{u}}(\mathbf{k}', t)\mathbf{z}(\mathbf{k}') \right\},$$

satisfies the unclosed variational-derivative equation

$$\frac{\partial}{\partial t}\Phi[t;\mathbf{z}] = \int d\mathbf{k}\, z_i(\mathbf{k}) \left\{ -\frac{1}{2}\int d\mathbf{k}_1 \int d\mathbf{k}_2 \Lambda_{i,\alpha\beta}(\mathbf{k}_1,\mathbf{k}_2,\mathbf{k})\frac{\delta^2}{\delta z_\alpha(\mathbf{k}_1)\delta z_\beta(\mathbf{k}_2)} \right.$$
$$\left. - \nu k^2 \frac{\delta}{\delta z_i(\mathbf{k})}\Phi[t;\mathbf{z}] \right\} + \left\langle \dot{\Theta}_t\left[t;\frac{\delta}{i\delta\mathbf{f}(\boldsymbol{\kappa},\tau)}\right]\varphi[t;\mathbf{z}] \right\rangle,$$

where

$$\dot{\Theta}_t\left[t;\boldsymbol{\psi}(\boldsymbol{\kappa},\tau)\right] = \frac{d}{dt}\ln\left\langle \exp\left\{ i\int_0^t d\tau \int d\boldsymbol{\kappa}\,\mathbf{f}(\boldsymbol{\kappa},\tau)\boldsymbol{\psi}(\boldsymbol{\kappa},\tau) \right\} \right\rangle$$

is the derivative of the characteristic functional logarithm of external forces $\hat{\mathbf{f}}(\mathbf{k},t)$.

Problem 37 *Assuming that $\mathbf{f}(\mathbf{x},t)$ is the homogeneous stationary random field delta-correlated in time, derive the equation for the one-time characteristic functional of the solution to stochastic nonlinear integro-differential equation (1.57), 29. Consider the case of the Gaussian field $\mathbf{f}(\mathbf{x},t)$ [3].*

Solution.

$$\frac{\partial}{\partial t}\Phi[t;\mathbf{z}] = \int d\mathbf{k}\, z_i(\mathbf{k}) \left\{ -\frac{1}{2}\int d\mathbf{k}_1 \int d\mathbf{k}_2 \Lambda_{i,\alpha\beta}(\mathbf{k}_1,\mathbf{k}_2,\mathbf{k})\frac{\delta^2}{\delta z_\alpha(\mathbf{k}_1)\delta z_\beta(\mathbf{k}_2)} \right.$$
$$\left. - \nu k^2 \frac{\delta}{\delta z_i(\mathbf{k})} \right\}\Phi[t;\mathbf{z}] + \dot{\Theta}_t\left[t;\mathbf{z}(\mathbf{k})\right]\Phi[t;\mathbf{z}], \tag{6.86}$$

where

$$\dot{\Theta}_t\left[t;\boldsymbol{\psi}(\boldsymbol{\kappa},\tau)\right] = \frac{d}{dt}\ln\left\langle \exp\left\{ i\int_0^t d\tau \int d\boldsymbol{\kappa}\,\hat{\mathbf{f}}(\boldsymbol{\kappa},\tau)\boldsymbol{\psi}(\boldsymbol{\kappa},\tau) \right\} \right\rangle$$

is the derivative of the characteristic functional logarithm of external forces $\hat{\mathbf{f}}(\mathbf{k},t)$.

If we assume now that $\mathbf{f}(\mathbf{x},t)$ is the Gaussian random field homogeneous and isotropic in space and stationary in time with the correlation tensor $B_{ij}(\mathbf{x}_1-\mathbf{x}_2,t_1-t_2) = \langle f_i(\mathbf{x}_1,t_1)f_j(\mathbf{x}_2,t_2)\rangle$, then the field $\hat{\mathbf{f}}(\mathbf{k},t)$ will also be the Gaussian stationary random field with the correlation tensor

$$\left\langle \hat{f}_i(\mathbf{k},t+\tau)\hat{f}_j(\mathbf{k}',t) \right\rangle = \frac{1}{2}F_{ij}(\mathbf{k},\tau)\delta(\mathbf{k}+\mathbf{k}),$$

where $F_{ij}(\mathbf{k},\tau)$ is the external force spatial spectrum given by the formula

$$F_{ij}(\mathbf{k},\tau) = 2(2\pi)^3 \int d\mathbf{x}\,B_{ij}(\mathbf{x},\tau)e^{-i\mathbf{k}\mathbf{x}}.$$

In view of the fact that forces are spatially isotropic, we have

$$F_{ij}(\mathbf{k},\tau) = F(k,\tau)\Delta_{ij}(\mathbf{k}).$$

As long as field $\hat{\mathbf{f}}(\mathbf{k},t)$ is delta-correlated in time, we have $F(k,\tau) = F(k)\delta(\tau)$, so that functional $\Theta\left[t;\boldsymbol{\psi}(\boldsymbol{\kappa},\tau)\right]$ is given by the formula

$$\Theta\left[t;\boldsymbol{\psi}(\boldsymbol{\kappa},\tau)\right] = -\frac{1}{4}\int_0^t d\tau \int d\boldsymbol{\kappa}\, F(\kappa)\Delta_{ij}(\boldsymbol{\kappa})\psi_i(\boldsymbol{\kappa},\tau)\psi_j(-\boldsymbol{\kappa},\tau),$$

and Eq. (6.86) assumes the closed form

$$\frac{\partial}{\partial t}\Phi[t;\mathbf{z}] = -\frac{1}{4}\int d\mathbf{k}F(k)\Delta_{ij}(\mathbf{k})z_i(\mathbf{k})z_j(-\mathbf{k})\Phi[t;\mathbf{z}]$$

$$-\int d\mathbf{k}z_i(\mathbf{k})\left\{\frac{1}{2}\int d\mathbf{k}_1\int d\mathbf{k}_2\Lambda_{i,\alpha\beta}(\mathbf{k}_1,\mathbf{k}_2,\mathbf{k})\frac{\delta^2}{\delta z_\alpha(\mathbf{k}_1)\delta z_\beta(\mathbf{k}_2)} + \nu k^2\frac{\delta}{\delta z_i(\mathbf{k})}\right\}\Phi[t;\mathbf{z}].$$

Problem 38 *Average the linear stochastic integral equation*

$$S(\mathbf{r},\mathbf{r}') = S_0(\mathbf{r},\mathbf{r}')$$
$$+ \int d\mathbf{r}_1\int d\mathbf{r}_2\int d\mathbf{r}_3 S_0(\mathbf{r},\mathbf{r}_1)\Lambda(\mathbf{r}_1,\mathbf{r}_2,\mathbf{r}_3)\left[\eta(\mathbf{r}_2) + f(\mathbf{r}_2)\right]S(\mathbf{r}_3,\mathbf{r}'),$$

Solution.

$$G[\mathbf{r},\mathbf{r}';\eta(\mathbf{r})] = S_0(\mathbf{r},\mathbf{r}')$$
$$+ \int d\mathbf{r}_1\int d\mathbf{r}_2\int d\mathbf{r}_3 S_0(\mathbf{r},\mathbf{r}_1)\Lambda(\mathbf{r}_1,\mathbf{r}_2,\mathbf{r}_3)\eta(\mathbf{r}_2)G[\mathbf{r}_3,\mathbf{r}';\eta(\mathbf{r})]$$
$$+ \int d\mathbf{r}_1\int d\mathbf{r}_2\int d\mathbf{r}_3 S_0(\mathbf{r},\mathbf{r}_1)\Lambda(\mathbf{r}_1,\mathbf{r}_2,\mathbf{r}_3)\Omega_{\mathbf{r}_2}\left[\frac{\delta}{i\delta\eta(\mathbf{r})}\right]G[\mathbf{r}_3,\mathbf{r}';\eta(\mathbf{r})].$$

Here,

$$G[\mathbf{r},\mathbf{r}';\eta(\mathbf{r})] = \langle S[\mathbf{r},\mathbf{r}';f(\mathbf{r}) + \eta(\mathbf{r})]\rangle$$

and functional

$$\Omega_{\mathbf{r}}[v(\mathbf{r})] = \frac{\delta}{i\delta v(\mathbf{r})}\Theta[v(\mathbf{r})], \quad \Theta[v(\mathbf{r})] = \ln\left\langle\exp\left\{i\int d\mathbf{r}f(\mathbf{r})v(\mathbf{r})\right\}\right\rangle.$$

Problem 39 *Assuming that random external force* $\mathbf{f}(\mathbf{r},t)$ *is the homogeneous stationary Gaussian field, derive the equation for the characteristic functional of space-time harmonics of turbulent velocity field (1.59), page 29 [39].*

Solution. Characteristic functional of the velocity field

$$\Phi[\mathbf{z}] = \left\langle\exp\left\{i\int d^4\mathbf{K}'\hat{\mathbf{u}}\left(\mathbf{K}'\right)\mathbf{z}\left(\mathbf{K}'\right)\right\}\right\rangle$$

and functional

$$G_{ij}\left[\mathbf{K},\mathbf{K}';\mathbf{z}\right] = \left\langle\frac{\delta u_i(\mathbf{K})}{\delta\widehat{f}_j(\mathbf{K}')}\varphi[\mathbf{z}]\right\rangle$$

satisfy the closed system of functional equations

$$(i\omega + \nu k^2)\frac{\delta}{\delta z_i(\mathbf{K})}\Phi[\mathbf{z}] = -\frac{1}{2}F_{ij}(\mathbf{K})\int d^4\mathbf{K}_1 z_\alpha\left(\mathbf{K}_1\right)G_{\alpha j}\left[\mathbf{K}_1,-\mathbf{K};\mathbf{z}\right]$$

$$-\frac{1}{2}\int d^4\mathbf{K}_1\int d^4\mathbf{K}_2\Lambda_i^{\alpha\beta}\left(\mathbf{K}_1,\mathbf{K}_2,\mathbf{K}\right)\frac{\delta^2\Phi[\mathbf{z}]}{\delta z_\alpha\left(\mathbf{K}_1\right)\delta z_\beta\left(\mathbf{K}_2\right)},$$

$$(i\omega + \nu k^2)G_{ij}\left[\mathbf{K},\mathbf{K}';\mathbf{z}\right] + \int d^4\mathbf{K}_1\int d^4\mathbf{K}_2\Lambda_i^{\alpha\beta}\left(\mathbf{K}_1,\mathbf{K}_2,\mathbf{K}\right)\frac{\delta}{\delta z_\alpha(\mathbf{K}_1)}G_{\beta j}\left[\mathbf{K}_2,\mathbf{K}';\mathbf{z}\right]$$

$$= \delta_{ij}\delta^4\left(\mathbf{K}-\mathbf{K}'\right)\Phi[\mathbf{z}],$$

where $\left\langle\widehat{f}_i(\mathbf{K}_1)\widehat{f}_j(\mathbf{K}_2)\right\rangle = \frac{1}{2}\delta^4\left(\mathbf{K}_1+\mathbf{K}_2\right)F_{ij}(\mathbf{K}_1)$, and $F_{ij}(\mathbf{K})$ is the space-time spectrum of external forces.

Remark. Expansion of functionals $\Phi[\mathbf{z}]$ and $G_{ij}\left[\mathbf{K},\mathbf{K}';\mathbf{z}\right]$ in the functional Taylor series in $\mathbf{z}\left(\mathbf{K}\right)$ determines velocity field cumulants and correlators of Green's function analog $S_{ij}\left(\mathbf{K},\mathbf{K}'\right) = \delta u_i\left(\mathbf{K}\right)/\delta f_j\left(\mathbf{K}'\right)$ with the velocity field.

Chapter 7

Stochastic equations with the Markovian fluctuations of parameters

In the preceding chapter, we dealt with the statistical description of dynamic systems at the level of general methods assuming that the characteristic functional of fluctuating parameters is known. However, this functional is unknown in most cases, and we are forced to resort either to assumptions on the model of parameter fluctuations, or to asymptotic approximations.

The methods based on approximating the fluctuating parameters with the Markovian random processes and fields with a finite temporal correlation radius are widely used. Such approximations can be found, for example, as solutions to the dynamic equations with delta-correlated parameter fluctuations (see the preceding chapter). Consider such methods in greater detail using the Markovian random processes as an example.

Consider stochastic equations of the form

$$\frac{d}{dt}\mathbf{x}(t) = f\left(t, \mathbf{x}, \mathbf{z}(t)\right), \quad \mathbf{x}(0) = \mathbf{x}_0, \tag{7.1}$$

where $f\left(t, \mathbf{x}, \mathbf{z}(t)\right)$ is the deterministic function of its arguments and $\mathbf{z}(t) = \{z_1(t), ..., z_n(t)\}$ is the Markovian vector process.

Our task consists in the determination of statistical characteristics of the solution to Eq. (7.1) from known statistical characteristics of process $\mathbf{z}(t)$.

In the general case of arbitrary Markovian process $\mathbf{z}(t)$, we cannot judge about process $\mathbf{x}(t)$. We can only assert that the joint process $\{\mathbf{x}(t), \mathbf{z}(t)\}$ is the Markovian process with joint probability density $P(\mathbf{x}, \mathbf{z}, t)$.

If we would able to solve equation for $P(\mathbf{x}, \mathbf{z}, t)$, then we could integrate the solution over $\mathbf{z}$ to obtain the probability density of the solution to Eq. (7.1), i.e., function $P(\mathbf{x}, t)$. In this case, process $\mathbf{x}(t)$ would not be the Markovian process.

There are several types of processes $\mathbf{z}(t)$ that allow obtaining equations for density $P(\mathbf{x}, t)$ without solving equation for $P(\mathbf{x}, \mathbf{z}, t)$. Among these processes, we mention first of all the telegrapher's and generalized telegrapher's processes, Markovian processes with finite number of states, and Gaussian Markovian processes. Below, we discuss the telegrapher's and Gaussian Markovian processes in more detail as examples of processes widely used in different applications.

111

7.1   Telegrapher's processes

Recall that telegrapher's random process $z(t)$ (the binary, or two-state process) is defined by the equality

$$z(t) = a(-1)^{n(0,t)},$$

where random quantity a assumes values $a = \pm a_0$ with probabilities $1/2$.

Telegrapher's process $z(t)$ is stationary in time and its correlation function

$$\langle z(t)z(t')\rangle = a_0^2 e^{-2\nu|t-t'|}$$

has the temporal correlation radius $\tau_0 = 1/(2\nu)$.

For the correlator between telegrapher's process $z(t)$ and arbitrary functional $R[t; z(\tau)]$, where $\tau \le t$, we obtained the relationship of correlation splitting (5.26), page 75

$$\langle z(t)R[t; z(\tau)]\rangle = a_0^2 \int\limits_0^t dt_1 e^{-2\nu(t-t_1)}\left\langle \frac{\delta}{\delta z(t_1)}\tilde{R}[t; z(\tau)]\right\rangle, \tag{7.2}$$

where functional $\tilde{R}[t; z(\tau)]$ is given by the formula

$$\tilde{R}[t,t_1; z(\tau)] = R[t; z(\tau)\theta(t_1 - \tau + 0)] \quad (t_1 < t). \tag{7.3}$$

Formula (7.2) is appropriate for analyzing stochastic equations linear in process $z(t)$. Let functional $R[t; z(\tau)]$ be the solution to a system of differential equations of the first order in time with initial conditions at $t = 0$. Functional $\tilde{R}[t,t_1; z(\tau)]$ will also satisfy the same system of equations with product $z(t)\theta(t_1 - t)$ instead of $z(t)$. Consequently, for all times $t > t_1$, functional $\tilde{R}[t,t_1; z(\tau)] = R[t; 0]$ and satisfies the same system of equations in the absence of fluctuations (i.e., at $z(t) = 0$) with the initial condition $\tilde{R}[t_1,t_1; z(\tau)] = R[t_1; z(\tau)]$.

Another formula convenient in the context of stochastic equations linear in random telegrapher's process $z(t)$ concerns the differentiation of a correlator of this process with arbitrary functional $R[t; z(\tau)]$ $(\tau \le t)$ (5.27)

$$\frac{d}{dt}\langle z(t)R[t; z(\tau)]\rangle = -2\nu\langle z(t)R[t; z(\tau)]\rangle + \left\langle z(t)\frac{d}{dt}R[t; z(\tau)]\right\rangle. \tag{7.4}$$

Formula (7.4) determines the rule of factoring the differentiation operation out of averaging brackets

$$\left\langle z(t)\frac{d^n}{dt^n}R[t; z(\tau)]\right\rangle = \left(\frac{d}{dt} + 2\nu\right)^n\langle z(t)R[t; z(\tau)]\rangle. \tag{7.5}$$

We consider some special examples to show the usability of these formulas. It is evident that both methods give the same result. However, the method based on the differentiation formula appears more practicable.

The first example concerns the system of linear operator equations

$$\frac{d}{dt}\mathbf{x}(t) = \hat{A}(t)\mathbf{x}(t) + z(t)\hat{B}(t)\mathbf{x}(t), \quad \mathbf{x}(0) = \mathbf{x}_0, \tag{7.6}$$

where $\hat{A}(t)$ and $\hat{B}(t)$ are certain differential operators (they may include differential operators with respect to auxiliary variables). If operators $\hat{A}(t)$ and $\hat{B}(t)$ are matrixes, then Eqs. (7.6) describe the linear dynamic system.

Average Eqs. (7.6) over an ensemble of random functions $z(t)$. The result will be the equation

$$\frac{d}{dt}\langle \mathbf{x}(t)\rangle = \hat{A}(t)\langle \mathbf{x}(t)\rangle + \hat{B}(t)\boldsymbol{\psi}(t), \quad \langle \mathbf{x}(0)\rangle = \mathbf{x}_0, \tag{7.7}$$

where we introduced new functions

$$\boldsymbol{\psi}(t) = \langle z(t)\mathbf{x}(t)\rangle.$$

We can use formula (7.4) for these functions; as a result, we obtain the equality

$$\frac{d}{dt}\boldsymbol{\psi}(t) = -2\nu\boldsymbol{\psi}(t) + \left\langle z(t)\frac{d}{dt}\mathbf{x}(t)\right\rangle. \tag{7.8}$$

Substituting now derivative $d\mathbf{x}/dt$ (7.6) in Eq. (7.8), we obtain the equation for the vector function $\boldsymbol{\psi}(t)$

$$\left(\frac{d}{dt} + 2\nu\right)\boldsymbol{\psi}(t) = \hat{A}(t)\boldsymbol{\psi}(t) + \hat{B}(t)\left\langle z^2(t)\mathbf{x}(t)\right\rangle. \tag{7.9}$$

Because $z^2(t) \equiv a_0^2$ for telegrapher's process, we obtain finally the closed system of linear equations for vectors $\langle \mathbf{x}(t)\rangle$ and $\boldsymbol{\psi}(t)$

$$\frac{d}{dt}\langle \mathbf{x}(t)\rangle = \hat{A}(t)\langle \mathbf{x}(t)\rangle + \hat{B}(t)\boldsymbol{\psi}(t), \quad \langle \mathbf{x}(0)\rangle = \mathbf{x}_0,$$
$$\left(\frac{d}{dt} + 2\nu\right)\boldsymbol{\psi}(t) = \hat{A}(t)\boldsymbol{\psi}(t) + a_0^2\hat{B}(t)\langle \mathbf{x}(t)\rangle, \quad \boldsymbol{\psi}(0) = (0). \tag{7.10}$$

If operators $\hat{A}(t)$ and $\hat{B}(t)$ are the time-independent matrixes A and B, we can solve system (7.10) using the Laplace transform. After the Laplace transform, system (7.10) becomes the algebraic system of equations

$$(pE - A)\langle \mathbf{x}\rangle_p - B\boldsymbol{\psi}_p = \mathbf{x}_0,$$
$$[(p + 2\nu)E - A]\boldsymbol{\psi}_p - a_0^2 B\langle \mathbf{x}\rangle_p = 0, \tag{7.11}$$

where E is the unit matrix. From this system, we obtain solution $\langle \mathbf{x}\rangle_p$ in the form

$$\langle \mathbf{x}\rangle_p = \left[(pE - A) - a_0^2 B\frac{1}{(p + 2\nu)E - A}B\right]^{-1}\mathbf{x}_0. \tag{7.12}$$

Let now operators $\hat{A}(t)$ and $\hat{B}(t)$ in Eq. (7.6) be the matrixes. If only one component is of our interest, we can obtain for it the operator equation

$$\hat{L}\left(\frac{d}{dt}\right)x(t) + \sum_{i+j=0}^{n-1} b_{ij}(t)\frac{d^i}{dt^i}z(t)\frac{d^j}{dt^j}x(t) = f(t), \tag{7.13}$$

where

$$\hat{L}\left(\frac{d}{dt}\right) = \sum_{i=0}^{n} a_i(t)\frac{d^i}{dt^i}$$

and n is the order of matrixes A and B in Eq. (7.6). The initial conditions for x are included in function $f(t)$ through the corresponding derivatives of delta function. Note that function $f(t)$ can depend additionally on derivatives of random process $z(t)$ at $t = 0$, i.e., $f(t)$ is also the random function statistically related to process $z(t)$.

Averaging Eq. (7.13) over an ensemble of realizations of process $z(t)$ with the use of formula (7.5), we obtain the equation

$$\hat{L}\left(\frac{d}{dt}\right)\langle x(t)\rangle + M\left[\frac{d}{dt},\frac{d}{dt}+2\nu\right]\langle z(t)x(t)\rangle = \langle f(t)\rangle,\qquad(7.14)$$

where

$$M[p,q] = \sum_{i+j=0}^{n-1} b_{ij}(t)p^i q^j.$$

However, Eq. (7.14) is unclosed because of the presence of correlation $\langle z(t)x(t)\rangle$. Multiplying Eq. (7.13) by $z(t)$ and averaging the result, we obtain the equation for correlation $\langle z(t)x(t)\rangle$

$$\hat{L}\left(\frac{d}{dt}+2\nu\right)\langle z(t)x(t)\rangle + a_0^2 M\left[\frac{d}{dt}+2\nu,\frac{d}{dt}\right]\langle x(t)\rangle = \langle z(t)f(t)\rangle.\qquad(7.15)$$

System of equations (7.14) and (7.15) is the closed system. If functions $a_i(t)$ and $b_{ij}(t)$ are independent of time, this system can be solved using Laplace transform. The solution is as follows:

$$\langle x\rangle_p = \frac{\hat{L}(p+2\nu)\langle f\rangle_p - M[p,p+2\nu]\langle zf\rangle_p}{\hat{L}(p)\hat{L}(p+2\nu) - a_0^2 M[p+2\nu,p]M[p,p+2\nu]}.\qquad(7.16)$$

7.2 Gaussian Markovian processes

Here, we consider several examples associated with the Gaussian Markovian processes. Define random process $z(t)$ by the formula

$$z(t) = z_1(t) + \dots + z_N(t),\qquad(7.17)$$

where $z_i(t)$ are statistically independent telegrapher's processes with correlation functions

$$\langle z_i(t)z_j(t')\rangle = \delta_{ij}\left\langle z^2\right\rangle e^{-\alpha|t-t'|}\quad(\alpha = 2\nu).$$

If we set $\langle z^2\rangle = \sigma^2/N$, then this process grades for $N \to \infty$ into the Gaussian Markovian process with correlation function

$$\langle z_i(t)z_j(t')\rangle = \left\langle z^2\right\rangle e^{-\alpha|t-t'|}.$$

Thus, process $z(t)$ (7.17) approximates the Gaussian Markovian process in terms of the Markovian process with a finite number of states.

It is evident that the differentiation formula and the rule of factoring the derivative out of averaging brackets for process $z(t)$ assume the forms

$$\left(\frac{d}{dt}+\alpha k\right)\langle z_1(t)\dots z_k(t)R[t;z(\tau)]\rangle = \left\langle z_1(t)\dots z_k(t)\frac{d}{dt}R[t;z(\tau)]\right\rangle,$$

$$\left\langle z_1(t)\dots z_k(t)\frac{d^n}{dt^n}R[t;z(\tau)]\right\rangle = \left(\frac{d}{dt}+\alpha k\right)^n\langle z_1(t)\dots z_k(t)R[t;z(\tau)]\rangle,$$

$$(7.18)$$

where $R[t;z(\tau)]$ is an arbitrary functional of process $z(t)$ $(\tau \le t)$.

Consider again Eq. (7.6), which we represent here in the form

$$\frac{d}{dt}\mathbf{x}(t) = \hat{A}(t)\mathbf{x}(t) + [z_1(t) + \dots + z_N(t)]\hat{B}(t)\mathbf{x}(t), \quad \mathbf{x}(0) = \mathbf{x}_0, \tag{7.19}$$

and introduce vector-function

$$\mathbf{X}_k(t) = \langle z_1(t)\dots z_k(t)\mathbf{x}(t)\rangle, \quad k = 1, \dots, N; \quad \mathbf{X}_0(t) = \langle \mathbf{x}(t)\rangle. \tag{7.20}$$

Using formula (7.18) for differentiating correlations (7.20) and Eq. (7.19), we obtain the recurrence equation for $\mathbf{x}_k(t), k = 0, 1, \dots, N$

$$\begin{aligned}
\frac{d}{dt}\mathbf{X}_k(t) &= \hat{A}(t)\mathbf{X}_k(t) + \left\langle z_1(t)\dots z_k(t)[z_1(t) + \dots + z_N(t)]\hat{B}(t)\mathbf{x}(t)\right\rangle \\
&= \hat{A}(t)\mathbf{X}_k(t) + k\left\langle z^2\right\rangle \hat{B}(t)\mathbf{X}_{k-1}(t) + (N-k)\hat{B}(t)\mathbf{X}_{k+1}(t),
\end{aligned} \tag{7.21}$$

with the initial condition

$$\mathbf{X}_k(0) = \mathbf{x}_0\delta_{k,0}.$$

Thus, the mean value of the solution to system (7.19) satisfies the closed system of $(N+1)$ vector equations. If operators $\hat{A}(t)$, $\hat{B}(t)$ are time-independent matrixes, system (7.21) can be easily solved using the Laplace transform. It is clear that such a solution will have the form of a finite segment of continued fraction. If we set $\langle z^2\rangle = \sigma^2/N$ and proceed to the limit $N \to \infty$, then random process $z(t) = z_1(t) + \dots + z_N(t)$, as was mentioned earlier, will grade into the Gaussian Markovian process and solution to system (7.21) will assume the form of the infinite continued fraction.

Problems

Problem 40 *Assuming that $z(t)$ is telegrapher's random process, find average potential energy of the stochastic oscillator described by the stochastic equation*

$$\frac{d^3}{dt^3}U(t) + 4\omega_0^2\frac{d}{dt}U(t) + 2\omega_0^2\left(z(t)\frac{d}{dt}U(t) + \frac{d}{dt}z(t)U(t)\right) = 0,$$

$$x(0) = 0, \quad \left.\frac{d}{dt}U(t)\right|_{t=0} = 0, \quad \left.\frac{d^2}{dt^2}U(t)\right|_{t=0} = 2y_0^2.$$

Solution.

$$\frac{d^3}{dt^3}\langle U(t)\rangle + 4\omega_0^2\frac{d}{dt}\langle U(t)\rangle + 4\omega_0^2\left(\frac{d}{dt} + 2\nu\right)\langle z(t)U(t)\rangle = 0,$$

$$\left(\frac{d}{dt} + 2\nu\right)^3\langle z(t)U(t)\rangle + 4\omega_0^2\left(\frac{d}{dt} + 2\nu\right)\langle z(t)U(t)\rangle + 4\omega_0^2 a_0^2\left(\frac{d}{dt} + 2\nu\right)\langle U(t)\rangle = 0. \tag{7.22}$$

Systems of equations (7.22) can be solved using Laplace transform, and we obtain the solution in the form

$$\begin{aligned}
\langle U\rangle_p &= 2y_0^2\frac{L(p + 2\nu)}{L(p)L(p + 2\nu) - a_0^2M^2(p)}, \\
L(p) &= p\left(p^2 + 4\omega_0^2\right), \quad M(p) = 4\omega_0^2\left(p^2 + \nu\right).
\end{aligned} \tag{7.23}$$

Remark. In the limiting case of great parameters ν and a_0^2, but finite ratio $a_0^2/2\nu = \sigma^2\tau_0$, we obtain from the second equation of system (7.22)

$$\langle z(t)U(t)\rangle = -\frac{\omega_0^2\sigma^2\tau_0}{\nu}\,\langle U(t)\rangle\,.$$

Consequently, average potential energy $\langle U(t)\rangle$ satisfies in this limiting case the closed third-order equation

$$\frac{d^3}{dt^3}\langle U(t)\rangle + 4\omega_0^2\frac{d}{dt}\langle U(t)\rangle - 4\omega_0^4\sigma^2\tau_0\langle U(t)\rangle = 0,$$

which coincides with Eq. (6.78), page 105 and corresponds to the Gaussian delta-correlated process $z(t)$.

Problem 41 *Assuming that $z(t)$ is telegrapher's random process, obtain the steady-state probability distribution of the solution to the stochastic equation*

$$\frac{d}{dt}x(t) = f(x) + z(t)g(x), \quad x(0) = x_0.$$

Solution. Under the condition that $|f(x)| < a_0|g(x)|$, we have

$$P(x) = \frac{C|g(x)|}{a_0^2g^2(x) - f^2(x)}\exp\left\{\frac{2\nu}{a_0^2}\int dx\frac{f(x)}{a_0^2g^2(x) - f^2(x)}\right\}, \tag{7.24}$$

where the positive constant C is determined from the normalization condition.

Remark. In the limit $\nu \to \infty$ and $a_0^2 \to \infty$ under the condition that $a_0^2\tau_0 = \text{const}$ $(\tau_0 = 1/(2\nu))$, probability distribution (7.24) grades into the expression

$$P(x) = \frac{C}{|g(x)|}\exp\left\{\frac{2\nu}{a_0^2}\int dx\frac{f(x)}{g^2(x)}\right\}$$

corresponding to the Gaussian delta-correlated process $z(t)$, i.e., the Gaussian process with correlation function $\langle z(t)z(t')\rangle = 2a_0^2\tau_0\delta(t - t')$.

Problem 42 *Assuming that $z(t)$ is telegrapher's random process, obtain the steady-state probability distribution of the solution to the stochastic equation*

$$\frac{d}{dt}x(t) = -x + z(t), \quad x(0) = x_0.$$

Solution.
$$P(x) = \frac{1}{B(\nu, 1/2)}(1 - x^2)^{\nu-1} \quad (|x| < 1), \tag{7.25}$$

where $B(\nu, 1/2)$ is the beta-function. This distribution has essentially different behaviors for $\nu > 1$, $\nu = 1$ and $\nu < 1$. One can easily see that this system resides mainly near state $x = 0$ if $\nu > 1$, and near states $x = \pm 1$ if $\nu < 1$. In the case $\nu = 1$, we obtain the uniform probability distribution on segment $[-1, 1]$. (Fig. 7.1).

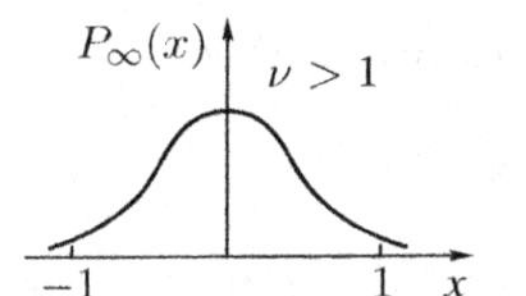

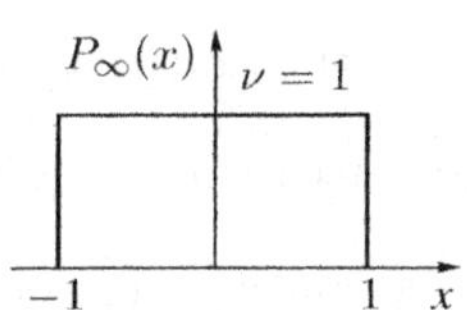

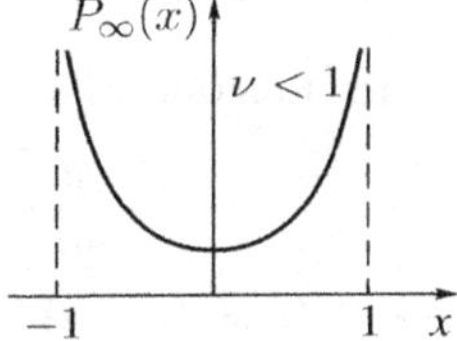

Figure 7.1: Steady-state probability distribution (7.25) versus parameter ν.

Chapter 8

Gaussian delta-correlated random field (ordinary differential equations)

In the foregoing chapters, we considered in detail general methods of analyzing stochastic equations. Here, we give an alternative and more detailed consideration of the approximation of the Gaussian random delta-correlated (in time) field in the context of stochastic equations and discuss the physical meaning of this widely used approximation.

8.1 The Fokker-Planck equation

Let vector function $\mathbf{x}(t) = \{x_1(t), x_2(t), ..., x_n(t)\}$ satisfies the dynamic equation

$$\frac{d}{dt}\mathbf{x}(t) = \mathbf{v}(\mathbf{x}, t) + \mathbf{f}(\mathbf{x}, t), \quad \mathbf{x}(t_0) = \mathbf{x}_0, \tag{8.1}$$

where $v_i(\mathbf{x}, t)$ $(i = 1, ..., n)$ are the deterministic functions and $f_i(\mathbf{x}, t)$ are the random functions of $(n + 1)$ variable that have the following properties:

(a) $f_i(\mathbf{x}, t)$ is the Gaussian random field in the $(n + 1)$-dimensional space $(\mathbf{x}, t)$;

(b) $\langle f_i(\mathbf{x}, t) \rangle = 0$.

For definiteness, we will assume that t is the temporal variable and $\mathbf{x}$ is the spatial variable.

Statistical characteristics of field $f_i(\mathbf{x}, t)$ are completely described by the correlation tensor

$$B_{ij}(\mathbf{x}, t; \mathbf{x}', t') = \langle f_i(\mathbf{x}, t) f_j(\mathbf{x}', t') \rangle.$$

Because Eq. (8.1) is the first-order equation with the initial condition, its solution satisfies the dynamic causality condition

$$\frac{\delta}{\delta f_j(\mathbf{x}', t')} x_i(t) = 0 \quad \text{for} \quad t' < t_0 \quad \text{and} \quad t' > t, \tag{8.2}$$

which means that solution $\mathbf{x}(t)$ depends only on values of function $f_j(\mathbf{x}, t')$ for times t' preceding time t, i.e., $t_0 \leq t' \leq t$. In addition, we have the following equality for the variational derivative

$$\frac{\delta}{\delta f_j(\mathbf{x}', t - 0)} x_i(t) = \delta_{ij}\delta(\mathbf{x}(t) - \mathbf{x}'). \tag{8.3}$$

Nevertheless, the statistical relationship between $\mathbf{x}(t)$ and function values $f_j(\mathbf{x}, t'')$ for consequent times $t'' > t$ can exist, because such function values $f_j(\mathbf{x}, t'')$ are correlated with values $f_j(\mathbf{x}, t')$ for $t' \le t$. It is obvious that the correlation between function $\mathbf{x}(t)$ and consequent values $f_j(\mathbf{x}, t'')$ is appreciable only for $t'' - t \le \tau_0$, where τ_0 is the correlation radius of field $\mathbf{f}(\mathbf{x}, t)$ with respect to variable t.

For many actual physical processes, characteristic temporal scale T of function $\mathbf{x}(t)$ significantly exceeds correlation radius τ_0 ($T \gg \tau_0$); in this case, the problem has the small parameter τ_0/T that can be used for constructing an approximate solution.

In the first approximation in this small parameter, one can consider the asymptotic solution for $\tau_0 \to 0$. In this case values of function $\mathbf{x}(t')$ for $t' < t$ will be independent of values $\mathbf{f}(\mathbf{x}, t'')$ for $t'' > t$ not only functionally, but also statistically. This approximation is equivalent to the replacement of correlation tensor B_{ij} with the effective tensor

$$B_{ij}^{\text{eff}}(\mathbf{x}, t; \mathbf{x}', t') = 2\delta(t - t') F_{ij}(\mathbf{x}, \mathbf{x}'; t). \tag{8.4}$$

Here, quantity $F_{ij}(\mathbf{x}, \mathbf{x}'; t)$ is determined from the condition that integrals of $B_{ij}(\mathbf{x}, t; \mathbf{x}', t')$ and $B_{ij}^{\text{eff}}(\mathbf{x}, t; \mathbf{x}', t')$ over t' must coincide

$$F_{ij}(\mathbf{x}, \mathbf{x}'; t) = \frac{1}{2} \int\limits_{-\infty}^{\infty} dt' B_{ij}(\mathbf{x}, t; \mathbf{x}', t'),$$

which just corresponds to the passage to the Gaussian random field delta-correlated in time t.

Introduce the indicator function

$$\varphi(\mathbf{x}, t) = \delta(\mathbf{x}(t) - \mathbf{x}), \tag{8.5}$$

where $\mathbf{x}(t)$ is the solution to Eq. (8.1), which satisfies the Liouville equation

$$\left(\frac{\partial}{\partial t} + \frac{\partial}{\partial \mathbf{x}} \mathbf{v}(\mathbf{x}, t) \right) \varphi(\mathbf{x}, t) = -\frac{\partial}{\partial \mathbf{x}} \mathbf{f}(\mathbf{x}, t) \varphi(\mathbf{x}, t) \tag{8.6}$$

and the equality

$$\frac{\delta}{\delta f_j(\mathbf{x}', t - 0)} \varphi(\mathbf{x}, t) = -\frac{\partial}{\partial x_j} \left\{ \delta(\mathbf{x} - \mathbf{x}') \varphi(\mathbf{x}, t) \right\}. \tag{8.7}$$

The equation for the probability density of the solution to Eq. (8.1)

$$P(\mathbf{x}, t) = \langle \varphi(\mathbf{x}, t) \rangle = \langle \delta(\mathbf{x}(t) - \mathbf{x}) \rangle$$

can be obtained by averaging Eq. (8.6) over an ensemble of realizations of field $\mathbf{f}(\mathbf{x}, t)$,

$$\left(\frac{\partial}{\partial t} + \frac{\partial}{\partial \mathbf{x}} \mathbf{v}(\mathbf{x}, t) \right) P(\mathbf{x}, t) = -\frac{\partial}{\partial \mathbf{x}} \langle \mathbf{f}(\mathbf{x}, t) \varphi(\mathbf{x}, t) \rangle. \tag{8.8}$$

We rewrite Eq. (8.8) in the form

$$\left(\frac{\partial}{\partial t} + \frac{\partial}{\partial \mathbf{x}} \mathbf{v}(\mathbf{x}, t) \right) P(\mathbf{x}, t) = -\frac{\partial}{\partial x_i} \int dx' \int\limits_{t_0}^{t} dt' B_{ij}(\mathbf{x}, t; \mathbf{x}', t') \left\langle \frac{\delta \varphi(\mathbf{x}, t)}{\delta f_j(\mathbf{x}', t')} \right\rangle, \tag{8.9}$$

where we used the Furutsu–Novikov formula

$$\langle f_k(\mathbf{x}, t) R[t; \mathbf{f}(\mathbf{y}, \tau)] \rangle = \int dx' \int dt' B_{kl}(\mathbf{x}, t; \mathbf{x}', t') \left\langle \frac{\delta R[t; \mathbf{f}(\mathbf{y}, \tau)]}{\delta f_l(\mathbf{x}', t')} \right\rangle \tag{8.10}$$

for the correlation between the Gaussian random field $\mathbf{f}(\mathbf{x}, t)$ and arbitrary functional $R[t; \mathbf{f}(\mathbf{y}, \tau)]$ of it and the dynamic causality condition (8.2).

Equation (8.9) shows that the one-time probability density of solution $\mathbf{x}(t)$ at instant t is governed by functional dependence of solution $\mathbf{x}(t)$ on field $\mathbf{f}(\mathbf{x}', t)$ for all times in the interval (t_0, t).

In the general case, there is no closed equation for the probability density $P(\mathbf{x}, t)$. However, if we use approximation (8.4) for the correlation function of field $\mathbf{f}(\mathbf{x}, t)$, there appear terms related to variational derivatives $\delta\varphi[\mathbf{x}, t; \mathbf{f}(\mathbf{y}, \tau)]/\delta f_j(\mathbf{x}', t')$ at coincident temporal arguments $t' = t - 0$,

$$\left(\frac{\partial}{\partial t} + \frac{\partial}{\partial \mathbf{x}}\mathbf{v}(\mathbf{x}, t)\right) P(\mathbf{x}, t) = -\frac{\partial}{\partial x_i} \int d\mathbf{x}' F_{ij}(\mathbf{x}, \mathbf{x}'; t) \left\langle \frac{\delta\varphi(\mathbf{x}, t)}{\delta f_j(\mathbf{x}', t - 0)} \right\rangle.$$

According to Eq. (8.7), these variational derivatives can be expressed immediately in terms of quantity $\varphi[\mathbf{x}, t; \mathbf{f}(\mathbf{y}, \tau)]$ itself. Thus we obtain the closed Fokker–Planck equation

$$\left(\frac{\partial}{\partial t} + \frac{\partial}{\partial x_k}\left[v_k(\mathbf{x}, t) + A_k(\mathbf{x}, t)\right]\right) P(\mathbf{x}, t) = \frac{\partial^2}{\partial x_k \partial x_l}\left[F_{kl}(\mathbf{x}, \mathbf{x}; t)P(\mathbf{x}, t)\right], \qquad (8.11)$$

where

$$A_k(\mathbf{x}, t) = \left.\frac{\partial}{\partial x_l'}F_{kl}(\mathbf{x}, \mathbf{x}'; t)\right|_{\mathbf{x}'=\mathbf{x}}.$$

Equation (8.11) should be solved with the initial condition $P(\mathbf{x}, t_0) = \delta(\mathbf{x} - \mathbf{x}_0)$, or with the more general initial condition $P(\mathbf{x}, t_0) = W_0(\mathbf{x})$ if the initial conditions are also random, but statistically independent of field $\mathbf{f}(\mathbf{x}, t)$.

The Fokker-Planck equation (8.11) is a partial differential equation and its further analysis essentially depends on boundary conditions with respect to $\mathbf{x}$ whose form can vary depending on a particular problem at hand.

Consider the quantities appeared in Eq. (8.11). In this equation, the terms containing $A_k(\mathbf{x}, t)$ and $F_{kl}(\mathbf{x}, \mathbf{x}'; t)$ are stipulated by fluctuations of field $\mathbf{f}(\mathbf{x}, t)$. If field $\mathbf{f}(\mathbf{x}, t)$ is stationary in time, quantities $A_k(\mathbf{x})$ and $F_{kl}(\mathbf{x}, \mathbf{x}')$ are independent of time. If field $\mathbf{f}(\mathbf{x}, t)$ is additionally homogeneous and isotropic in all spatial coordinates, then $F_{kl}(\mathbf{x}, \mathbf{x}, t) = \text{const}$, which corresponds to the constant tensor of diffusion coefficients, and $A_k(\mathbf{x}, t) = 0$ (note however that quantities $F_{kl}(\mathbf{x}, \mathbf{x}'; t)$ and $A_k(\mathbf{x}, t)$ can depend on $\mathbf{x}$ because of the use of curvilinear coordinate systems).

8.2   Transition probability distributions

Turn back to dynamic system (8.1) and consider the m-time probability density

$$P_m(\mathbf{x}_1, t_1; ...; \mathbf{x}_m, t_m) = \langle \delta(\mathbf{x}(t_1) - \mathbf{x}_1)...\delta(\mathbf{x}(t_m) - \mathbf{x}_m)\rangle \qquad (8.12)$$

for m different instants $t_1 < t_2 < ... < t_m$. Differentiating Eq. (8.12) with respect to time t_m and using then dynamic equation (8.1), dynamic causality condition (8.2), definition of function $F_{kl}(\mathbf{x}, \mathbf{x}'; t)$, and the Furutsu–Novikov formula (8.10), one can obtain the equation similar to the Fokker–Planck equation (8.11),

$$\frac{\partial}{\partial t_m} P_m(\mathbf{x}_1, t_1; ...; \mathbf{x}_m, t_m)$$

$$+ \sum_{k=1}^{n} \frac{\partial}{\partial x_{mk}} \left[v_k(\mathbf{x}_m, t_m) + A_k(\mathbf{x}_m, t_m) \right] P_m(\mathbf{x}_1, t_1; ...; \mathbf{x}_m, t_m)$$

$$= \sum_{k=1}^{n} \sum_{l=1}^{n} \frac{\partial^2}{\partial x_{mk} \partial x_{ml}} \left[F_{kl}(\mathbf{x}_m, \mathbf{x}_m; t_m) P_m(\mathbf{x}_1, t_1; ...; \mathbf{x}_m, t_m)) \right]. \tag{8.13}$$

No summation over index m is performed here. The initial condition to Eq. (8.13) can be found from Eq. (8.12). Setting $t_m = t_{m-1}$ in (8.12), we obtain

$$P_m(\mathbf{x}_1, t_1; ...; \mathbf{x}_m, t_{m-1}) = \delta(\mathbf{x}_m - \mathbf{x}_{m-1}) P_{m-1}(\mathbf{x}_1, t_1; ...; \mathbf{x}_{m-1}, t_{m-1}). \tag{8.14}$$

Equation (8.13) assumes the solution of the form

$$P_m(\mathbf{x}_1, t_1; ...; \mathbf{x}_m, t_m) = p(\mathbf{x}_m, t_m | \mathbf{x}_{m-1}, t_{m-1}) P_{m-1}(\mathbf{x}_1, t_1; ...; \mathbf{x}_{m-1}, t_{m-1}). \tag{8.15}$$

Because all differential operations in Eq. (8.13) concern only t_m and $\mathbf{x}_m$, we can find the equation for the *transition probability density* by substituting Eq. (8.15) in Eqs. (8.13) and (8.14):

$$\left(\frac{\partial}{\partial t} + \frac{\partial}{\partial x_k} \left[v_k(\mathbf{x}, t) + A_k(\mathbf{x}, t) \right] \right) p(\mathbf{x}, t | \mathbf{x}_0, t_0) = \frac{\partial^2}{\partial x_k \partial x_l} \left[F_{kl}(\mathbf{x}, \mathbf{x}; t) p(\mathbf{x}, t | \mathbf{x}_0, t_0) \right],$$

$$p(\mathbf{x}, t | \mathbf{x}_0, t_0) |_{t \to t_0} = \delta(\mathbf{x} - \mathbf{x}_0), \tag{8.16}$$

where

$$p(\mathbf{x}, t | \mathbf{x}_0, t_0) = \langle \delta(\mathbf{x}(t) - \mathbf{x}) | \mathbf{x}(t_0) = \mathbf{x}_0 \rangle.$$

In Eq. (8.16) we denoted variables $\mathbf{x}_m$ and t_m as $\mathbf{x}$ and t, and variables $\mathbf{x}_{m-1}$ and t_{m-1} as $\mathbf{x}_0$ and t_0.

Using formula (8.15) $(m-1)$ times, we obtain the relationship

$$P_m(\mathbf{x}_1, t_1; ...; \mathbf{x}_m, t_m) = p(\mathbf{x}_m, t_m | \mathbf{x}_{m-1}, t_{m-1}) \ ... \ p(\mathbf{x}_2, t_2 | \mathbf{x}_1, t_1) P(\mathbf{x}_1, t_1), \tag{8.17}$$

where $P(\mathbf{x}_1, t_1)$ is the probability density governed by Eq. (8.11) for one instant t_1. Equality (8.17) expresses the multi-time probability density as the product of transition probability densities, which means that random process $\mathbf{x}(t)$ is the Markovian process.

Equation (8.11) is usually called the *forward Fokker–Planck equation*. The *backward Fokker–Planck equation* (it describes the transition probability density as a function of initial parameters t_0 and $\mathbf{x}_0$) can also be easily derived.

Indeed, in Chapter 3 (page 43), we obtained the backward Liouville equation (3.4) for indicator function

$$\left(\frac{\partial}{\partial t_0} + \mathbf{v}(\mathbf{x}_0, t_0) \frac{\partial}{\partial \mathbf{x}_0} \right) \varphi(\mathbf{x}, t | \mathbf{x}_0, t_0) = -\mathbf{f}(\mathbf{x}_0, t_0) \frac{\partial}{\partial \mathbf{x}_0} \varphi(\mathbf{x}, t | \mathbf{x}_0, t_0), \tag{8.18}$$

with the initial condition

$$\varphi(\mathbf{x}, t | \mathbf{x}_0, t) = \delta(\mathbf{x} - \mathbf{x}_0).$$

This equation describes the dynamic system evolution in terms of initial parameters t_0 and $\mathbf{x}_0$. From Eq. (8.18) follows the equality similar to Eq. (8.7),

$$\frac{\delta}{\delta f_j(\mathbf{x}', t_0 + 0)} \varphi(\mathbf{x}, t | \mathbf{x}_0, t_0) = \delta(\mathbf{x} - \mathbf{x}') \frac{\partial}{\partial x_{0j}} \varphi(\mathbf{x}, t | \mathbf{x}_0, t_0). \qquad (8.19)$$

Averaging now the backward Liouville equation (8.18) over an ensemble of realizations of random field $\mathbf{f}(\mathbf{x}, t)$ with the effective correlation tensor (8.4), using the Furutsu–Novikov formula (8.10), and relationship (8.19) for the variational derivative, we obtain the backward Fokker–Planck equation

$$\left(\frac{\partial}{\partial t_0} + [v_k(\mathbf{x}_0, t_0) + A_k(\mathbf{x}_0, t_0)] \frac{\partial}{\partial x_{0k}} \right) p(\mathbf{x}, t | \mathbf{x}_0, t_0)$$

$$= -F_{kl}(\mathbf{x}_0, \mathbf{x}_0; t_0) \frac{\partial^2}{\partial x_{0k} \partial x_{0l}} p(\mathbf{x}, t | \mathbf{x}_0, t_0), \quad p(\mathbf{x}, t | \mathbf{x}_0, t) = \delta(\mathbf{x} - \mathbf{x}_0).$$

The forward and backward Fokker–Planck equations are equivalent. The forward equation is more convenient for analyzing the temporal behavior of statistical characteristics of the solution to Eq. (8.1). The backward equation appears more convenient for studying statistical characteristics related to initial conditions, such as the time during which process $\mathbf{x}(t)$ resides in certain spatial region and the time at which the process arrives at region boundary. In this case, the probability of the fact that random process $\mathbf{x}(t)$ resides in spatial region V is given by the integral

$$G(t; \mathbf{x}_0, t_0) = \int\limits_V d\mathbf{x} p(\mathbf{x}, t | \mathbf{x}_0, t_0),$$

which satisfies the closed equation

$$\left(\frac{\partial}{\partial t_0} + [v_k(\mathbf{x}_0, t_0) + A_k(\mathbf{x}_0, t_0)] \frac{\partial}{\partial x_{0k}} \right) G(t; \mathbf{x}_0, t_0)$$

$$= -F_{kl}(\mathbf{x}_0, \mathbf{x}_0; t_0) \frac{\partial^2}{\partial x_{0k} \partial x_{0l}} G(t; \mathbf{x}_0, t_0), \quad G(t; \mathbf{x}_0, t_0) = \begin{cases} 1 & (\mathbf{x}_0 \in V), \\ 0 & (\mathbf{x}_0 \notin V). \end{cases} \qquad (8.20)$$

For Eq. (8.20), we must formulate additional boundary conditions, which depend on characteristics of both region V and its boundaries.

8.3 Applicability range of the Fokker–Planck equation

To estimate the applicability range of the Fokker–Planck equation, we must include into consideration the finite-valued correlation radius τ_0 of field $\mathbf{f}(\mathbf{x}, t)$ with respect to time. Thus, smallness of parameter τ_0/T is the necessary but generally not sufficient condition in order that one can describe statistical characteristics of the solution to Eq. (8.1) using the approximation of the delta-correlated random field of which a consequence is the Fokker–Planck equation. Every particular problem requires more detailed investigations. Below, we give a more physical method called the *diffusion approximation*. This method also leads to the Markovian property of the solution to Eq. (8.1); moreover, it offers certain possibility of considerating the finite-valued temporal correlation radius.

Here, we emphasize that the approximation of the delta-correlated random field cannot be reduced to the formal replacement of random field $\mathbf{f}(\mathbf{x}, t)$ in Eq. (8.1) with the random field with correlation function (8.4). This approximation corresponds in fact to the construction of an asymptotic expansion in terms of the temporal correlation radius τ_0 of filed $\mathbf{f}(\mathbf{x}, t)$ approaching to zero. It is in such limit process that exact average quantities like

$$\langle \mathbf{f}(\mathbf{x}, t) R[t; \mathbf{f}(\mathbf{x}', \tau)] \rangle$$

grade into the expressions obtained by the formal replacement of the correlation tensor of field $\mathbf{f}(\mathbf{x}, t)$ with the effective tensor (8.4).

8.3.1 Langevin equation

We illustrate the above material by the example of the *Langevin equation* that allows an exhaustive statistical analysis. This equation has the form

$$\frac{d}{dt} x(t) = -\lambda x(t) + f(t), \quad x(t_0) = 0, \tag{8.21}$$

and assumes that the sufficiently fine smooth function $f(t)$ is the stationary Gaussian process with zero-valued average and correlation function

$$\langle f(t) f(t') \rangle = B_f(t - t').$$

For any particular realization of random force $f(t)$, the solution to Eq. (8.21) has the form

$$x(t) = \int_{t_0}^{t} d\tau f(\tau) e^{-\lambda(t-\tau)}.$$

Consequently, this solution $x(t)$ is also the Gaussian process with the parameters

$$\langle x(t) \rangle = 0, \quad \langle x(t) x(t') \rangle = \int_{t_0}^{t} d\tau_1 \int_{t_0}^{t'} d\tau_2 B_f(\tau_1 - \tau_2) e^{-\lambda(t+t'-\tau_1-\tau_2)}.$$

In addition, we have, for example,

$$\langle f(t) x(t) \rangle = \int_{0}^{t-t_0} d\tau B_f(\tau) e^{-\lambda\tau}.$$

Note that the one-point probability density $P(x, t) = \langle \delta(x(t) - x) \rangle$ for Eq. (8.21) satisfies the equation

$$\left(\frac{\partial}{\partial t} - \lambda \frac{\partial}{\partial x} x \right) P(x, t) = \int_{0}^{t-t_0} d\tau B_f(\tau) e^{-\lambda\tau} \frac{\partial^2}{\partial x^2} P(x, t), \quad P(x, t_0) = \delta(x),$$

which rigorously follows from Eq. (6.24). As a consequence, we can obtain

$$\frac{d}{dt} \left\langle x^2(t) \right\rangle = -2\lambda \left\langle x^2(t) \right\rangle + 2 \int_{0}^{t-t_0} d\tau B_f(\tau) e^{-\lambda\tau}.$$

For $t_0 \to -\infty$, process $x(t)$ grades into the steady-state Gaussian process with the following one-time statistical parameters

$$\langle x(t) \rangle = 0, \quad \sigma_x^2 = \left\langle x^2(t) \right\rangle = \frac{1}{\lambda} \int_{0}^{\infty} d\tau B_f(\tau) e^{-\lambda\tau}, \quad \langle f(t) x(t) \rangle = \int_{0}^{\infty} d\tau B_f(\tau) e^{-\lambda\tau}.$$

In particular, for exponential correlation function $B_f(t)$,

$$B_f(t) = \sigma_f^2 e^{-|\tau|/\tau_0},$$

we obtain the expressions

$$\langle x(t) \rangle = 0, \quad \left\langle x^2(t) \right\rangle = \frac{\sigma_f^2 \tau_0}{\lambda(1 + \lambda\tau_0)}, \quad \langle f(t)x(t) \rangle = \frac{\sigma_f^2 \tau_0}{1 + \lambda\tau_0}, \tag{8.22}$$

which grade into the asymptotic expressions

$$\left\langle x^2(t) \right\rangle = \frac{\sigma_f^2 \tau_0}{\lambda}, \quad \langle f(t)x(t) \rangle = \sigma_f^2 \tau_0 \tag{8.23}$$

for $\tau_0 \to 0$.

Multiply now Eq. (8.21) by $x(t)$. Assuming that function $x(t)$ is sufficiently fine function, we obtain the equality

$$x(t)\frac{d}{dt}x(t) = \frac{1}{2}\frac{d}{dt}x^2(t) = -\lambda x^2(t) + f(t)x(t).$$

Averaging this equation over an ensemble of realizations of function $f(t)$, we obtain the equation

$$\frac{1}{2}\frac{d}{dt}\left\langle x^2(t) \right\rangle = -\lambda\left\langle x^2(t) \right\rangle + \langle f(t)x(t) \rangle, \tag{8.24}$$

whose steady-state solution (it corresponds to the limit process $t_0 \to -\infty$ and $\tau_0 \to 0$)

$$\left\langle x^2(t) \right\rangle = \frac{1}{\lambda}\langle f(t)x(t) \rangle$$

coincides with Eqs. (8.22) and (8.23).

Taking into account the fact that $\delta x(t)/\delta f(t-0) = 1$, we obtain the same result for correlation $\langle f(t)x(t) \rangle$ by using the formula

$$\langle f(t)x(t) \rangle = \int\limits_{-\infty}^{t} d\tau\, B_f(t - \tau)\left\langle \frac{\delta}{\delta f(\tau)}x(t) \right\rangle \tag{8.25}$$

with the effective correlation function

$$B_f^{\text{eff}}(t) = 2\sigma_f^2 \tau_0 \delta(t).$$

Earlier, we mentioned that statistical characteristics of solutions to dynamic problems in the approximation of the delta-correlated random process (field) coincide with the statistical characteristics of the Markovian processes. However, one should clearly understand that this is the case only for statistical averages and equations for these averages. In particular, realizations of process $x(t)$ satisfying the Langevin equation (8.21) drastically differ from realizations of the corresponding Markovian process. The latter satisfies Eq. (8.21) in which function $f(t)$ in the right-hand side is the ideal white noise with the correlation function $B_f(t) = 2\sigma_f^2 \tau_0 \delta(t)$; moreover, this equation must be treated in the sense of generalized functions, because the Markovian processes are not differentiable in the ordinary sense. At the same time, process $x(t)$ — whose statistical characteristics coincide with characteristics of the Markovian process — behaves as sufficiently fine function and is differentiable in the ordinary sense. For example,

$$x(t)\frac{d}{dt}x(t) = \frac{1}{2}\frac{d}{dt}x^2(t),$$

and we have for $t_0 \to -\infty$ in particular

$$\left\langle x(t)\frac{d}{dt}x(t) \right\rangle = 0. \tag{8.26}$$

On the other hand, in the case of the ideal Markovian process $x(t)$ satisfying (in the sense of generalized functions) the Langevin equation (8.21) with the white noise in the right-hand side, Eq. (8.26) makes no sense at all, and the meaning of the relationship

$$\left\langle x(t)\frac{d}{dt}x(t) \right\rangle = -\lambda\left\langle x^2(t) \right\rangle + \langle f(t)x(t)\rangle \tag{8.27}$$

depends on the definition of averages. Indeed, if we will mean Eq. (8.27) as the limit of the equality

$$\left\langle x(t+\Delta)\frac{d}{dt}x(t) \right\rangle = -\lambda\langle x(t)x(t+\Delta)\rangle + \langle f(t)x(t+\Delta)\rangle \tag{8.28}$$

for $\Delta \to 0$, the result will be essentially different depending on whether we use limit processes $\Delta \to +0$, or $\Delta \to -0$. For limit process $\Delta \to +0$, we have

$$\lim_{\Delta \to +0} \langle f(t)x(t+\Delta)\rangle = 2\sigma_f^2\tau_0,$$

and, taking into account Eq. (8.25), we can rewrite Eq. (8.28) in the form

$$\left\langle x(t+0)\frac{d}{dt}x(t) \right\rangle = \sigma_f^2\tau_0. \tag{8.29}$$

On the contrary, for limit process $\Delta \to -0$, we have $\langle f(t)x(t-0)\rangle = 0$ because of the dynamic causality condition, and Eq. (8.28) assumes the form

$$\left\langle x(t-0)\frac{d}{dt}x(t) \right\rangle = -\sigma_f^2\tau_0. \tag{8.30}$$

Comparing Eq. (8.26) with Eqs. (8.29) and (8.30), we see that, for the ideal Markovian process described by the solution to the Langevin equation with the white noise in the right-hand side and commonly called the *Ohrnstein–Ulenbeck process*, we have

$$\left\langle x(t+0)\frac{d}{dt}x(t) \right\rangle \neq \left\langle x(t-0)\frac{d}{dt}x(t) \right\rangle \neq \frac{1}{2}\frac{d}{dt}\left\langle x^2(t) \right\rangle.$$

Note that equalities (8.29) and (8.30) can also be obtained from the correlation function

$$\langle x(t)x(t+\tau)\rangle = \frac{\sigma_f^2\tau_0}{\lambda}e^{-\lambda|\tau|}$$

of process $x(t)$.

To conclude with the discussion of the approximation of the delta-correlated random process (field), we emphasize that, in all further examples, we will treat the sentence like 'dynamic system (equation) with the delta-correlated parameter fluctuations' as asymptotic limit in which these parameters have temporal correlation radii small in comparison with all characteristic temporal scales of the problem under consideration.

8.3.2   Diffusion approximation

Applicability of the approximation of the delta-correlated random field $\mathbf{f}(\mathbf{x}, t)$ (i.e., applicability of the Fokker–Planck equation) is restricted by the smallness of the temporal correlation radius τ_0 of random field $\mathbf{f}(\mathbf{x}, t)$ with respect to all temporal scales of the problem under consideration. The effect of the finite-valued temporal correlation radius of random field $\mathbf{f}(\mathbf{x}, t)$ can be considered within the framework of the diffusion approximation. The diffusion approximation appears more obvious and physical than the formal mathematical derivation of the approximation of the delta-correlated random field. This approximation also holds for sufficiently weak parameter fluctuations of the stochastic dynamic system and allows describing new physical effects caused by the finite-valued temporal correlation radius of random parameters, rather than only obtaining the applicability range of the delta-correlated approximation. The diffusion approximation assumes that the effect of random actions is insignificant during temporal scales about τ_0, i.e., the system behaves during these times as the free system.

Again, let vector function $\mathbf{x}(t)$ satisfies the dynamic equation (8.1)

$$\frac{d}{dt}\mathbf{x}(t) = \mathbf{v}(\mathbf{x}, t) + \mathbf{f}(\mathbf{x}, t), \quad \mathbf{x}(t_0) = \mathbf{x}_0, \tag{8.31}$$

where $\mathbf{v}(\mathbf{x}, t)$ is the vector deterministic function and $\mathbf{f}(\mathbf{x}, t)$ is the random statistically homogeneous and stationary Gaussian vector field with the statistical characteristics

$$\langle f(\mathbf{x}, t)\rangle = 0, \quad B_{ij}(\mathbf{x}, t; \mathbf{x}', t') = B_{ij}(\mathbf{x} - \mathbf{x}'; t - t') = \langle f_i(\mathbf{x}, t) f_j(\mathbf{x}', t')\rangle.$$

Introduce the indicator function

$$\varphi(\mathbf{x}, t) = \delta(\mathbf{x}(t) - \mathbf{x}), \tag{8.32}$$

($\mathbf{x}(t)$ is the solution to Eq. (8.31)) satisfying the Liouville equation (8.6)

$$\left(\frac{\partial}{\partial t} + \frac{\partial}{\partial \mathbf{x}}\mathbf{v}(\mathbf{x}, t)\right)\varphi(\mathbf{x}, t) = -\frac{\partial}{\partial \mathbf{x}}\mathbf{f}(\mathbf{x}, t)\varphi(\mathbf{x}, t). \tag{8.33}$$

As earlier, we obtain the equation for the probability density of the solution to Eq. (8.31)

$$P(\mathbf{x}(t) = \langle\varphi(\mathbf{x}, t)\rangle = \langle\delta(\mathbf{x}(t) - \mathbf{x})\rangle$$

by averaging Eq. (8.33) over an ensemble of realizations of field $\mathbf{f}(\mathbf{x}, t)$

$$\left(\frac{\partial}{\partial t} + \frac{\partial}{\partial \mathbf{x}}\mathbf{v}(\mathbf{x}, t)\right)P(\mathbf{x}, t) = -\frac{\partial}{\partial \mathbf{x}}\langle\mathbf{f}(\mathbf{x}, t)\varphi(\mathbf{x}, t)\rangle, \quad P(\mathbf{x}, t_0) = \delta(\mathbf{x} - \mathbf{x}_0). \tag{8.34}$$

Using the Furutsu–Novikov formula (8.10) we can rewrite Eq. (8.34) in the form

$$\left(\frac{\partial}{\partial t} + \frac{\partial}{\partial \mathbf{x}}\mathbf{v}(\mathbf{x}, t)\right)P(\mathbf{x}, t) = -\frac{\partial}{\partial x_i}\int d\mathbf{x}' \int_{t_0}^{t} dt' B_{ij}(\mathbf{x}, t; \mathbf{x}', t')\left\langle\frac{\delta}{\delta f_j(\mathbf{x}', t')}\varphi(\mathbf{x}, t)\right\rangle. \tag{8.35}$$

In the diffusion approximation, Eq. (8.35) is the exact equation, and the variational derivative and indicator function satisfy, within temporal scales of about temporal correlation radius τ_0 of random field $\mathbf{f}(\mathbf{x}, t)$, the system of dynamic equations

$$\frac{\partial}{\partial t} \frac{\delta \varphi(\mathbf{x}, t)}{\delta f_i(\mathbf{x}', t')} = -\frac{\partial}{\partial \mathbf{x}} \left\{ \mathbf{v}(\mathbf{x}, t) \frac{\delta \varphi(\mathbf{x}, t)}{\delta f_i(\mathbf{x}', t')} \right\},$$

$$\frac{\delta \varphi(\mathbf{x}, t)}{\delta f_i(\mathbf{x}', t')} \bigg|_{t=t'} = -\frac{\partial}{\partial x_i} \left\{ \delta(\mathbf{x} - \mathbf{x}') \varphi(\mathbf{x}, t') \right\},$$

$$\frac{\partial}{\partial t} \varphi(\mathbf{x}, t) = -\frac{\partial}{\partial \mathbf{x}} \left\{ \mathbf{v}(\mathbf{x}, t) \varphi(\mathbf{x}, t) \right\}, \quad \varphi(\mathbf{x}, t)|_{t=t'} = \varphi(\mathbf{x}, t'). \tag{8.36}$$

The solution to problem (8.35), (8.36) holds for all times t. In this case, the solution $\mathbf{x}(t)$ to problem (8.31) cannot be considered as the vector Markovian random process because its multi-time probability density cannot be factorized in terms of the transition probability density. However, in asymptotic limit $t \gg \tau_0$, the diffusion-approximation solution to the initial dynamic system (8.31) will be the Markovian random process, and the corresponding conditions of applicability are formulated as smallness of all statistical effects within temporal scales of about temporal correlation radius τ_0.

Thus, the diffusion approximation lifts the basic restriction on smallness of the temporal correlation radius τ_0 remaining within the framework of the Markovian processes.

Problems

Problem 43 *Assuming that $\varepsilon(x, \mathbf{R})$ is the homogeneous isotropic delta-correlated Gaussian field with zero-valued mean and correlation function*

$$B_\varepsilon(x - x', \mathbf{R} - \mathbf{R}') = \langle \varepsilon(x, \mathbf{R}) \varepsilon(x', \mathbf{R}') \rangle = A(\mathbf{R} - \mathbf{R}') \delta(x - x'),$$

derive the Fokker–Planck equation for the diffusion of rays satisfying system of equations (1.52), page 28

$$\frac{d}{dx} \mathbf{R}(x) = \mathbf{p}(x), \quad \frac{d}{dx} \mathbf{p}(x) = \frac{1}{2} \boldsymbol{\nabla}_{\mathbf{R}} \varepsilon(x, \mathbf{R}). \tag{8.37}$$

Solution. Function $P(x; \mathbf{R}, \mathbf{p}) = \langle \delta(\mathbf{R}(x) - \mathbf{R}) \delta(\mathbf{p}(x) - \mathbf{p}) \rangle$ satisfies the Fokker–Planck equation

$$\left(\frac{\partial}{\partial x} + \mathbf{p} \frac{\partial}{\partial \mathbf{R}} \right) P(x; \mathbf{R}, \mathbf{p}) = D \Delta_{\mathbf{R}} P(x; \mathbf{R}, \mathbf{p})$$

with the diffusion coefficient

$$D = -\frac{1}{8} \Delta_{\mathbf{R}} A(\mathbf{R})|_{\mathbf{R}=0} = \pi^2 \int\limits_0^\infty d\kappa \, \kappa^3 \Phi_\varepsilon(0, \kappa).$$

Here,

$$\Phi_\varepsilon(q, \boldsymbol{\kappa}) = \frac{1}{(2\pi)^3} \int\limits_{-\infty}^\infty dx \int d\mathbf{R} B_\varepsilon(x, \mathbf{R}) e^{-i(qx + \boldsymbol{\kappa}\mathbf{R})}, \quad A(\mathbf{R}) = 2\pi \int d\boldsymbol{\kappa} \Phi_\varepsilon(0, \boldsymbol{\kappa}) e^{i\boldsymbol{\kappa}\mathbf{R}},$$

is the three-dimensional spectral density of random field $\varepsilon(x, \mathbf{R})$.

Solution corresponding to the initial condition $P(0; \mathbf{R}, \mathbf{p}) = \delta(\mathbf{R}) \delta(\mathbf{p})$ is the Gaussian probability density with the parameters

$$\langle R_j(x) R_k(x) \rangle = \frac{2}{3} D \delta_{jk} x^3, \quad \langle R_j(x) p_k(x) \rangle = D \delta_{jk} x^2, \quad \langle p_j(x) p_k(x) \rangle = 2 D \delta_{jk} x.$$

Problem 44 *Assuming that $\varepsilon(x, \mathbf{R})$ is the homogeneous isotropic delta-correlated Gaussian field with zero-valued mean and correlation function*

$$B_\varepsilon(x - x', \mathbf{R} - \mathbf{R}') = \langle \varepsilon(x, \mathbf{R})\varepsilon(x', \mathbf{R}') \rangle = A(\mathbf{R} - \mathbf{R}')\delta(x - x'),$$

derive, starting from Eqs. (8.37), the longitudinal correlation function of ray displacement.

Solution.
$$\langle \mathbf{R}(x)\mathbf{R}(x') \rangle = 2D(x')^2 \left(x - \frac{1}{3}x' \right).$$

Problem 45 *Consider the diffusion of two rays described by the system of equations*

$$\frac{d}{dx}\mathbf{R}_\nu(x) = \mathbf{p}_\nu(x), \quad \frac{d}{dx}\mathbf{p}_\nu(x) = \frac{1}{2}\nabla_{\mathbf{R}_\nu}\varepsilon(x, \mathbf{R}_\nu),$$

where index $\nu = 1, 2$ denotes the number of the corresponding ray.

Solution. The joint probability density

$$\begin{aligned}
P(x; \mathbf{R}_1, \mathbf{p}_1, \mathbf{R}_2, \mathbf{p}_2) \\
= \langle \delta(\mathbf{R}_1(x) - \mathbf{R}_1)\delta(\mathbf{p}_1(x) - \mathbf{p}_1)\delta(\mathbf{R}_2(x) - \mathbf{R}_2)\delta(\mathbf{p}_2(x) - \mathbf{p}_2) \rangle
\end{aligned}$$

satisfies the Fokker–Planck equation

$$\begin{aligned}
\left(\frac{\partial}{\partial x} + \mathbf{p}_1 \frac{\partial}{\partial \mathbf{R}_1} + \mathbf{p}_2 \frac{\partial}{\partial \mathbf{R}_2} \right) P(x; \mathbf{R}_1, \mathbf{p}_1, \mathbf{R}_2, \mathbf{p}) \\
= \widehat{L}\left(\frac{\partial}{\partial \mathbf{p}_1}, \frac{\partial}{\partial \mathbf{p}_2} \right) P(x; \mathbf{R}_1, \mathbf{p}_1, \mathbf{R}_2, \mathbf{p}),
\end{aligned}$$

where operator

$$\begin{aligned}
\widehat{L}\left(\frac{\partial}{\partial \mathbf{p}_1}, \frac{\partial}{\partial \mathbf{p}_2} \right) = \frac{\pi}{4} \int d\boldsymbol{\kappa} \Phi_\varepsilon(0, \boldsymbol{\kappa}) \left[\left(\boldsymbol{\kappa}\frac{\partial}{\partial \mathbf{p}_1} \right)^2 + \left(\boldsymbol{\kappa}\frac{\partial}{\partial \mathbf{p}_2} \right)^2 \right. \\
\left. + 2\cos\left[\boldsymbol{\kappa}(\mathbf{R}_1 - \mathbf{R}_2)\right] \left(\boldsymbol{\kappa}\frac{\partial}{\partial \mathbf{p}_1} \right) \left(\boldsymbol{\kappa}\frac{\partial}{\partial \mathbf{p}_1} \right) \right].
\end{aligned}$$

Problem 46 *Consider the relative diffusion of two rays.*

Solution. Probability density of relative diffusion of two rays

$$P(x; \mathbf{R}, \mathbf{p}) = \langle \delta(\mathbf{R}_1(x) - \mathbf{R}_2(x) - \mathbf{R})\delta(\mathbf{p}_1(x) - \mathbf{p}_2(x) - \mathbf{p}) \rangle,$$

satisfies the Fokker–Planck equation

$$\left(\frac{\partial}{\partial x} + \mathbf{p}\frac{\partial}{\partial \mathbf{R}} \right) P(x; \mathbf{R}, \mathbf{p}) = D_{\alpha\beta}(\mathbf{R})\frac{\partial^2}{\partial p_\alpha \partial p_\beta} P(x; \mathbf{R}, \mathbf{p}), \tag{8.38}$$

where

$$D_{\alpha\beta}(\mathbf{R}) = 2\pi \int d\boldsymbol{\kappa} \left[1 - \cos\left(\boldsymbol{\kappa}\mathbf{R}\right)\right] \kappa_\alpha \kappa_\beta \Phi_\varepsilon(0, \boldsymbol{\kappa}).$$

Remark. If l_0 is the correlation radius of random field $\varepsilon(x, \mathbf{R})$, than under the condition $R \gg l_0$ we have

$$D_{\alpha\beta}(\mathbf{R}) = 2D\delta_{\alpha\beta},$$

i.e., relative diffusion of rays is characterized by the diffusion coefficient exceeding the diffusion coefficient of a single ray by a factor of 2, which corresponds to statistically independent rays. In this case, the joint probability density of relative diffusion in Gaussian.

In the general case, Eq. (8.38) hardly can be solved in the analytic form. The only clear point is that the solution corresponding to variable diffusion coefficient is not the Gaussian distribution.

Asymptotic case $R \ll l_0$ can be analyzed in more detail. In this case, we can expand function $\{1 - \cos(\boldsymbol{\kappa}\mathbf{R})\}$ in the Taylor series to obtain the diffusion matrix in the form

$$D_{\alpha\beta}(\mathbf{R}) = \pi B \left(\mathbf{R}^2 \delta_{a\beta} + 2 R_\alpha R_\beta \right), \tag{8.39}$$

where

$$B = \frac{\pi}{4} \int\limits_0^\infty d\kappa \, \kappa^5 \Phi_\varepsilon(0, \kappa).$$

From Eq. (8.38) with coefficients (8.39) follow two equations for moment functions

$$\frac{d}{dx} \left\langle \mathbf{p}^2(x) \right\rangle = 8\pi B \left\langle \mathbf{R}^2(x) \right\rangle, \quad \frac{d}{dx} \left\langle \mathbf{R}^2(x) \right\rangle = 2 \left\langle \mathbf{R}(x)\mathbf{p}(x) \right\rangle,$$
$$\frac{d}{dx} \left\langle \mathbf{R}(x)\mathbf{p}(x) \right\rangle = \left\langle \mathbf{p}^2(x) \right\rangle.$$

These equations can be easily solved. From the solution follows that, for x from the interval in which $\alpha x \gg 1$ ($\alpha = (16\pi B)^{1/3}$), but $R_0^2 e^{\alpha x} \ll l_0^2$ (such an interval always exists for sufficiently small distances R_0 between the rays under consideration), quantities $\left\langle \mathbf{R}^2(x) \right\rangle$, $\left\langle \mathbf{R}(x)\mathbf{p}(x) \right\rangle$, and $\left\langle \mathbf{p}^2(x) \right\rangle$ are exponentially increasing functions of x.

Problem 47 *Starting from parabolic equation (1.49), page 26*

$$\frac{\partial}{\partial x} u(x, \mathbf{R}) = \frac{i}{2k} \Delta_\mathbf{R} u(x, \mathbf{R}) + i\frac{k}{2}\varepsilon(x, \mathbf{R}) u(x, \mathbf{R}), \quad u(0, \mathbf{R}) = u_0(\mathbf{R}), \tag{8.40}$$

derive the equations for average field $\langle u(x, \mathbf{R}) \rangle$ and second-order coherence function

$$\Gamma_2(x; \mathbf{R}, \boldsymbol{\rho}) = \left\langle u\left(x, \mathbf{R} + \frac{1}{2}\boldsymbol{\rho}\right) u^*\left(x, \mathbf{R} - \frac{1}{2}\boldsymbol{\rho}\right) \right\rangle$$

in the approximation of the homogeneous isotropic delta-correlated Gaussian field $\varepsilon(x, \mathbf{R})$ with the parameters

$$\langle \varepsilon(x, \mathbf{R}) \rangle = 0, \quad B_\varepsilon(x, \mathbf{R}) = B_\varepsilon^{e\!f\!f}(x, \mathbf{R}) = A(\mathbf{R})\delta(x).$$

Solution.

$$\frac{\partial}{\partial x} \langle u(x, \mathbf{R}) \rangle = \frac{i}{2k} \Delta_\mathbf{R} \langle u(x, \mathbf{R}) \rangle - \frac{k^2}{8} A(0) \langle u(x, \mathbf{R}) \rangle, \quad \langle u(0, \mathbf{R}) \rangle = u_0(\mathbf{R}), \tag{8.41}$$

$$\left(\frac{\partial}{\partial x} - \frac{i}{k} \boldsymbol{\nabla}_\mathbf{R} \boldsymbol{\nabla}_\rho \right) \Gamma_2(x, \mathbf{R}, \boldsymbol{\rho}) = -\frac{k^2}{4} D(\boldsymbol{\rho}) \Gamma_2(x, \mathbf{R}, \boldsymbol{\rho}),$$
$$\Gamma_2(0, \mathbf{R}, \boldsymbol{\rho}) = u_0\left(\mathbf{R} + \frac{1}{2}\boldsymbol{\rho}\right) u_0^*\left(\mathbf{R}' - \frac{1}{2}\boldsymbol{\rho}\right), \tag{8.42}$$

where $D(\boldsymbol{\rho}) = A(0) - A(\boldsymbol{\rho})$.

Remark. The solution to Eq. (8.41) has the form

$$\langle u(x, \mathbf{R})\rangle = u_0(x, \mathbf{R})e^{-\frac{\gamma}{2}x}, \tag{8.43}$$

where $u_0(x, \mathbf{R})$ is the solution to the problem with absent fluctuations of medium parameters and quantity $\gamma = \frac{k^2}{4}A(0)$ is called the *extinction coefficient*. The corresponding solution for the second-order coherence function is

$$\Gamma_2(x, \mathbf{R}, \boldsymbol{\rho}) = \int d\mathbf{q}\gamma_0\left(\mathbf{q}, \boldsymbol{\rho} - \mathbf{q}\frac{x}{k}\right)\exp\left\{i\mathbf{q}\mathbf{R} - \frac{k^2}{4}\int\limits_0^x d\xi D\left(\boldsymbol{\rho} - \mathbf{q}\frac{\xi}{k}\right)\right\}, \tag{8.44}$$

where

$$\gamma_0\left(\mathbf{q}, \boldsymbol{\rho}\right) = \frac{1}{(2\pi)^2}\int d\mathbf{R}\Gamma_2(0, \mathbf{R}, \boldsymbol{\rho})e^{-i\mathbf{q}\mathbf{R}}.$$

In the case of the plane incident wave, we have

$$u_0(\mathbf{R}) = u_0 = \text{const}, \quad \Gamma_2(0, \mathbf{R}, \boldsymbol{\rho}) = |u_0|^2, \quad \gamma_0\left(\mathbf{q}, \boldsymbol{\rho}\right) = |u_0|^2\delta(\mathbf{q}).$$

Expressions (8.43) and (8.44) become significantly simpler

$$\langle u(x, \mathbf{R})\rangle = u_0 e^{-\frac{1}{2}\gamma x}, \quad \Gamma_2(x, \mathbf{R}, \boldsymbol{\rho}) = |u_0|^2 e^{-\frac{1}{4}k^2 x D(\boldsymbol{\rho})}$$

and appear independent of the effect of plane wave diffraction in random medium.

Problem 48 *Find the second-order coherence function and average intensity at the axis of parabolic waveguide in the problem formulated in terms of the dynamic equation*

$$\frac{\partial}{\partial x}u(x, \mathbf{R}) = \frac{i}{2k}\Delta_{\mathbf{R}}u(x, \mathbf{R}) + \frac{ik}{2}\left[-\alpha^2\mathbf{R}^2 + \varepsilon(x, \mathbf{R})\right]u(x, \mathbf{R})$$

assuming that $\varepsilon(x, \mathbf{R})$ is the homogeneous isotropic Gaussian field delta-correlated in x and characterized by the parameters

$$\langle\varepsilon(x, \mathbf{R})\rangle = 0, \quad B_\varepsilon(x, \mathbf{R}) = B_\varepsilon^{eff}(x, \mathbf{R}) = A(\mathbf{R})\delta(x).$$

Solution.

$$\Gamma_2(x, \mathbf{R}, \boldsymbol{\rho}) = \frac{e^{-i\alpha\mathbf{R}\boldsymbol{\rho}\tan(\alpha x)}}{\cos^2(\alpha x)}\int d\mathbf{q}\gamma_0\left(\mathbf{q}, \frac{1}{\cos(\alpha x)}\boldsymbol{\rho} - \frac{\mathbf{q}}{\alpha k}\tan(\alpha x)\right)$$

$$\times \exp\left\{i\frac{1}{\cos(\alpha x)}\mathbf{q}\mathbf{R} - \frac{k^2}{4}\int\limits_0^x d\xi D\left(\frac{\cos(\alpha\xi)}{\cos(\alpha x)}\boldsymbol{\rho} - \frac{1}{\alpha k}\frac{\sin\left[\alpha\left(x - \xi\right)\right]}{\cos(\alpha x)}\mathbf{q}\right)\right\},$$

where

$$\gamma_0\left(\mathbf{q}, \boldsymbol{\rho}\right) = \frac{1}{(2\pi)^2}\int d\mathbf{R}\Gamma_2(0, \mathbf{R}, \boldsymbol{\rho})e^{-i\mathbf{q}\mathbf{R}}$$

and $D(\boldsymbol{\rho}) = A(0) - A(\boldsymbol{\rho})$.

Setting $\mathbf{R} = 0$ and $\boldsymbol{\rho} = 0$, we obtain the average intensity at waveguide axis as a function of longitudinal coordinate x

$$\langle I(x, \mathbf{0})\rangle = \frac{1}{\cos^2(\alpha x)}\int d\mathbf{q}\gamma_0\left(\mathbf{q}, -\frac{\mathbf{q}}{\alpha k}\tan(\alpha x)\right)\exp\left\{-\frac{k^2}{4}\int\limits_0^x d\xi D\left(\frac{1}{\alpha k}\frac{\sin\left(\alpha\xi\right)}{\cos(\alpha x)}\mathbf{q}\right)\right\}.$$

Problem 49 *Construct the diffusion approximation for the problem of the dynamics of a particle with linear friction under random forces, which is described by the stochastic system (1.11), page 14*

$$\frac{d}{dt}\mathbf{r}(t) = \mathbf{v}(t), \quad \frac{d}{dt}\mathbf{v}(t) = -\lambda\mathbf{v}(t) + \mathbf{f}(\mathbf{r}, t),$$
$$\mathbf{r}(0) = \mathbf{r}_0, \quad \mathbf{v}(0) = \mathbf{v}_0.$$

Solution. The one-time probability density of particle position and velocity

$$P(\mathbf{r}, \mathbf{v}, t) = \langle \delta(\mathbf{r}(t) - \mathbf{r})\delta(\mathbf{v}(t) - \mathbf{v}) \rangle$$

satisfies the Fokker–Planck equation

$$\left(\frac{\partial}{\partial t} + \mathbf{v}\frac{\partial}{\partial \mathbf{r}} - \lambda\frac{\partial}{\partial \mathbf{v}}\mathbf{v} \right) P(\mathbf{r}, \mathbf{v}, t)$$
$$= \frac{\partial}{\partial v_i} \left\{ D_{ij}^{(1)}(\mathbf{v})\frac{\partial}{\partial v_j} + D_{ij}^{(2)}(\mathbf{v})\frac{\partial}{\partial r_j} \right\} P(\mathbf{r}, \mathbf{v}, t). \tag{8.45}$$

with the diffusion coefficients

$$D_{ij}^{(1)}(\mathbf{v}) = \int_0^\infty d\tau e^{-\lambda\tau} B_{ij}\left(\frac{1}{\lambda}\left[e^{\lambda\tau} - 1 \right]\mathbf{v}, \tau \right),$$
$$D_{ij}^{(2)}(\mathbf{v}) = \frac{1}{\lambda}\int_0^\infty d\tau \left[1 - e^{-\lambda\tau} \right] B_{ij}\left(\frac{1}{\lambda}\left[e^{\lambda\tau} - 1 \right]\mathbf{v}, \tau \right). \tag{8.46}$$

Problem 50 *In the diffusion approximation, find the steady-state probability distribution of the velocity of a particle described by dynamic system (1.11), page 14 in the one-dimensional case assuming that the force correlation function has the form*

$$B_f(x, t) = \sigma_f^2 \exp\left\{ -\frac{|x|}{l_0} - \frac{|t|}{\tau_0} \right\}.$$

Solution. From Eq. (8.45) follows that steady-state probability density of particle's velocity (i.e., probability density in limit $t \to \infty$) satisfies the equation

$$-\lambda v P(v) = D(v)\frac{\partial}{\partial v}P(v), \tag{8.47}$$

where, in accordance with Eq. (8.46),

$$D(v) = \int_0^\infty d\tau e^{-\lambda\tau} B\left(\frac{1}{\lambda}\left[e^{\lambda\tau} - 1 \right]v, \tau \right).$$

For a sufficiently small friction ($\lambda\tau_0 \ll 1$), the solution to Eq. (8.47) has the form

$$P(v) = C\exp\left\{ -\frac{\lambda v^2}{2\sigma_f^2\tau_0}\left[1 + \frac{2}{3}\frac{|v|\tau_0}{l_0} \right] \right\}. \tag{8.48}$$

For small particle velocity $|v|\tau_0 \ll l_0$, probability distribution (8.48) grades into the Gaussian distribution corresponding to the approximation of the delta-correlated (in time) random field $f(x, t)$. However, in the opposite limiting case $|v|\tau_0 \gg l_0$, probability distribution (8.48) decreases significantly faster than in the case of the approximation of the delta-correlated (in time) random field $f(x, t)$, namely,

$$P(v) = C \exp \left\{ -\frac{\lambda v^2 |v|}{3\sigma_f^2 l_0} \right\}, \tag{8.49}$$

which corresponds to the diffusion coefficient decreasing according to the law $D(v) \sim 1/|v|$ for great particle velocities. Physically, it means that the effect of random force $f(x, t)$ on faster particles is significantly smaller than on slower ones.

Problem 51 *Within the framework of the diffusion approximation, derive equations for average field $\langle u(x, \mathbf{R}) \rangle$ and second-order coherence function*

$$\Gamma_2(x; \mathbf{R}, \boldsymbol{\rho}) = \left\langle u \left(x, \mathbf{R} + \frac{1}{2}\boldsymbol{\rho} \right) u^* \left(x, \mathbf{R} - \frac{1}{2}\boldsymbol{\rho} \right) \right\rangle$$

starting from parabolic equation (8.40) [41, 42].

Solution.

$$\left(\frac{\partial}{\partial x} - \frac{i}{2k}\Delta_{\mathbf{R}} \right) \langle u(x, \mathbf{R}) \rangle$$
$$= -\frac{k^2}{4} \int\limits_0^x dx' \int d\mathbf{R}' B_\varepsilon(x', \mathbf{R} - \mathbf{R}') e^{\frac{ix'}{2k}\Delta_{\mathbf{R}}} \left[\delta(\mathbf{R} - \mathbf{R}') e^{-\frac{ix'}{2k}\Delta_{\mathbf{R}}} \langle u(x, \mathbf{R}) \rangle \right],$$

$$\left(\frac{\partial}{\partial x} - \frac{i}{k}\boldsymbol{\nabla}_{\mathbf{R}}\boldsymbol{\nabla}_{\boldsymbol{\rho}} \right) \Gamma_2(x, \mathbf{R}, \boldsymbol{\rho})$$
$$= -\frac{k^2}{4} \int\limits_0^x dx_1 \int d\mathbf{R}_1 \left[B_\varepsilon \left(x_1, \mathbf{R} - \mathbf{R}_1 + \frac{1}{2}\boldsymbol{\rho} \right) - B_\varepsilon \left(x_1, \mathbf{R} - \mathbf{R}_1 - \frac{1}{2}\boldsymbol{\rho} \right) \right]$$
$$\times e^{\frac{i}{k}x_1 \boldsymbol{\nabla}_{\mathbf{R}}\boldsymbol{\nabla}_{\boldsymbol{\rho}}} \left[\delta \left(\mathbf{R} - \mathbf{R}_1 + \frac{1}{2}\boldsymbol{\rho} \right) - \delta \left(\mathbf{R} - \mathbf{R}_1 - \frac{1}{2}\boldsymbol{\rho} \right) \right] e^{-\frac{i}{k}x_1 \boldsymbol{\nabla}_{\mathbf{R}}\boldsymbol{\nabla}_{\boldsymbol{\rho}}} \Gamma_2(x - x_1, \mathbf{R}, \boldsymbol{\rho}).$$

If we introduce now the two-dimensional spectral density of inhomogeneities

$$B_\varepsilon(x, \mathbf{R}) = \int d\mathbf{q}\Phi_\varepsilon^{(2)}(x, \mathbf{q})e^{i\mathbf{q}\mathbf{R}}, \quad \Phi_\varepsilon^{(2)}(x, \mathbf{q}) = \frac{1}{(2\pi)^2} \int d\mathbf{R}B_\varepsilon(x, \mathbf{R})e^{-i\mathbf{q}\mathbf{R}}$$

and Fourier transforms of wave field $u(x, \mathbf{R})$ and coherence function $\Gamma_2(x, \mathbf{R}, \boldsymbol{\rho})$ with respect to transverse coordinates

$$u(x, \mathbf{R}) = \int d\mathbf{q}\tilde{u}(x, \mathbf{q})e^{i\mathbf{q}\mathbf{R}}, \quad \tilde{u}(x, \mathbf{q}) = \frac{1}{(2\pi)^2} \int d\mathbf{R}u(x, \mathbf{R})e^{-i\mathbf{q}\mathbf{R}},$$

$$\Gamma_2(x, \mathbf{R}, \boldsymbol{\rho}) = \int d\mathbf{q}\tilde{\Gamma}_2(x, \mathbf{q}, \boldsymbol{\rho})e^{i\mathbf{q}\mathbf{R}}, \quad \tilde{\Gamma}_2(x, \mathbf{q}, \boldsymbol{\rho}) = \frac{1}{(2\pi)^2} \int d\mathbf{R}\Gamma_2(x, \mathbf{R}, \boldsymbol{\rho})e^{-i\mathbf{q}\mathbf{R}},$$

then we obtain that average field is given by the expression

$$\langle u(x, \mathbf{R}) \rangle = \frac{1}{(2\pi)^2} \int d\mathbf{q} \int d\mathbf{R}'u_0(\mathbf{R}') \exp \left\{ i\mathbf{q}(\mathbf{R} - \mathbf{R}') - i\frac{\mathbf{q}^2 x}{2} - \frac{k^2}{2} \int\limits_0^x dx'D(x', \mathbf{q}) \right\},$$

where

$$D(x, \mathbf{q}) = \int\limits_0^x d\xi \int d\mathbf{q}' \Phi_\varepsilon^{(2)}(\xi, \mathbf{q}') \exp\left\{ -\frac{i\xi}{2k}\left(\mathbf{q}'^2 - 2\mathbf{q}'\mathbf{q}\right) \right\},$$

and function $\tilde{\Gamma}_2(x, \mathbf{q}, \boldsymbol{\rho})$ satisfies the integro-differential equation

$$\left(\frac{\partial}{\partial x} + \frac{1}{k}\mathbf{q}\boldsymbol{\nabla}_\rho \right) \tilde{\Gamma}_2(x, \mathbf{q}, \boldsymbol{\rho}) = -\frac{k^2}{4} \int\limits_0^x dx_1 \int d\mathbf{q}_1 \Phi_\varepsilon^{(2)}\left(x_1, \mathbf{q}_1\right)$$

$$\times \left\{ \cos\left[\frac{x_1}{2k}\mathbf{q}_1(\mathbf{q}_1 - \mathbf{q}) \right] - \cos\left[\mathbf{q}_1\boldsymbol{\rho} - \frac{x_1}{2k}\mathbf{q}_1(\mathbf{q}_1 - \mathbf{q}) \right] \right\} \tilde{\Gamma}_2(x, \mathbf{q}_1, \boldsymbol{\rho}).$$

If distances a wave passes in medium satisfy the condition $x \gg l_\parallel$, where $l_\parallel$ is the longitudinal correlation radius of field $\varepsilon(x, \mathbf{R})$, then we obtain

$$\langle u(x, \mathbf{R}) \rangle = \frac{1}{(2\pi)^2} \int d\mathbf{q} \int d\mathbf{R}' u_0(\mathbf{R}') \exp\left\{ i\mathbf{q}(\mathbf{R} - \mathbf{R}') - i\frac{\mathbf{q}^2 x}{2} - \frac{k^2}{2}xD(\mathbf{q}) \right\},$$

where

$$D(\mathbf{q}) = \int\limits_0^\infty d\xi \int d\mathbf{q}' \Phi_\varepsilon^{(2)}(\xi, \mathbf{q}') \exp\left\{ -\frac{i\xi}{2k}\left(\mathbf{q}'^2 - 2\mathbf{q}'\mathbf{q}\right) \right\}.$$

For the plane incident wave, we have $u_0(\mathbf{R}) = 1$, and, consequently, average intensity is independent of $\mathbf{R}$,

$$\langle u(x, \mathbf{R}) \rangle = e^{-\frac{1}{2}k^2 x D(0)}.$$

The applicability range of this expression is described by the condition

$$\frac{k^2}{2}D(0)l_\parallel \ll 1.$$

Chapter 9

Methods for solving and analyzing the Fokker-Planck equation

The Fokker–Planck equations for the one-point probability density (8.11) and for the transition probability density (8.15) are the partial differential equations of parabolic type, so that we can use methods of the theory of mathematical physics equations to solve them. In this context, the basic methods are such as the method of separation of variables, the Fourier transformation with respect to spatial coordinates, and other integral transformations.

However, there are only few Fokker–Planck equations that allow an exact solution. First of all, among them are the Fokker–Planck equations corresponding to the stochastic equations that are themselves solvable in the analytic form. Such problems often allow determination of not only the one-point and transition probability densities, but also the characteristic functional and other statistical characteristics important for practice.

The simplest special case is the equation that defines the *Wiener random process*. In view of the significant role that such processes plays in physics (for example, they describe the *Brownian motion of particles*), we consider the Wiener process in detail.

9.1 Wiener random process

The Wiener random process is defined as the solution to the stochastic equation

$$\frac{d}{dt}w(t) = z(t), \quad w(0) = 0,$$

where $z(t)$ is the Gaussian process delta-correlated in time and described by the parameters

$$\langle z(t)\rangle = 0, \quad \langle z(t)z(t')\rangle = 2\sigma^2\tau_0\delta(t - t').$$

The solution to this equation

$$w(t) = \int\limits_{0}^{t} d\tau\, z(\tau)$$

is the continuous Gaussian nonstationary random process with the parameters

$$\langle w(t)\rangle = 0, \quad \langle w(t)w(t')\rangle = 2\sigma^2\tau_0 \min(t, t').$$

Consider a more general process that includes additionally the drift dependent on parameter α

$$w(t;\alpha) = -\alpha t + w(t), \quad \alpha > 0.$$

Process $w(t;\alpha)$ is the Markovian process, and its probability density

$$P(w,t;\alpha) = \langle \delta(w(t;\alpha) - w)\rangle$$

satisfies the Fokker–Planck equation

$$\left(\frac{\partial}{\partial t} - \alpha\frac{\partial}{\partial w}\right) P(w,t;\alpha) = D\frac{\partial^2}{\partial w^2}P(w,t;\alpha), \quad P(w,0;\alpha) = \delta(w), \tag{9.1}$$

where $D = \sigma^2\tau_0$ is the diffusion coefficient. The solution to this equation has the form of the Gaussian distribution

$$P(w,t;\alpha) = \frac{1}{2\sqrt{\pi Dt}}\exp\left\{-\frac{(w+\alpha t)^2}{4Dt}\right\}. \tag{9.2}$$

The corresponding integral distribution function defined as probability of the event that $w(t;\alpha) < w$ is given by the formula

$$F(w,t;\alpha) = \int_{-\infty}^{w} dw P(w,t;\alpha) = \Phi\left(\frac{w}{\sqrt{2Dt}} + \alpha\sqrt{\frac{t}{2D}}\right), \tag{9.3}$$

where

$$\Phi(z) = \frac{1}{\sqrt{2\pi}}\int_{-\infty}^{z} dy\exp\left\{-\frac{y^2}{2}\right\} \tag{9.4}$$

is the error function.

In addition to the initial condition, supplement Eq. (9.1) with the boundary condition

$$P(w,t;\alpha)|_{w=h} = 0, \quad (t > 0). \tag{9.5}$$

This condition breaks down realizations of process $w(t;\alpha)$ at the instant they reach boundary h. For $w < h$, the solution to the boundary-value problem (9.1), (9.5) (we denote it as $P(w,t;\alpha,h)$) describes the probability distribution of those realizations of process $w(t;\alpha)$ that survived instant t, i.e., never reached boundary h during the whole temporal interval. Correspondingly, the norm of the probability density appears not unity, but the probability of the event that $t < t^*$, where t^* is the instant at which process $w(t;\alpha)$ reaches boundary h for the first time

$$\int_{-\infty}^{h} dw P(w,t;\alpha,h) = P(t < t^*). \tag{9.6}$$

Introduce the integral distribution function and probability density of random instant at which the process reaches boundary h

$$F(t;\alpha,h) = P(t^* < t) = 1 - P(t < t^*) = 1 - \int_{-\infty}^{h} dw P(w,t;\alpha,h),$$

$$P(t;\alpha,h) = \frac{\partial}{\partial t}F(t;\alpha,h) = -D\left.\frac{\partial}{\partial w}P(w,t;\alpha,h)\right|_{w=h}. \tag{9.7}$$

If $\alpha > 0$, process $w(t; \alpha)$ moves on average out of boundary h; as a result, probability $P(t < t^*)$ (9.6) tends for $t \to \infty$ to the probability of the event that process $w(t; \alpha)$ never reaches boundary h. In other words, limit

$$\lim_{t \to \infty} \int_{-\infty}^{h} dw P(w, t; \alpha, h) = P\left(w_{\max}(\alpha) < h\right) \tag{9.8}$$

is equal to the probability of the event that the process absolute maximum

$$w_{\max}(\alpha) = \max_{t \in (0, \infty)} w(t; \alpha)$$

is less than h. Thus, from Eq. (9.8) follows that the integral distribution function of the absolute maximum $w_{\max}(\alpha)$ is given by the formula

$$F(h; \alpha) = P\left(w_{\max}(\alpha) < h\right) = \lim_{t \to \infty} \int_{-\infty}^{h} dw P(w, t; \alpha, h). \tag{9.9}$$

After we solve boundary-value problem (9.1), (9.5) by using, for example, the reflection method, we obtain

$$P(w, t; \alpha, h) = \frac{1}{2\sqrt{\pi D t}} \left\{ \exp\left[-\frac{(w + \alpha t)^2}{4Dt} \right] - \exp\left[-\frac{h\alpha}{D} - \frac{(w - 2h + \alpha t)^2}{4Dt} \right] \right\}. \tag{9.10}$$

Substituting this expression in Eq. (9.7), we obtain the probability density of instant t^* at which process $w(t; \alpha)$ reaches boundary h for the first time

$$P(t; \alpha, h) = \frac{1}{2Dt\sqrt{\pi D t}} \exp\left\{ -\frac{(h + \alpha t)^2}{4Dt} \right\}.$$

Finally, integrating Eq. (9.10) over w and setting $t \to \infty$, we obtain, in accordance with Eq. (9.9), the integral distribution function of absolute maximum $w_{\max}(\alpha)$ of process $w(t; \alpha)$ in the form

$$F(h; \alpha) = P\left(w_{\max}(\alpha) < h\right) = 1 - \exp\left\{ -\frac{h\alpha}{D} \right\}. \tag{9.11}$$

Consequently, the absolute maximum of the Wiener process has the exponential probability density

$$P(h; \alpha) = \langle \delta\left(w_{\max}(\alpha) - h\right) \rangle = \frac{\alpha}{D} \exp\left\{ -\frac{h\alpha}{D} \right\}.$$

The Wiener random process offers a possibility of constructing different other processes convenient for modeling different physical phenomena. In the case of positive quantities, the simplest approximation of such kind is the logarithmic-normal (lognormal) process. Consider this process in greater detail.

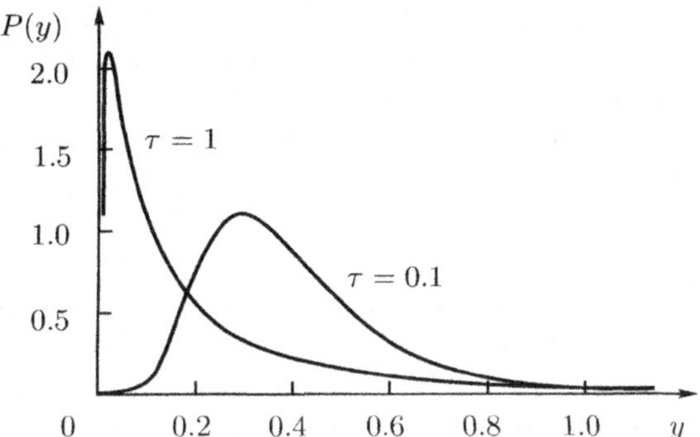

Figure 9.1: Logarithmic-normal probability density (9.13) for $\alpha/D = 1$ and dimensionless times $\tau = 0.1$ and 1.

9.2 Logarithmic-normal random process

We define the lognormal process by the formula

$$y(t;\alpha) = e^{w(t;\alpha)} = \exp\left\{ -\alpha t + \int\limits_0^t d\tau\, z(\tau) \right\}, \qquad (9.12)$$

where $z(t)$ is the Gaussian white noise process with the parameters

$$\langle z(t)\rangle = 0, \quad \langle z(t)z(t')\rangle = 2\sigma^2 \tau_0 \delta(t - t').$$

The lognormal process satisfies the stochastic equation

$$\frac{d}{dt} y(t;\alpha) = \{-\alpha + z(t)\}\, y(t;\alpha), \qquad y(0;\alpha) = 1.$$

The one-time probability density of the lognormal process is given by the formula

$$P(y,t;\alpha) = \left\langle \delta\left(e^{w(t;\alpha)} - y\right)\right\rangle = \frac{1}{y}\, P(w,t;\alpha)|_{w=\ln y},$$

where $P(w,t;\alpha)$ is the one-time probability density of the Wiener process with a drift (9.2), so that

$$P(y,t;\alpha) = \frac{1}{2y\sqrt{\pi Dt}} \exp\left\{ -\frac{(\ln y + \alpha t)^2}{4Dt} \right\} = \frac{1}{2y\sqrt{\pi Dt}} \exp\left\{ -\frac{\ln^2\left(ye^{\alpha t}\right)}{4Dt} \right\}, \qquad (9.13)$$

where $D = \sigma^2 \tau_0$.

Figure 9.1 shows the curves of the lognormal probability density (9.13) for $\alpha/D = 1$ and dimensionless times $\tau = Dt = 0.1$ and 1. One can see the long flat *tail* that appears for the curve at $\tau = 1$; this tail increases the role of high peaks of process $y(t;\alpha)$ in the formation of the one-time statistics. Correspondingly, the integral distribution function is given, in accordance with Eqs. (9.3), (9.4), by the expression

$$F(y,t;\alpha) = P\left(y(t;\alpha) < y\right) = \Phi\left(\frac{1}{\sqrt{2Dt}} \ln\left(ye^{\alpha t}\right)\right). \qquad (9.14)$$

Having only the one-point statistical characteristics of process $y(t; \alpha)$, one can obtain a deeper insight into the behavior of realizations of process $y(t; \alpha)$ on the whole interval of times $(0, \infty)$ [2, 8]. In particular,

1. From the integral distribution function, one can calculate the typical realization curve of the lognormal process $y(t; \alpha)$ (see Chapter 4); this curve appears the exponentially decaying curve

$$y^*(t; \alpha) = e^{-\alpha t}. \tag{9.15}$$

2. The lognormal process $y(t; \alpha)$ is the Markovian process and its one-time probability density (9.13) satisfies the Fokker–Planck equation

$$\left(\frac{\partial}{\partial t} - \alpha \frac{\partial}{\partial y} y \right) P(y, t; \alpha) = D \frac{\partial}{\partial y} y \frac{\partial}{\partial y} y P(y, t; \alpha), \quad P(y, 0; \alpha) = \delta(y - 1). \tag{9.16}$$

From Eq. (9.16), one can easily derive the equations for moment functions of process $y(t; \alpha)$; solutions to these equations are given by the formulas

$$\langle y^n(t; \alpha) \rangle = e^{n(n - \alpha/D)Dt}, \quad \left\langle \frac{1}{y^n(t; \alpha)} \right\rangle = e^{n(n + \alpha/D)Dt}, \quad n = 1, 2, \ldots \tag{9.17}$$

from which follows that moments exponentially grow with time. Consequently, the exponential increase of moments is caused by deviations of process $y(t; \alpha)$ from the curve of typical realization curve $y^*(t; \alpha)$ towards both large and small values of y.

At $\alpha/D = 1$, the average value of process $y(t; D)$ is independent of time and is equal to unity. Despite this fact, according to Eq. (9.14), the probability of the event that $y < 1$ for $Dt \gg 1$ rapidly approaches the unity by the law

$$P\left(y(t; D) < 1\right) = \Phi \left(\sqrt{\frac{Dt}{2}} \right) = 1 - \frac{1}{\sqrt{\pi Dt}} e^{-Dt/4},$$

i.e., the curves of process realizations run mainly below the level of the process average $\langle y(t; D) \rangle = 1$ though large process peaks exactly govern the behavior of statistical moments of process $y(t; D)$.

Here, we have a clear contradiction between the behavior of statistical characteristics of process $y(t; \alpha)$ and the behavior of process realizations.

3. The behavior of realizations of process $y(t; \alpha)$ on the whole temporal interval can also be evaluated with the use of the p-majorant curves $M_p(t, \alpha)$ whose definition is as follows. We call the majorant curve the curve $M_p(t, \alpha)$ for which inequality $y(t; \alpha) < M_p(t, \alpha)$ is satisfied for all times t with probability p, i.e.,

$$P\left\{ y(t; \alpha) < M_p(t, \alpha) \quad \text{for all} \quad t \in (0, \infty) \right\} = p.$$

The above statistics (9.11) of the absolute maximum of the Wiener process with a drift $w(t; \alpha)$ makes it possible to outline a wide enough class of the majorant curves. Indeed, let p be the probability of the event that the absolute maximum $w_{\max}(\beta)$ of the auxiliary process $w(t; \beta)$ with arbitrary parameter β in the interval $0 < \beta < \alpha$ satisfies inequality $w(t; \beta) < h = \ln A$. It is clear that the whole realization of process $y(t; \alpha)$ will run in this case below the majorant curve

$$M_p(t, \alpha, \beta) = A e^{(\beta - \alpha)t}. \tag{9.18}$$

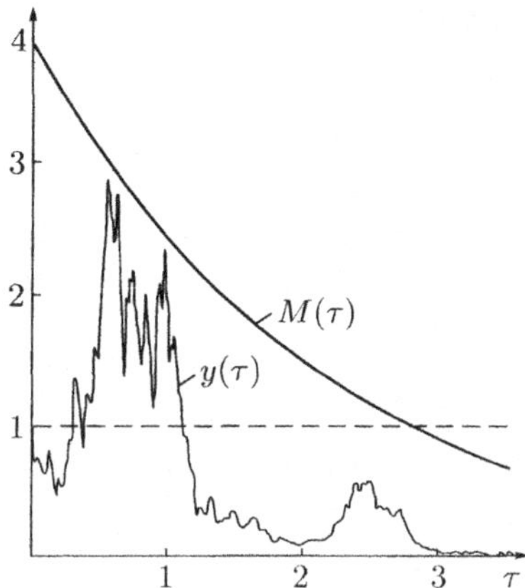

Figure 9.2: Schematic behaviors of a realization of process $y(t; D)$ and majorant curve $M(\tau)$ (9.20).

with the same probability p. As may be seen from Eq. (9.11), the probability of the event that process $y(t; \alpha)$ never exceeds majorant curve (9.18) depends on this curve parameters according to the formula

$$p = 1 - A^{-\beta/D}.$$

This means that we derived the one-parameter class of exponentially decaying majorant curves

$$M_p(t, \alpha, \beta) = \frac{1}{(1-p)^{D/\beta}} e^{(\beta-\alpha)t}. \tag{9.19}$$

Notice the remarkable fact that, despite statistical average $\langle y(t; D) \rangle$ remains constant ($\langle y(t; D) \rangle = 1$) and higher-order moments of process $y(t; D)$ are exponentially increasing functions, one can always select an exponentially decreasing majorant curve (9.19) such that realizations of process $y(t; D)$ will run below it with arbitrary predetermined probability $p < 1$. In particular, inequality ($\tau = Dt$)

$$y(t; D) < M_{1/2}(t, D, D/2) = M(\tau) = 4e^{-\tau/2} \tag{9.20}$$

is satisfied with probability $p = 1/2$ for any instant t from interval $(0, \infty)$.

Figure 9.2 schematically shows the behaviors of a realization of process $y(t; D)$ and the majorant curve $M(\tau)$ (9.20). This schematic is an additional fact in favor of our conclusion that the exponential growth of moments of process $y(t; D)$ with time is the purely statistical effect caused by averaging over the whole ensemble of realizations.

Note that the area below the exponentially decaying majorant curves has a finite value. Consequently, high peaks of process $y(t; \alpha)$, which are the reason of the exponential growth of higher moments, only insignificantly contribute to the area below realizations; this area appears finite for almost all realizations, which means that the peaks of the lognormal process $y(t; \alpha)$ are sufficiently narrow.

4. In this connection, it is of interest to investigate the statistics of random area below the powers of realizations of process $y(t; \alpha)$

$$S_n(t; \alpha) = \int\limits_0^t d\tau \, y^n(\tau; \alpha). \qquad (9.21)$$

One can easily show that random integrals $S_n(\alpha) = \int_0^\infty d\tau \, y^n(\tau; \alpha)$, which are the limits of areas (9.21) for $t \to \infty$, are distributed according to the steady-state probability density

$$P(S_n; \alpha) = \frac{1}{(n^2 D)^{\alpha/nD} \, \Gamma\left(\frac{\alpha}{D}\right) S_n^{1+\alpha/D}} \exp\left\{-\frac{1}{n^2 D S_n}\right\},$$

where $\Gamma(z)$ is the *gamma function*. In the special case $n = 1$, quantity

$$S(\alpha) = S_1(\alpha) = \int\limits_0^\infty d\tau \, y(\tau; \alpha)$$

has the following probability density

$$P(S; \alpha) = \frac{1}{D^{\alpha/D} \Gamma\left(\frac{\alpha}{D}\right) S^{1+\alpha/D}} \exp\left\{-\frac{1}{DS}\right\}. \qquad (9.22)$$

If we set now $\alpha = D$, then the steady-state probability density and the corresponding integral distribution function have the form

$$P(S; D) = \frac{1}{DS^2} \exp\left\{-\frac{1}{DS}\right\}, \quad F(S; D) = \exp\left\{-\frac{1}{DS}\right\}. \qquad (9.23)$$

All these properties of lognormal processes manifest themselves in the dynamics of concrete physical systems as coherent phenomena such as localization and clustering.

9.3　Integral transformations

Integral transformations are very practicable for solving the Fokker Planck equation. Indeed, earlier we mentioned the convenience of the Fourier transformation in (8.11) if the diffusion coefficient tensor $F_{kl}(\mathbf{x}, \mathbf{x}; t)$ is independent of $\mathbf{x}$. Different integral transformations related to eigenfunctions of the diffusion operator

$$\hat{L} = \frac{\partial^2}{\partial x_k \partial x_l} F_{kl}(\mathbf{x}, \mathbf{x}; t)$$

can be used in other situations.

For example, in the case of the Legendre operator

$$\hat{L} = \frac{\partial}{\partial x}(x^2 - 1)\frac{\partial}{\partial x},$$

it is quite natural to use the integral transformation related to the *Legendre functions*. This transformation is called the *Meller Fock* transform and is defined by the formula

$$F(\mu) = \int\limits_1^\infty dx \, f(x) P_{-1/2+i\mu}(x) \quad (\mu > 0), \qquad (9.24)$$

where $P_{-1/2+i\mu}(x)$ is the *complex index Legendre function of the first kind*, which satisfies the equation

$$\frac{d}{dx}(x^2-1)\frac{d}{dx}P_{-1/2+i\mu}(x) = -\left(\mu^2+\frac{1}{4}\right)P_{-1/2+i\mu}(x). \qquad (9.25)$$

The inversion of the transform (9.24) has the form

$$f(x) = \int_0^\infty d\mu\,\mu\tanh(\pi\mu)F(\mu)P_{-1/2+i\mu}(x) \quad (1 \le x \le \infty), \qquad (9.26)$$

where $F(\mu)$ is given by formula (9.24).

Another integral transformation called the Kantorovich–Lebedev transform, is related to diffusion operator

$$\hat{L} = \frac{\partial}{\partial x}x^2\frac{\partial}{\partial x}$$

and has the form

$$F(\tau) = \int_0^\infty dx\,f(x)K_{i\tau}(x) \quad (\tau > 0),$$

where $K_{i\tau}(x)$ is the *imaginary index McDonalds function of the first kind*, which satisfies the equations

$$\left(x^2\frac{d^2}{dx^2} + x\frac{d}{dx} - x^2 + \tau^2\right)K_{i\tau}(x) = 0,$$

$$\left(\frac{d}{dx}x^2\frac{d}{dx} - x\frac{d}{dx}\right)K_{i\tau}(x) = \left(x^2 - \tau^2\right)K_{i\tau}(x). \qquad (9.27)$$

The corresponding inversion has the form

$$f(x) = \frac{2}{\pi^2 x}\int_0^\infty d\tau\,\sinh(\pi\tau)F(\tau)K_{i\tau}(x). \qquad (9.28)$$

9.4   Steady-state solutions of the Fokker–Planck equation

In previous sections, we discussed general methods of solving the Fokker–Planck equation for both transition and one-point probability densities. However, the problem on the one-point probability density can have peculiar features caused by possible existence of the steady-state solution; in a number of cases, such a solution can be obtained immediately. The steady-state solution, if it exists, is independent of the initial conditions and is the solution of the Fokker–Planck equation in the limit $t \to \infty$.

There are two classes of problems for which the steady-state solution of the Fokker–Planck equation can be easily found. These classes deal with the one-dimensional differential equations and with the Hamiltonian systems of equations. Consider these cases in greater detail.

9.4.1   One-dimensional nonlinear differential equation

The one-dimensional nonlinear systems are described by the stochastic equation

$$\frac{d}{dt}x(t) = f(x) + z(t)g(x), \quad x(0) = x_0, \tag{9.29}$$

where $z(t)$ is, as earlier, the Gaussian delta-correlated process with the parameters

$$\langle z(t) \rangle = 0, \quad \langle z(t)z(t') \rangle = 2D\delta(t - t') \quad (D = \sigma_z^2 \tau_0).$$

The corresponding Fokker–Planck equation has the form

$$\left(\frac{\partial}{\partial t} + \frac{\partial}{\partial x}f(x) \right) P(x,t) = D\frac{\partial}{\partial x}g(x)\frac{\partial}{\partial x}g(x)P(x,t). \tag{9.30}$$

The steady-state probability distribution $P(x)$, if it exists, satisfies the equation

$$f(x)P(x) = Dg(x)\frac{d}{dx}g(x)P(x) \tag{9.31}$$

(we assume that $P(x)$ is distributed over the whole space, i.e., for $-\infty < x < \infty$) whose solution is as follows

$$P(x) = \frac{C}{|g(x)|} \exp\left\{ \frac{1}{D} \int dx \frac{f(x)}{g^2(x)} \right\}, \tag{9.32}$$

where constant C is determined from the normalization condition

$$\int\limits_{-\infty}^{\infty} dx P(x) = 1.$$

In the special case of the Langevin equation (8.21), page 123 ($f(x) = -\lambda x$, $g(x) = 1$), Eq. (9.32) grades into the Gaussian probability distribution

$$P(x) = \sqrt{\frac{\lambda}{2\pi D}} \exp\left\{ -\frac{\lambda}{2D}x^2 \right\}. \tag{9.33}$$

9.4.2   Hamiltonian systems

Another type of dynamic systems that allow obtaining the steady-state probability distribution is described by the Hamiltonian system with linear friction

$$\begin{aligned}
\frac{d}{dt}\mathbf{r}_i(t) &= \frac{\partial}{\partial \mathbf{p}_i}H(\{\mathbf{r}_i\}, \{\mathbf{p}_i\}), \\
\frac{d}{dt}\mathbf{p}_i(t) &= -\frac{\partial}{\partial \mathbf{r}_i}H(\{\mathbf{r}_i\}, \{\mathbf{p}_i\}) - \lambda\mathbf{p}_i + \mathbf{f}_i(t),
\end{aligned} \tag{9.34}$$

where $i = 1, 2, ..., N$,

$$H(\{\mathbf{r}_i\}, \{\mathbf{p}_i\}) = \frac{\mathbf{p}_i^2}{2} + U(\mathbf{r}_1, \ldots, \mathbf{r}_N),$$

is the Hamiltonian, λ is a constant coefficient (friction), and random forces $\mathbf{f}_i(t)$ are the Gaussian delta-correlated random vector functions with the correlation tensor

$$\left\langle f_i^\alpha(t)f_j^\beta(t') \right\rangle = 2D\delta_{ij}\delta_{\alpha\beta}\delta(t - t'), \quad D = \sigma_f^2\tau_0. \tag{9.35}$$

Here, α and β are the vector indexes.

System of equations (9.34) describes the Brownian motion of the system of N interacting particles. The Fokker–Planck equation for the joint probability density of the solution to system (9.34) has the form

$$\frac{\partial}{\partial t}P(\{\mathbf{r}_i\},\{\mathbf{p}_i\},t) + \sum_{k=1}^{N}\{H,P\}_{(k)} - \lambda\sum_{k=1}^{N}\frac{\partial}{\partial \mathbf{p}_k}\{\mathbf{p}_k P\}$$
$$= D\sum_{k=1}^{N}\frac{\partial^2}{\partial \mathbf{p}_k^2}P(\{\mathbf{r}_i\},\{\mathbf{p}_i\},t), \tag{9.36}$$

where

$$\{\varphi,\psi\}_{(k)} = \frac{\partial\varphi}{\partial\mathbf{p}_k}\frac{\partial\psi}{\partial\mathbf{r}_k} - \frac{\partial\psi}{\partial\mathbf{p}_k}\frac{\partial\varphi}{\partial\mathbf{r}_k}$$

is the Poisson bracket for the k-th particle.

One can easily check that the steady-state solution to Eq. (9.36) is the *Gibbs canonical distribution*

$$P(\{\mathbf{r}_i\},\{\mathbf{p}_i\}) = C\exp\left\{-\frac{\lambda}{D}H(\{\mathbf{r}_i\},\{\mathbf{p}_i\})\right\}. \tag{9.37}$$

The specificity of this distribution consists in the Gaussian behavior with respect to momenta and statistical independence of particle coordinates and momenta.

Integrating Eq. (9.37) over all $\mathbf{r}$, we can obtain the *Maxwell distribution* that describes velocity fluctuations of the Brownian particles. The case $U(\mathbf{r}_1, \ldots, \mathbf{r}_N) = 0$ corresponds to the Brownian motion of a system of free particles (9.33).

If we integrate probability distribution (9.37) over momenta (velocities), we obtain the *Boltzmann distribution* for particle coordinates

$$P(\{\mathbf{r}_i\}) = C\exp\left\{-\frac{\lambda}{D}U(\{\mathbf{r}_i\})\right\}. \tag{9.38}$$

In the case of sufficiently strong friction, the equilibrium distribution (9.37) is formed in two stages. First, the Gaussian momentum distribution (the Maxwell distribution) is formed relatively quickly and then, the spatial distribution (the Boltzmann distribution) is formed at far slower rate. The latter stage is described by the Fokker–Planck equation

$$\frac{\partial}{\partial t}P(\{\mathbf{r}_i\},t) = \frac{1}{\lambda}\sum_{k=1}^{N}\frac{\partial}{\partial\mathbf{r}_k}\left(\frac{\partial U(\{\mathbf{r}_i\})}{\partial\mathbf{r}_k}P(\{\mathbf{r}_i\},t)\right) + \frac{D}{\lambda^2}\sum_{k=1}^{N}\frac{\partial^2}{\partial\mathbf{r}_k^2}P(\{\mathbf{r}_i\},t), \tag{9.39}$$

which is usually called the *Einstein–Smolukhovsky equation*. Derivation of Eq. (9.39) from the Fokker–Planck equation (9.36) is called the *Kramers problem*. Note that Eq. (9.39) statistically corresponds to the stochastic equation

$$\frac{d}{dt}\mathbf{r}_i(t) = -\frac{1}{\lambda}\frac{\partial}{\partial\mathbf{r}_i}U(\{\mathbf{r}_i\}) + \frac{1}{\lambda}\mathbf{f}_i(t),$$

which, nevertheless, cannot be considered as the limit of Eq. (9.34) for $\lambda \to \infty$.

In the one-dimensional case, Eqs. (9.34) are simplified and assume the form of the system of two equations

$$\frac{d}{dt}x(t) = y(t), \quad \frac{d}{dt}y(t) = -\frac{\partial}{\partial x}U(x) - \lambda y(t) + f(t). \tag{9.40}$$

The corresponding steady-state probability distribution has the form

$$P(x, y) = C \exp\left\{-\frac{\lambda}{D} H(x, y)\right\}, \quad H(x, y) = \frac{y^2}{2} + U(x). \tag{9.41}$$

9.5 Boundary-value problems for the Fokker-Planck equation (transfer phenomena)

The Fokker–Planck equations are the partial differential equations and they generally require boundary conditions whose particular form depends on the problem at hand. We can proceed from both forward and backward Fokker–Planck equations, which are equivalent. Consider one typical example.

Consider the nonlinear oscillator with friction described by the equation

$$\frac{d^2}{dt^2}x(t) + \lambda\frac{d}{dt}x(t) + \omega_0^2 x(t) + \beta x^3(t) = f(t) \quad (\beta, \lambda > 0) \tag{9.42}$$

and assume that random force $f(t)$ is the delta-correlated random function with the parameters

$$\langle f(t)\rangle = 0, \quad \langle f(t)f(t')\rangle = 2D\delta(t - t') \quad (D = \sigma_f^2 \tau_0).$$

At $\lambda = 0$ and $f(t) = 0$, this equation is called the *Duffing equation*.

We can rewrite Eq. (9.42) in the standard form of the Hamiltonian system in functions $x(t)$ and $v(t) = \frac{d}{dt}x(t)$,

$$\frac{d}{dt}x(t) = \frac{\partial}{\partial v} H(x, v), \quad \frac{d}{dt}v(t) = -\frac{\partial}{\partial x} H(x, v) - \lambda v + f(t),$$

where

$$H(x, v) = \frac{v^2}{2} + U(x), \quad U(x) = \frac{\omega_0^2 x^2}{2} + \beta\frac{x^4}{4}$$

is the Hamiltonian.

According to Eq. (9.41), the steady-state solution to the corresponding Fokker–Planck equation has the form

$$P(x, v) = C \exp\left\{-\frac{\lambda}{D} H(x, v)\right\}. \tag{9.43}$$

It is clear that this distribution is the product of two independent distributions, of which one — the steady-state probability distribution of quantity $v(t)$ — is the Gaussian distribution and the other — the steady-state probability distribution of quantity $x(t)$ — is the non-Gaussian distribution. Integrating Eq. (9.43) over v, we obtain the steady-state probability distribution of $x(t)$

$$P(x, v) = C \exp\left\{-\frac{\lambda}{D}\left(\frac{\omega_0^2 x^2}{2} + \beta\frac{x^4}{4}\right)\right\}.$$

This distribution is maximum at the stable equilibrium point $x = 0$.

Consider now the equation

$$\frac{d^2}{dt^2}x(t) + \lambda\frac{d}{dt}x(t) - \omega_0^2 x(t) + \beta x^3(t) = f(t) \quad (\beta, \lambda > 0). \tag{9.44}$$

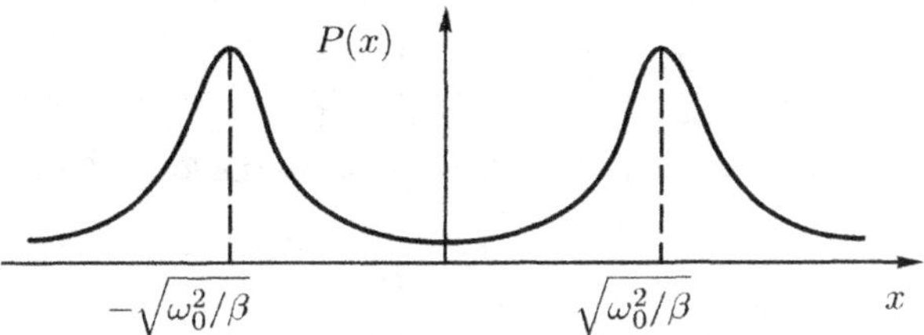

Figure 9.3: Probability distribution (9.45).

In this case again, the steady-state probability distribution has the form (9.43), where now

$$H(x,v) = \frac{v^2}{2} + U(x), \quad U(x) = -\frac{\omega_0^2 x^2}{2} + \beta\frac{x^4}{4}.$$

The steady-state probability distribution of $x(t)$ assumes now the form

$$P(x,v) = C\exp\left\{-\frac{\lambda}{D}\left(-\frac{\omega_0^2 x^2}{2} + \beta\frac{x^4}{4}\right)\right\} \tag{9.45}$$

and has maxima at points $x = \pm\sqrt{\omega_0^2/\beta}$ and a minimum at point $x = 0$; the maxima correspond to the stable equilibrium points of problem (9.44) for $f(t) = 0$ and the minimum, to the instable equilibrium point. Figure 9.3 shows the behavior of the probability distribution (9.45).

As we mentioned earlier, the process of establishing distribution (9.45) is described by the Einstein–Smolukhovsky equation (9.39), which has in this case the form

$$\frac{\partial}{\partial t}P(x,t) = \frac{1}{\lambda}\frac{\partial}{\partial x}\left(\frac{\partial U(x)}{\partial x}P(x,t)\right) + \frac{1}{\lambda^2}\frac{\partial^2}{\partial x^2}P(x,t). \tag{9.46}$$

This equation is statistically equivalent to the dynamic equation

$$\frac{d}{dt}x(t) = -\frac{1}{\lambda}\frac{\partial U(x)}{\partial x} + \frac{1}{\lambda}f(t). \tag{9.47}$$

Probability distribution (9.45) corresponds to averaging over an ensemble of realizations of random process $f(t)$. If we deal with a single realization, the system arrives with probability $1/2$ at one of states corresponding to the distribution maxima. In this case, averaging over time will form the probability distribution around the maximum position. However, after a lapse of certain time T (the longer, the smaller D), the system will be transferred in the vicinity of the other maximum due to the fact that function $f(t)$ can assume sufficiently large values. For this reason, temporal averaging will form probability distribution (9.46) only if averaging time $t \gg T$.

Introducing dimensionless coordinate $x \to \sqrt{\frac{\omega_0^2}{\beta}}x$ and time $t \to \frac{\lambda}{\omega_0^2}t$, we can rewrite Eq. (9.46) in the form

$$\frac{\partial}{\partial t}P(x,t) = \frac{\partial}{\partial x}\left(\frac{\partial U(x)}{\partial x}P(x,t)\right) + \mu\frac{\partial^2}{\partial x^2}P(x,t), \tag{9.48}$$

where

$$\mu = \frac{\beta D}{\lambda \omega_0^4}, \quad U(x) = -\frac{x^2}{2} + \frac{x^4}{4}.$$

In this case, the equivalent stochastic equation (9.47) assumes the form of Eq. (1.10), page 14 of Chapter 1

$$\frac{d}{dt}x(t) = -\frac{\partial U(x)}{\partial x} + f(t). \tag{9.49}$$

Estimate the time required for the system to switch from a most probable state $x = -1$ to the other $x = 1$.

Let the system described by stochastic equation (9.49) was at a point from the interval (a, b) at instant t_0. The corresponding probability for the system to leave this interval

$$G(t; x_0, t_0) = 1 - \int_a^b dx\, p(x, t | x_0, t_0)$$

satisfies Eq. (8.20) following from the backward Fokker–Planck equation, i.e., the equation

$$\frac{\partial}{\partial t_0} G(t; x_0, t_0) = \frac{\partial U(x_0)}{\partial x_0} \frac{\partial}{\partial x_0} G(t; x_0, t_0) - \mu \frac{\partial^2}{\partial x_0^2} G(t; x_0, t_0)$$

with the boundary conditions

$$G(t; x_0, t) = 0, \quad G(t; a, t_0) = G(t; b, t_0) = 1.$$

Taking into account the fact that $G(t; x_0, t_0) = G(t - t_0; x_0)$ in our problem, we can denote $(t - t_0) = \tau$ and rewrite the boundary-value problem in the form

$$\frac{\partial}{\partial \tau} G(\tau; x_0) = \frac{\partial U(x_0)}{\partial x_0} \frac{\partial}{\partial x_0} G(\tau; x_0) - \mu \frac{\partial^2}{\partial x_0^2} G(\tau; x_0),$$

$$G(0; x_0) = 0, \; G(\tau; a,) = G(\tau; b) = 1 \quad \left(\lim_{\tau \to \infty} G(\tau; x_0) = 0 \right). \tag{9.50}$$

From Eq. (9.50), one can easily see that average time required for the system to leave interval (a, b)

$$T(x_0) = \int_0^\infty d\tau\, \tau \frac{\partial G(\tau; x_0)}{\partial \tau}$$

satisfies the boundary-value problem

$$\mu \frac{d^2 T(x_0)}{dx_0^2} - \frac{dU(x_0)}{dx_0} \frac{dT(x_0)}{dx_0} = -1, \quad T(a) = T(b) = 0. \tag{9.51}$$

Equation (9.51) can be easily solved, and we obtain that the average time required for the system under random force to switch its state from $x_0 = -1$ to $x_0 = 1$ (this time is usually called the *Kramers time*)

$$T = \frac{1}{\mu} \int_{-1}^1 d\xi \int_{-\infty}^\xi d\eta \exp\left\{ \frac{1}{\mu} [U(\xi) - U(\eta)] \right\} = \frac{C(\mu)}{\mu} \int_0^1 d\xi \exp\left\{ \frac{1}{\mu} U(\xi) \right\}, \tag{9.52}$$

where $C(\mu) = \int_{-\infty}^\infty d\xi\, e^{\frac{1}{\mu} U(\xi)}$. For $\mu \ll 1$, we obtain

$$T \approx \sqrt{2}\pi e^{\frac{1}{4\mu}},$$

i.e., the average switching time increases exponentially with decreasing the intensity of fluctuations of the force.

9.6 Method of fast oscillation averaging

If fluctuations of dynamic system parameters are sufficiently small, we can analyze the Fokker–Planck equation using different asymptotic and approximate techniques. Here, we consider in greater detail the method that is most often used in the statistical analysis.

Let a stochastic system be described by the dynamic equations

$$
\begin{aligned}
\frac{d}{dt}x(t) &= A(x,\widetilde{\phi}) + z(t)B(x,\widetilde{\phi}), \\
\frac{d}{dt}\phi(t) &= C(x,\widetilde{\phi}) + z(t)D(x,\widetilde{\phi}),
\end{aligned}
\tag{9.53}
$$

where

$$
\widetilde{\phi}(t) = \omega_0 t + \phi(t),
$$

functions $A(x,\widetilde{\phi})$, $B(x,\widetilde{\phi})$, $C(x,\widetilde{\phi})$, and $D(x,\widetilde{\phi})$ are the periodic functions of variable $\widetilde{\phi}$, and $z(t)$ is the Gaussian delta-correlated process with the parameters

$$
\langle z(t)\rangle = 0, \quad \langle z(t)z(t')\rangle = 2D\delta(t - t'), \quad D = \sigma^2 \tau_0.
$$

Variables $x(t)$ and $\phi(t)$ can be, for example, the vector module and phase, respectively. The Fokker–Planck equation corresponding to system of equations (9.53) has the form

$$
\begin{aligned}
\frac{\partial}{\partial t}P(x,\phi,t) &= -\frac{\partial}{\partial x}A(x,\widetilde{\phi})P(x,\phi,t) - \frac{\partial}{\partial \phi}C(x,\widetilde{\phi})P(x,\phi,t) \\
&+ D\left[\frac{\partial}{\partial x}B(x,\widetilde{\phi}) + \frac{\partial}{\partial \phi}D(x,\widetilde{\phi})\right]^2 P(x,\phi,t).
\end{aligned}
\tag{9.54}
$$

Generally, Eq. (9.54) is very complicated to immediately analyze the joint probability density. We rewrite this equation in the form

$$
\begin{aligned}
\frac{\partial}{\partial t}P(x,\phi,t) &= -\frac{\partial}{\partial x}A(x,\widetilde{\phi})P(x,\phi,t) - \frac{\partial}{\partial \phi}C(x,\widetilde{\phi})P(x,\phi,t) \\
&- D\frac{\partial}{\partial x}\left(\frac{\partial B^2(x,\widetilde{\phi})}{2\partial x} + \frac{\partial B(x,\widetilde{\phi})}{\partial \widetilde{\phi}}D(x,\widetilde{\phi})\right)P(x,\phi,t) \\
&- D\frac{\partial}{\partial \phi}\left(\frac{\partial D(x,\widetilde{\phi})}{\partial x}B(x,\widetilde{\phi}) + \frac{\partial D^2(x,\widetilde{\phi})}{2\partial\widetilde{\phi}}\right)P(x,\phi,t) \\
&+ D\left\{\frac{\partial^2}{\partial x^2}B^2(x,\widetilde{\phi}) + 2\frac{\partial^2}{\partial x\partial\phi}B(x,\widetilde{\phi})D(x,\widetilde{\phi}) + \frac{\partial^2}{\partial\phi^2}D^2(x,\widetilde{\phi})\right\}P(x,\phi,t).
\end{aligned}
\tag{9.55}
$$

Now, we assume that functions $A(x,\widetilde{\phi})$ and $C(x,\widetilde{\phi})$ are sufficiently small and fluctuation intensity of process $z(t)$ is also small. In this case, statistical characteristics of system of equations (9.53) only slightly vary during times $\sim 1/\omega_0$. To study these small variations (accumulated effects), we can average Eq. (9.55) over the period of all oscillating functions. Assuming that function $P(x,\phi,t)$ remains intact under averaging, we obtain the equation

$$
\begin{aligned}
\frac{\partial}{\partial t}\overline{P(x,\phi,t)} &= -\frac{\partial}{\partial x}\overline{A(x,\widetilde{\phi})}P(x,\phi,t) - \frac{\partial}{\partial \phi}\overline{C(x,\widetilde{\phi})}P(x,\phi,t) \\
&- D\left\{\frac{\partial}{\partial x}\overline{\left(\frac{\partial B^2(x,\widetilde{\phi})}{2\partial x} + \frac{\partial B(x,\widetilde{\phi})}{\partial\widetilde{\phi}}D(x,\widetilde{\phi})\right)} + \frac{\partial}{\partial \phi}\overline{\frac{\partial D(x,\widetilde{\phi})}{\partial x}B(x,\widetilde{\phi})}\right\}\overline{P(x,\phi,t)} \\
&+ D\left\{\frac{\partial^2}{\partial x^2}\overline{B^2(x,\widetilde{\phi})} + 2\frac{\partial^2}{\partial x\partial\phi}\overline{B(x,\widetilde{\phi})D(x,\widetilde{\phi})} + \frac{\partial^2}{\partial\phi^2}\overline{D^2(x,\widetilde{\phi})}\right\}\overline{P(x,\phi,t)},
\end{aligned}
\tag{9.56}
$$

where the overbar denotes quantities averaged over the oscillation period.

Integrating Eq. (9.56) over ϕ, we obtain the Fokker–Planck equation for function $\overline{P(x,t)}$

$$\frac{\partial}{\partial t}\overline{P(x,\phi,t)} = -\frac{\partial}{\partial x}\overline{A(x,\widetilde{\phi})\,P(x,\phi,t)}$$
$$+D\frac{\partial}{\partial x}\left(\frac{\overline{\partial B^2(x,\widetilde{\phi})}}{2\partial x} + \overline{\frac{\partial B(x,\widetilde{\phi})}{\partial\widetilde{\phi}}D(x,\widetilde{\phi})}\right)\overline{P(x,\phi,t)} + D\frac{\partial}{\partial x}\overline{B^2(x,\widetilde{\phi})}\frac{\partial}{\partial x}\overline{P(x,\phi,t)}.$$

$$(9.57)$$

Note that quantity $x(t)$ appears the one-dimensional Markovian random process in this approximation.

If we assume that

$$\overline{B(x,\widetilde{\phi})D(x,\widetilde{\phi})} = \overline{\frac{\partial D(x,\widetilde{\phi})}{\partial x}B(x,\widetilde{\phi})} = 0, \text{ and } \overline{C(x,\widetilde{\phi})} = \text{const}, \ \overline{D^2(x,\widetilde{\phi})} = \text{const}$$

in Eq. (9.56), then processes $x(t)$ and $\phi(t)$ become statistically independent, and process $\phi(t)$ becomes the Markovian Gaussian process whose variance is the linear increasing function of time t. This means that probability distribution of quantity $\phi(t)$ over segment $[0, 2\pi]$ becomes uniform for large t (at $\overline{C(x,\widetilde{\phi})} = 0$).

Problems

Problem 52 *Derive the equation for the one-time probability density that describes the statistics of the area*

$$S_n(t;\alpha) = \int\limits_0^t d\tau\, y^n(\tau;\alpha)$$

below realizations of lognormal process $y(t;\alpha)$ (9.12)

Instruction. Show that, in the context of one-time statistics, quantity $S_n(t;\alpha)$ is statistically equivalent to the quantity

$$S_n(t;\alpha) = \int\limits_0^t d\tau \exp\left\{-n\alpha(t-\tau) + n\int\limits_0^{t-\tau} d\tau_1 z(\tau+\tau_1)\right\}$$

that satisfies the following stochastic equation of the first order

$$\frac{d}{dt}S_n(t;\alpha) = 1 - n\{\alpha - z(t)\}S_n(t;\alpha), \quad S_n(0;\alpha) = 0.$$

Solution.

$$\frac{\partial}{\partial t}P(S_n,t;\alpha) = \left\{-\frac{\partial}{\partial S_n} + n\alpha\frac{\partial}{\partial S_n}S_n + n^2 D\frac{\partial}{\partial S_n}S_n\frac{\partial}{\partial S_n}S_n\right\}P(S_n,y,t;\alpha),$$
$$P(S_n,0;\alpha) = \delta(S_n).$$

Problem 53 *Show that steady-state probability density of random integrals $S_n(\alpha) = \int_0^\infty d\tau\, y^n(\tau;\alpha)$ exists for $t \to \infty$ and is given by Eq. (9.22).*

Problem 54 *Find the expressions for probability density and integral distribution function of stochastic quantity*

$$\tilde{S}(t,\alpha) = \int\limits_{t}^{\infty} d\tau\, y(\tau;\alpha), \tag{9.58}$$

where $y(t;\alpha)$ is the lognormal random process.

Instruction. Represent integral (9.58) in the form

$$\tilde{S}(t,\alpha) = y(t;\alpha) \int\limits_{0}^{\infty} d\tau \exp\left\{ -\alpha\tau + \int\limits_{0}^{\tau} d\tau_1 z(\tau_1 + t) \right\}, \tag{9.59}$$

from which follows that random process $y(t;\alpha)$ is statistically independent of the integral in the right-hand side of Eq. (9.59), because they are functionals of random process $z(\tau)$ for nonoverlapping intervals of variable τ and the integral by itself is statistically equivalent to random quantity $S(\alpha)$.

Solution.

$$P(\tilde{S},t;\alpha) = \int\limits_{0}^{\infty}\int\limits_{0}^{\infty} dy\,dS\,\delta(yS - \tilde{S})P(y,t;\alpha) = \int\limits_{0}^{\infty} \frac{dy}{y} P(y,t;\alpha)P(\tilde{S}/y;\alpha),$$

where $P(y,t;\alpha)$ is the one-time probability density (9.13) of lognormal process $y(t;\alpha)$ and $P(\tilde{S}/y;\alpha)$ is the area probability density (9.22) at $n = 1$.

Problem 55 *Solve the Fokker–Planck equation*

$$\frac{\partial}{\partial t}p(x,t|x_0,t_0) = D\frac{\partial}{\partial x}(x^2 - 1)\frac{\partial}{\partial x}p(x,t|x_0,t_0) \quad (x \geq 1),$$

$$p(x,t_0|x_0,t_0) = \delta(x - x_0).$$

Instruction. Use the integral Meller–Fock transform.
Solution.

$$p(x,t|x_0,t_0)$$
$$= \int\limits_{0}^{\infty} d\mu\, \mu \tanh(\pi\mu) \exp\left\{ -D\left(\mu^2 + \frac{1}{4}\right)(t - t_0) \right\} P_{-1/2+i\mu}(x)P_{-1/2+i\mu}(x_0).$$

If $x_0 = 1$ at the initial instant $t_0 = 0$, then

$$P(x,t) = \int\limits_{0}^{\infty} d\mu\, \mu \tanh(\pi\mu) \exp\left\{ -D\left(\mu^2 + \frac{1}{4}\right)(t - t_0) \right\} P_{-1/2+i\mu}(x).$$

Problem 56 *In the context of singular stochastic problem (1.13), page 15 for $\lambda = 1$*

$$\frac{d}{dt}x(t) = -x^2(t) + f(t), \quad x(0) = x_0, \tag{9.60}$$

where $f(t)$ is the Gaussian delta-correlated process with the parameters

$$\langle f(t)\rangle = 0, \quad \langle f(t)f(t')\rangle = 2D\delta(t - t') \quad (D = \sigma_f^2\tau_0),$$

estimate average time required for the system to switch from state x_0 to state $(-\infty)$ and average time between two singularities.

Solution.

$$\langle T(x_0)\rangle = \int\limits_{-\infty}^{x_0} d\xi \int\limits_{\xi}^{\infty} d\eta \exp\left\{\frac{1}{3}\left(\xi^3 - \eta^3\right)\right\}, \quad \langle T(\infty)\rangle = \sqrt{\pi}\frac{12^{1/6}}{3}\Gamma\left(\frac{1}{6}\right) \approx 4.976.$$

Problem 57 *Find the steady-state probability distribution for stochastic problem (9.60).*

Instruction. Solve the Fokker–Planck equation under the condition that function $x(t)$ is discontinuous and defined for all times t in such a way that its value of $-\infty$ at instant $t \to t_0 - 0$ is immediately followed by its value of ∞ at instant $t \to t_0 + 0$. This behavior corresponds to the boundary condition of the Fokker–Planck equation formulated as continuity of probability flux density.

Solution.

$$P(x) = J \int\limits_{-\infty}^{x} d\xi \exp\left\{\frac{1}{3}\left(\xi^3 - x^3\right)\right\}, \tag{9.61}$$

where

$$J = \frac{1}{\langle T(\infty)\rangle}$$

is the steady-state probability flux density.

Remark. From Eq. (9.61) follows the asymptotic formula

$$P(x) \approx \frac{1}{\langle T(\infty)\rangle\, x^2} \tag{9.62}$$

for great x.

Problem 58 *Show that asymptotic formula (9.62) is formed by discontinuities of function $x(t)$.*

Instruction. Near discontinuous points t_k, represent the solution $x(t)$ in the form

$$x(t) = \frac{1}{t - t_k}.$$

Problem 59 *Derive the Fokker Planck equation for slowly varying amplitude and phase of the solution to the problem*

$$\frac{d}{dt}x(t) = y(t), \quad \frac{d}{dt}y(t) = -2\gamma y(t) - \omega_0^2[1 + z(t)]x(t), \tag{9.63}$$

where $z(t)$ is the Gaussian process with the parameters

$$\langle z(t)\rangle = 0, \quad \langle z(t)z(t')\rangle = 2\sigma^2\tau_0\delta(t - t').$$

Instruction. Replace functions $x(t)$ and $y(t)$ with new variables (oscillation amplitude and phase) according to the equalities

$$x(t) = A(t)\sin\left(\omega_0 t + \phi(t)\right), \quad y(t) = \omega_0 A(t)\cos\left(\omega_0 t + \phi(t)\right)$$

and represent amplitude $A(t)$ in the form $A(t) = e^{u(t)}$.

Solution.

$$\frac{\partial}{\partial t}\overline{P(t;u,\phi)} = \gamma\frac{\partial}{\partial u}\overline{P(t;u,\phi)} - \frac{D}{4}\frac{\partial}{\partial u}\overline{P(t;u,\phi)}$$
$$+ \frac{D}{8}\frac{\partial^2}{\partial u^2}\overline{P(t;u,\phi)} + \frac{3D}{8}\frac{\partial^2}{\partial\phi^2}\overline{P(t;u,\phi)}, \tag{9.64}$$

where $D = \sigma^2\tau_0\omega_0^2$. From Eq. (9.64) follows that statistical characteristics of oscillation amplitude and phase (averaged over the oscillation period) are statistically independent and have the Gaussian probability densities

$$\overline{P(t;u)} = \frac{1}{\sqrt{2\pi\sigma_u^2(t)}}\exp\left\{-\frac{(u-\langle u(t)\rangle)^2}{2\sigma_u^2(t)}\right\},$$
$$\overline{P(t;\phi)} = \frac{1}{\sqrt{2\pi\sigma_\phi^2(t)}}\exp\left\{-\frac{(\phi-\phi_0)^2}{2\sigma_\phi^2(t)}\right\}, \tag{9.65}$$

where

$$\langle u(t)\rangle = u_0 - \gamma t + \frac{D}{4}t, \quad \sigma_u^2(t) = \frac{D}{4}t,$$
$$\langle\phi(t)\rangle = \phi_0, \quad \sigma_\phi^2(t) = \frac{3D}{4}t.$$

Problem 60 *Using probability distributions (9.65), calculate quantities $\langle x(t)\rangle$ and $\langle x^2(t)\rangle$. Derive the condition of stochastic excitation of the second moment.*

Solution. Average value
$$\langle x(t)\rangle = e^{-\gamma t}\sin(\omega_0 t)$$

coincides with the problem solution in the absence of fluctuations.

Quantity
$$\left\langle x^2(t)\right\rangle = \frac{1}{2}e^{(D-2\gamma)t}\left\{1 - e^{-3Dt/2}\cos(2\omega_0 t)\right\}, \tag{9.66}$$

coincides in the absence of absorption with Eq. (6.77) to terms about $D/\omega_0 \ll 1$.

Problem 61 *Find the condition of stochastic parametric excitation of function $\langle A^n(t)\rangle$.*

Solution. We have

$$\langle A^n(t)\rangle = \left\langle e^{nu(t)}\right\rangle = A_0^n\exp\left\{-n\gamma t + \frac{1}{8}n(n+2)Dt\right\},$$

and, consequently, the stochastic dynamic system is parametrically excited beginning from moment function of order n if the condition

$$8\gamma < (n+2)D$$

is satisfied.

Remark. The typical realization curve of random amplitude has the form

$$A^*(t) = A_0\exp\left\{-\left(\gamma - \frac{D}{4}\right)t\right\}$$

and exponentially decays in time if absorption is sufficiently small, namely, if

$$1 < 4\frac{\gamma}{D} < 1 + \frac{1}{2}n,$$

which always holds for sufficiently great n, whereas all moment functions of random amplitude $A(t)$ of order n and higher exponentially grow in time. This means that statistics of random amplitude $A(t)$ is formed mainly by large spikes over the exponentially decaying typical realization curve. This fact follows from lognormal property of random amplitude $A(t)$.

Chapter 10

Gaussian delta-correlated random field (causal integral equations)

In problems discussed in Chapter 9, we succeeded in deriving the closed statistical description in the approximation of the delta-correlated random field due to the fact that every of these problems corresponded to a system of the first-order (in temporal coordinate) differential equations with given initial conditions at $t = 0$. Such systems possess the dynamic causality property, which means that the solution at instant t depends only on system parameter fluctuations for preceding times and is independent of fluctuations for consequent times.

However, problems described in terms of integral equations that generally cannot be reduced to a system of differential equations also can possess the causality property. For short, we illustrate this fact by the simplest example of the one-dimensional causal equation $(t > t')$

$$G(t;t') = g(t;t') + \Lambda \int_{t'}^{t} d\tau\, g(t;\tau) z(\tau) G(\tau;t'), \tag{10.1}$$

where function $g(t;t')$ is Green's function for the problem with absent parameter fluctuations, i.e., for $z(t) = 0$, and we assume that $z(t)$ is the Gaussian delta-correlated random function with the parameters

$$\langle z(t) \rangle = 0, \quad \langle z(t) z(t') \rangle = 2D\delta(t - t') \quad (D = \sigma_z^2 \tau_0).$$

Averaging then Eq. (10.1) over an ensemble of realizations of random function $z(t)$, we obtain the equation

$$\langle G(t;t') \rangle = g(t;t') + \Lambda \int_{t'}^{t} d\tau\, g(t;\tau) \langle z(\tau) G(\tau;t') \rangle. \tag{10.2}$$

Taking into account equality

$$\frac{\delta}{\delta z(t)} G(t;t') = g(t;t) \Lambda G(t;t') \tag{10.3}$$

following from Eq. (10.1), we can rewrite the correlator in the right-hand side of Eq. (10.2) in the from

$$\langle z(\tau) G(\tau;t') \rangle = D \left\langle \frac{\delta}{\delta z(\tau)} G(\tau;t') \right\rangle = \Lambda D g(\tau;\tau) \langle G(\tau;t') \rangle.$$

As a consequence, Eq. (10.2) grades into the closed integral equation for average Green's function

$$\langle G(t;t')\rangle = g(t;t') + \Lambda^2 D \int_{t'}^{t} d\tau\, g(t;\tau) g(\tau;\tau) \langle G(\tau;t')\rangle, \tag{10.4}$$

which has the form of the *Dyson equation* (in terminology of the quantum field theory)

$$\langle G(t;t')\rangle = g(t;t') + \Lambda \int_{t'}^{t} d\tau\, g(t;\tau) \int_{t'}^{\tau} d\tau'\, Q(\tau;\tau') \langle G(\tau';t')\rangle,$$

$$\langle G(t;t')\rangle = g(t;t') + \Lambda \int_{t'}^{t} d\tau\, \langle G(t;\tau)\rangle \int_{t'}^{\tau} d\tau'\, Q(\tau;\tau') g(\tau';t'), \tag{10.5}$$

with the *mass function*

$$Q(\tau;\tau') = \Lambda^2 D g(\tau;\tau)\delta(\tau - \tau').$$

Derive now the equation for the correlation function

$$\Gamma(t,t';t_1,t_1') = \langle G(t;t')G^*(t_1;t_1')\rangle \quad (t > t', \quad t_1 > t_1'),$$

where $G^*(t;t')$ is complex conjugated Green's function. With this goal in view, we multiply Eq. (10.1) by $G^*(t_1;t_1')$ and average the result over an ensemble of realizations of random function $z(t)$. The result is the equation that can be symbolically represented as

$$\Gamma = g\langle G^*\rangle + \Lambda g \langle zGG^*\rangle. \tag{10.6}$$

Taking into account the Dyson equation (10.5)

$$\langle G\rangle = \{1 + \langle G\rangle Q\}g,$$

we apply operator $\{1 + \langle G\rangle Q\}$ to Eq. (10.6). As a result, we obtain the symbolic-form equation

$$\Gamma = \langle G\rangle\langle G^*\rangle + \langle G\rangle\Lambda\{\langle zGG^*\rangle - Q\Gamma\},$$

which can be represented in common variables as

$$\Gamma(t,t';t_1,t_1') = \langle G(t;t')\rangle\langle G^*(t_1;t_1')\rangle$$

$$+\Lambda D \int_{0}^{t} d\tau\, \langle G(t;\tau)\rangle \left[\left\langle \frac{\delta G(\tau;t')}{\delta z(\tau)} G^*(t_1;t_1') + 2G(\tau;t')\frac{\delta G^*(t_1;t_1')}{\delta z(\tau)} \right\rangle\right]$$

$$-\Lambda^2 D \int_{0}^{t} d\tau\, \langle G(t;\tau)\rangle\, g(\tau;\tau)\Gamma(\tau,t';t_1,t_1'). \tag{10.7}$$

Deriving Eq. (10.7), we used additionally Eq. (5.31), page 76 for splitting correlations between the Gaussian delta-correlated process $z(t)$ and functionals of this process

$$\langle z(t')R[t;z(\tau)]\rangle = \begin{cases} D\left\langle \frac{\delta}{\delta z(t)}R[t;z(\tau)]\right\rangle & (t' = t, \quad \tau < t), \\[2ex] 2D\left\langle \frac{\delta}{\delta z(t')}R[t;z(\tau)]\right\rangle & (t' < t, \quad \tau < t). \end{cases}$$

Taking into account formulas (10.3) and equality

$$\frac{\delta}{\delta z(\tau)} G^*(t_1; t_1') = \Lambda G^*(t_1; \tau) G^*(\tau; t_1'),$$

following from Eq. (10.1) we can rewrite Eq. (10.7) as

$$\Gamma(t, t'; t_1, t_1') = \langle G(t; t') \rangle \langle G^*(t_1; t_1') \rangle$$
$$+ 2|\Lambda|^2 D \int\limits_0^t d\tau \, \langle G(t; \tau) \rangle \langle G^*(t_1; \tau) G(\tau; t') G^*(\tau; t_1') \rangle. \tag{10.8}$$

Now, we take into account the fact that function $G^*(t_1; \tau)$ functionally depends on random process $z(\tilde{\tau})$ for $\tilde{\tau} \geq \tau$ while functions $G(\tau; t')$ and $G^*(\tau; t_1')$ depend on it for $\tilde{\tau} \leq \tau$. Consequently, these functions are statistically independent in the case of the delta-correlated process $z(\tilde{\tau})$, and we can rewrite Eq. (10.8) in the form of the closed equation $(t_1 \geq t)$

$$\Gamma(t, t'; t_1, t_1') = \langle G(t; t') \rangle \langle G^*(t_1; t_1') \rangle + 2|\Lambda|^2 D \int\limits_0^t d\tau \, \langle G(t; \tau) \rangle \langle G^*(t_1; \tau) \rangle \Gamma(\tau; t'; \tau; t_1').$$
$$\tag{10.9}$$

Problems

Problem 62 *Show that integral equation (10.1) is equivalent to the variational derivative equation*

$$\frac{\delta}{\delta z(\tau)} G(t, t') = G(t, \tau) \Lambda G(\tau, t') \quad (t' \leq \tau \leq t)$$

with the initial condition

$$G(t, t')\big|_{z(\tau)=0} = g(t, t').$$

Part III

Examples of coherent phenomena in stochastic dynamic systems

Chapter 11

Passive tracer clustering and diffusion in random hydrodynamic flows

The evolution of the density (concentration) of a conservative passive tracer moving in velocity field $\mathbf{U}(\mathbf{r}, t)$ is described by the equation

$$\left(\frac{\partial}{\partial t} + \frac{\partial}{\partial \mathbf{r}}\mathbf{U}(\mathbf{r}, t)\right)\rho(\mathbf{r}, t) = \mu\Delta\rho(\mathbf{r}, t), \quad \rho(\mathbf{r}, 0) = \rho_0(\mathbf{r}). \tag{11.1}$$

where $\mathbf{U}(\mathbf{r}, t) = \mathbf{u}_0(\mathbf{r}, t) + \mathbf{u}(\mathbf{r}, t)$, $\mathbf{u}_0(\mathbf{r}, t)$ is the deterministic component of the velocity field (mean flow), and $\mathbf{u}(\mathbf{r}, t)$ is the random component. In the general case, random field $\mathbf{u}(\mathbf{r}, t)$ can be composed of both solenoidal (for which $\operatorname{div}\mathbf{u}(\mathbf{r}, t) = 0$) and potential (for which $\operatorname{div}\mathbf{u}(\mathbf{r}, t) \neq 0$) components. The right-hand side of Eq. (11.1) takes into account the molecular diffusion with diffusion coefficient μ; it is assumed that the total tracer mass is conserved during the evolution process, i.e.,

$$M = M(t) = \int d\mathbf{r}\rho(\mathbf{r}, t) = \int d\mathbf{r}\rho_0(\mathbf{r}) = \text{const}.$$

The effect of molecular diffusion can be neglected during the initial stages of diffusion development. In this case, Eq. (11.1) becomes simpler and assumes the form

$$\left(\frac{\partial}{\partial t} + \mathbf{U}(\mathbf{r}, t)\frac{\partial}{\partial \mathbf{r}}\right)\rho(\mathbf{r}, t) + \frac{\partial\mathbf{U}(\mathbf{r}, t)}{\partial \mathbf{r}}\rho(\mathbf{r}, t) = 0. \tag{11.2}$$

The above equations correspond to the *Eulerian description* of the concentration evolution.

Equation (11.2) is the first-order partial differential equation and can be solved by the method of characteristics. Introducing characteristic curves $\mathbf{r}(t)$ satisfying the equations of particle motion

$$\frac{d}{dt}\mathbf{r}(t) = \mathbf{U}(\mathbf{r}, t), \quad \mathbf{r}(0) = \mathbf{r}_0, \tag{11.3}$$

we can change over from Eq. (11.2) to the ordinary differential equation

$$\frac{d}{dt}\rho(t) = -\frac{\partial\mathbf{U}(\mathbf{r}, t)}{\partial \mathbf{r}}\rho(t), \quad \rho(0) = \rho_0(\mathbf{r}_0). \tag{11.4}$$

Solutions to Eqs. (11.3) and (11.4) have an obvious geometric interpretation. They describe the concentration behavior around a fixed tracer particle moving along trajectory $\mathbf{r} = \mathbf{r}(t)$. As may be seen from Eq. (11.4), the concentration in divergent flows varies: it increases in regions where medium is dense and decreases in regions where medium is rarefied.

Solutions to system (11.3), (11.4) depend on characteristic parameter $\mathbf{r}_0$ (the initial coordinate of the particle)

$$\mathbf{r}(t) = \mathbf{r}(t|\mathbf{r}_0), \qquad \rho(t) = \rho(t|\mathbf{r}_0), \tag{11.5}$$

which we will separate by the bar. Components of vector $\mathbf{r}_0$ are called the *Lagrangian coordinates* of the particle; they unambiguously specify the position of arbitrary particle. Equations (11.3), (11.4) correspond in this case to the *Lagrangian description* of concentration evolution. The first of equalities (11.5) specify the relationship between the Eulerian and Lagrangian descriptions. Solving it in $\mathbf{r}_0$, we obtain the relationship that expresses the Lagrangian coordinates in terms of the Eulerian ones

$$\mathbf{r}_0 = \mathbf{r}_0(\mathbf{r}, t). \tag{11.6}$$

Then, using Eq. (11.6), to eliminate $\mathbf{r}_0$ in the last equality in (11.5), we turn back to the concentration in the *Eulerian description*

$$\rho(\mathbf{r}, t) = \rho(t|\mathbf{r}_0(\mathbf{r}, t)) = \int d\mathbf{r}_0 \rho(t|\mathbf{r}_0) j(t|\mathbf{r}_0) \delta\left(\mathbf{r}(t|\mathbf{r}_0) - \mathbf{r}\right), \tag{11.7}$$

where we introduced new function called *divergence*

$$j(t|\mathbf{r}_0) = \det\|j_{ik}(t|\mathbf{r}_0)\| = \det\left\|\frac{\partial r_i(t|\mathbf{r}_0)}{\partial r_{0k}}\right\|,$$

which is the quantitative measure of the degree of compression (extension) of physically infinitely small liquid particles. One can easily obtain that it satisfies the equation

$$\frac{d}{dt} j(t|\mathbf{r}_0) = \frac{\partial \mathbf{U}(\mathbf{r}, t)}{\partial \mathbf{r}} j(t|\mathbf{r}_0), \qquad j(0|\mathbf{r}_0) = 1. \tag{11.8}$$

Comparing Eq. (11.4) with Eq. (11.8), we see that

$$\rho(t|\mathbf{r}_0) = \frac{\rho_0(\mathbf{r}_0)}{j(t|\mathbf{r}_0)}. \tag{11.9}$$

Thus, we can rewrite Eq. (11.7) as the equality

$$\rho(\mathbf{r}, t) = \int d\mathbf{r}_0 \rho_0(\mathbf{r}_0) \delta\left(\mathbf{r}(t|\mathbf{r}_0) - \mathbf{r}\right) \tag{11.10}$$

specifying the relationship between the Lagrangian and Eulerian characteristics. Delta function in the right-hand side of Eq. (11.10) is the *indicator function* for the position of the Lagrangian particle; as a consequence, after averaging Eq. (11.10) over an ensemble of realizations of random velocity field, we arrive at the well-known relationship between the average concentration in the Eulerian description and the one-time probability density

$$P(t, \mathbf{r}|\mathbf{r}_0) = \langle \delta\left(\mathbf{r}(t|\mathbf{r}_0) - \mathbf{r}\right)\rangle$$

of the Lagrangian particle

$$\langle \rho(\mathbf{r}, t) \rangle = \int d\mathbf{r}_0 \rho_0(\mathbf{r}_0) P(t, \mathbf{r}|\mathbf{r}_0).$$

The relationship between the spatial correlation function of the concentration field in the Eulerian description

$$\Gamma(\mathbf{r}_1, \mathbf{r}_2, t) = \langle \rho(\mathbf{r}_1, t) \rho(\mathbf{r}_2, t) \rangle$$

and the joint probability density function of positions of two particles

$$P(t, \mathbf{r}_1, \mathbf{r}_2 | \mathbf{r}_{01}, \mathbf{r}_{02}) = \langle \delta \left(\mathbf{r}_1(t|\mathbf{r}_{01}) - \mathbf{r}_1 \right) \delta \left(\mathbf{r}_2(t|\mathbf{r}_{02}) - \mathbf{r}_2 \right) \rangle$$

can be obtained similarly

$$\Gamma(\mathbf{r}_1, \mathbf{r}_2, t) = \int d\mathbf{r}_{01} \int d\mathbf{r}_{02} \rho_0(\mathbf{r}_{01}) \rho_0(\mathbf{r}_{02}) P(t, \mathbf{r}_1, \mathbf{r}_2 | \mathbf{r}_{01}, \mathbf{r}_{02}).$$

For the divergence-free velocity field ($\operatorname{div} \mathbf{U}(\mathbf{r}, t) = 0$), both particle divergence and particle concentration are invariant, i.e.,

$$j(t|\mathbf{r}_0) = 1, \qquad \rho(t|\mathbf{r}_0) = \rho_0(\mathbf{r}_0).$$

We assume in the general case that the random component of the velocity field is the divergent ($\operatorname{div} \mathbf{u}(\mathbf{r}, t) \neq 0$), statistically homogeneous and isotropic, stationary Gaussian random field whose average is zero-valued ($\langle \mathbf{u}(\mathbf{r}, t) \rangle = 0$) and correlation and spectral tensors are given by the formulas

$$\langle u_i(\mathbf{r}, t) u_j(\mathbf{r}', t') \rangle = B_{ij}(\mathbf{r} - \mathbf{r}', t - t') = \int d\mathbf{k} E_{ij}(\mathbf{k}, t - t') e^{i\mathbf{k}(\mathbf{r}-\mathbf{r}')},$$

$$B_{ij}(\mathbf{r}, t) = B_{ij}^{\mathrm{s}}(\mathbf{r}, t) + B_{ij}^{\mathrm{p}}(\mathbf{r}, t), \quad E_{ij}(\mathbf{k}, t) = E_{ij}^{\mathrm{s}}(\mathbf{k}, t) + E_{ij}^{\mathrm{p}}(\mathbf{k}, t)$$

$$E_{ij}^{\mathrm{s}}(\mathbf{k}, t) = \frac{1}{(2\pi)^d} \int d\mathbf{r} B_{ij}^{\mathrm{s}}(\mathbf{r}, t) e^{-i\mathbf{k}\mathbf{r}}, \quad E_{ij}^{\mathrm{p}}(\mathbf{k}, t) = \frac{1}{(2\pi)^d} \int d\mathbf{r} B_{ij}^{\mathrm{p}}(\mathbf{r}, t) e^{-i\mathbf{k}\mathbf{r}},$$

$$(11.11)$$

where d is the dimension of space and the spectral tensor of the velocity field assumes the following structure

$$E_{ij}^{\mathrm{s}}(\mathbf{k}, t) = E^{\mathrm{s}}(k, t) \left(\delta_{ij} - \frac{k_i k_j}{k^2} \right), \quad E_{ij}^{\mathrm{p}}(\mathbf{k}, t) = E^{\mathrm{p}}(k, t) \frac{k_i k_j}{k^2}. \qquad (11.12)$$

Here, $E^{\mathrm{s}}(k, t)$ and $E^{\mathrm{p}}(k, t)$ are the solenoidal and potential components of the spectral density of the velocity field, respectively.

Calculating statistical properties of the concentration field and its gradient, we will approximate the velocity field $\mathbf{u}(\mathbf{r}, t)$ by the random delta-correlated (in time) process. In the framework of this approximation, correlation tensor (11.11) is approximated by the expression

$$B_{ij}(\mathbf{r}, t) = 2 B_{ij}^{\mathrm{eff}}(\mathbf{r}) \delta(t), \qquad (11.13)$$

where

$$B_{ij}^{\mathrm{eff}}(\mathbf{r}) = \frac{1}{2} \int\limits_{-\infty}^{\infty} dt B_{ij}(\mathbf{r}, t) = \int\limits_{0}^{\infty} dt B_{ij}(\mathbf{r}, t).$$

In view of homogeneity and isotropy of the velocity field $\mathbf{u}(\mathbf{r}, t)$, we have the equalities

$$B_{kl}^{\text{eff}}(0) = D_0 \delta_{kl}, \quad \frac{\partial}{\partial r_i} B_{kl}^{\text{eff}}(0) = 0,$$

$$-\frac{\partial^2}{\partial r_i \partial r_j} B_{kl}^{\text{eff}}(0) = \frac{D^{\text{s}}}{d(d+2)} \left[(d+1)\delta_{kl}\delta_{ij} - \delta_{ki}\delta_{lj} - \delta_{kj}\delta_{li}\right]$$

$$+ \frac{D^{\text{p}}}{d(d+2)} \left[\delta_{kl}\delta_{ij} + \delta_{ki}\delta_{lj} + \delta_{kj}\delta_{li}\right], \tag{11.14}$$

where we introduced the following notation:

$$D_0 = \frac{1}{d} \int\limits_0^\infty dt \int d\mathbf{k} \left[(d-1)E^{\text{s}}(k,t) + E^{\text{p}}(k,t)\right],$$

$$D^{\text{s}} = \int\limits_0^\infty dt \int d\mathbf{k}\, k^2 E^{\text{s}}(k,t), \quad D^{\text{p}} = \int\limits_0^\infty dt \int d\mathbf{k}\, k^2 E^{\text{p}}(k,t). \tag{11.15}$$

11.1 Lagrangian description (particle diffusion)

11.1.1 One-point statistical characteristics

Thus, in the Lagrangian representation, the behavior of passive tracer is described in terms of ordinary differential equations (11.3), (11.4), (11.8). We can easily pass on from these equations to the linear Liouville equation in the corresponding phase space. With this goal in view, introduce the indicator function

$$\varphi_{\text{Lag}}(t; \mathbf{r}, \rho, j | \mathbf{r}_0) = \delta(\mathbf{r}(t|\mathbf{r}_0) - \mathbf{r})\delta(\rho(t|\mathbf{r}_0) - \rho)\delta(j(t|\mathbf{r}_0) - j), \tag{11.16}$$

where we explicitly emphasized the fact that the solution to the initial dynamic equations depends on the Lagrangian coordinates $\mathbf{r}_0$. Differentiating Eq. (11.16) with respect to time and using Eqs. (11.3), (11.4), and (11.8), we arrive at the Liouville equation equivalent to the initial problem

$$\left(\frac{\partial}{\partial t} + \frac{\partial}{\partial \mathbf{r}} \mathbf{U}(\mathbf{r}, t)\right) \varphi_{\text{Lag}}(t; \mathbf{r}, \rho, j | \mathbf{r}_0) = \frac{\partial \mathbf{U}(\mathbf{r}, t)}{\partial \mathbf{r}} \left(\frac{\partial}{\partial \rho}\rho - \frac{\partial}{\partial j}j\right) \varphi_{\text{Lag}}(t; \mathbf{r}, \rho, j | \mathbf{r}_0),$$

$$\varphi_{\text{Lag}}(0; \mathbf{r}, \rho, j | \mathbf{r}_0) = \delta(\mathbf{r}_0 - \mathbf{r})\delta(\rho_0(\mathbf{r}_0) - \rho)\delta(j - 1). \tag{11.17}$$

The one-time probability density of the solutions to dynamic problems (11.3), (11.4), and (11.8) coincides with the indicator function averaged over an ensemble of realizations

$$P(t; \mathbf{r}, \rho, j | \mathbf{r}_0) = \left\langle \varphi_{\text{Lag}}(t; \mathbf{r}, \rho, j | \mathbf{r}_0) \right\rangle.$$

Averaging Eq. (11.17) over an ensemble of realizations of random field $\mathbf{u}(\mathbf{r}, t)$, using the Furutsu–Novikov formula (8.10), and taking into account the equality

$$\frac{\delta}{\delta u_\beta(\mathbf{r}', t-0)} \varphi_{\text{Lag}}(t; \mathbf{r}, \rho, j | \mathbf{r}_0)$$

$$= \left\{ -\frac{\partial}{\partial r_\beta}\delta(\mathbf{r} - \mathbf{r}') + \frac{\partial \delta(\mathbf{r} - \mathbf{r}')}{\partial r_\beta}\left(\frac{\partial}{\partial \rho}\rho - \frac{\partial}{\partial j}j\right) \right\} \varphi_{\text{Lag}}(t; \mathbf{r}, \rho, j | \mathbf{r}_0),$$

and relationships (11.14), we arrive at the Fokker–Planck equation for the one-time Lagrangian probability density

$$\left(\frac{\partial}{\partial t} - D_0\Delta\right) P(t; \mathbf{r}, \rho, j|\mathbf{r}_0) = D^{\mathrm{p}}\left(\frac{\partial}{\partial \rho}\rho^2\frac{\partial}{\partial \rho} - 2\frac{\partial^2}{\partial \rho \partial j}\rho j + \frac{\partial^2}{\partial j^2}j^2\right) P(t; \mathbf{r}, \rho, j|\mathbf{r}_0),$$

$$P(0; \mathbf{r}, \rho, j|\mathbf{r}_0) = \delta(\mathbf{r} - \mathbf{r}_0)\delta(\rho_0(\mathbf{r}_0) - \rho)\delta(j - 1). \tag{11.18}$$

The solution to Eq. (11.18) is as follows

$$P(t; \mathbf{r}, \rho, j|\mathbf{r}_0) = P(t; \mathbf{r}|\mathbf{r}_0)P(t; j|\mathbf{r}_0)\delta\left(\rho - \frac{\rho_0(\mathbf{r}_0)}{j}\right), \tag{11.19}$$

where

$$P(t; \mathbf{r}|\mathbf{r}') = e^{D_0 t\Delta}\delta(\mathbf{r} - \mathbf{r}') = \frac{1}{(4\pi D_0 dt)^{d/2}}\exp\left\{\frac{(\mathbf{r} - \mathbf{r}')^2}{4D_0 t}\right\} \tag{11.20}$$

is the probability distribution of coordinates of passive tracer particle and

$$P(t; j|\mathbf{r}_0) = e^{D^{\mathrm{p}}t\frac{\partial^2}{\partial j^2}j^2}\delta(j - 1) = \frac{1}{2j\sqrt{\pi\tau}}\exp\left\{-\frac{\ln^2(je^\tau)}{4\tau}\right\} \tag{11.21}$$

is the probability distribution of the divergence field in the vicinity of this particle. In Eq. (11.21) and below, we use the dimensionless time $\tau = D^{\mathrm{p}}t$. We emphasize that solution (11.19) expresses the fact that coordinates $\mathbf{r}(t|\mathbf{r}_0)$ and divergence $j(t|\mathbf{r}_0)$ are statistically independent in the vicinity of the particle with the Lagrangian coordinates $\mathbf{r}_0$. From the lognormal distribution (11.21) follows that quantity $\chi(t|\mathbf{r}_0) = \ln j(t|\mathbf{r}_0)$ is distributed according to the Gaussian law with the parameters

$$\langle\chi(t|\mathbf{r}_0)\rangle = -\tau, \quad \sigma_\chi^2(t) = 2\tau. \tag{11.22}$$

In particular, Eq. (11.21) (and immediately Eq. (11.18), too) yields the following expressions for moments of the random divergence field

$$\langle j^n(t|\mathbf{r}_0)\rangle = e^{n(n-1)\tau}, \quad n = \pm 1, \pm 2, \dots . \tag{11.23}$$

We emphasize that average divergence remains constant $\langle j(t|\mathbf{r}_0)\rangle = 1$, while its higher moments exponentially grow in time.

In addition, we note that, according to Eqs. (11.9) and (11.23), the Lagrangian moments of concentration can be represented in the form

$$\langle\rho^n(t|\mathbf{r}_0)\rangle = \rho_0^n(\mathbf{r}_0)e^{n(n+1)\tau},$$

from which follows that both average concentration and its higher moments are exponentially increasing functions in the Lagrangian representation. In this case, the probability density of particle concentration has the form

$$P(t; \rho|\mathbf{r}_0) = \frac{1}{2\rho\sqrt{\pi\tau}}\exp\left\{-\frac{\ln^2\left(\rho e^{-\tau}/\rho_0(\mathbf{r}_0)\right)}{4\tau}\right\}. \tag{11.24}$$

This probability density can be obtained also as the solution to the Fokker–Planck equation following from Eq. (11.18)

$$\frac{\partial}{\partial t}P(t; \rho|\mathbf{r}_0) = D^{\mathrm{p}}\frac{\partial}{\partial \rho}\rho^2\frac{\partial}{\partial \rho}P(t; \rho|\mathbf{r}_0), \quad P(0; \rho|\mathbf{r}_0) = \delta(\rho_0(\mathbf{r}_0) - \rho).$$

The above paradoxial behavior of statistical characteristics of the divergence and concentration (simultaneous growth of all moment functions in time) is a consequence of the lognormal probability distribution. Indeed, the typical realization curve of random divergence is the exponentially decaying curve

$$j^*(t) = e^{-\tau}.$$

Moreover, realizations of the lognormal process satisfy certain majorant estimates. For example, with probability $p = 1/2$, we have

$$j(t|\mathbf{r}_0) < 4e^{-\tau/2}$$

throughout the whole temporal interval $t \in (t_1, t_2)$.

Similarly, the typical realization curve of concentration and its minorant estimate have the following form

$$\rho^*(t) = \rho_0 e^\tau, \quad \rho(t|\mathbf{r}_0) > \frac{\rho_0}{4} e^{\tau/2}.$$

We emphasize that the above Lagrangian statistical properties of a particle in flows containing the potential random component are qualitatively different from the statistical properties of a particle in divergence-free flows where $j(t|\mathbf{r}_0) \equiv 1$ and particle concentration remains invariant in the vicinity of a fixed particle $\rho(t|\mathbf{r}_0) = \rho_0(\mathbf{r}_0) = \text{const}$. The above statistical estimates are indicative of the fact that the statistics of random processes $j(t|\mathbf{r}_0)$ and $\rho(t|\mathbf{r}_0)$ is formed by the realization spikes relative typical realization curve.

At the same time, probability distributions of particle coordinates in essence coincide for both divergent and divergence-free velocity fields.

11.1.2   Two-point statistical characteristics

Consider now the joint dynamics of two particles in the absence of mean flow. In this case, the indicator function

$$\varphi(t; \mathbf{r}_1, \mathbf{r}_2) = \delta\left(\mathbf{r}_1(t) - \mathbf{r}_1\right) \delta\left(\mathbf{r}_2(t) - \mathbf{r}_2\right)$$

satisfies the Liouville equation

$$\frac{\partial}{\partial t}\varphi(t; \mathbf{r}_1, \mathbf{r}_2) = -\left[\frac{\partial}{\partial \mathbf{r}_1}\mathbf{u}_1(\mathbf{r}, t) + \frac{\partial}{\partial \mathbf{r}_2}\mathbf{u}_2(\mathbf{r}, t)\right]\varphi(t; \mathbf{r}_1, \mathbf{r}_2).$$

If we average the indicator function over an ensemble of realizations of field $\mathbf{u}(\mathbf{r}, t)$, use the Furutsu–Novikov formula (8.10), and the equality

$$\frac{\delta}{\delta u_j(\mathbf{r}', t - 0)}\varphi(t; \mathbf{r}_1, \mathbf{r}_2) = -\left[\frac{\partial}{\partial r_{1j}}\delta(\mathbf{r}_1 - \mathbf{r}') + \frac{\partial}{\partial r_{2j}}\delta(\mathbf{r}_2 - \mathbf{r}')\right]\varphi(t; \mathbf{r}_1, \mathbf{r}_2),$$

then we obtain that the joint probability density of positions of two particles

$$P(t; \mathbf{r}_1, \mathbf{r}_2) = \langle\varphi(t; \mathbf{r}_1, \mathbf{r}_2)\rangle$$

satisfies the Fokker–Planck equation

$$\frac{\partial}{\partial t}P(t; \mathbf{r}_1, \mathbf{r}_2) = \left[\frac{\partial^2}{\partial_{1i}\partial r_{1j}} + \frac{\partial^2}{\partial r_{2i}\partial r_{2j}}\right]B_{ij}^{\text{eff}}(0)P(t; \mathbf{r}_1, \mathbf{r}_2)$$

$$+2\frac{\partial^2}{\partial_{1i}\partial r_{2j}}B_{ij}^{\text{eff}}(\mathbf{r}_1 - \mathbf{r}_2)P(t; \mathbf{r}_1, \mathbf{r}_2). \tag{11.25}$$

Multiplying now Eq. (11.25) by function $\delta(\mathbf{r}_1 - \mathbf{r}_2 - \mathbf{l})$ and integrating over $\mathbf{r}_1$ and $\mathbf{r}_2$, we obtain that the probability density of relative diffusion of two particles

$$P(t;\mathbf{l}) = \langle \delta(\mathbf{r}_1(t) - \mathbf{r}_2(t) - \mathbf{l}) \rangle$$

satisfies the Fokker–Planck equation

$$\frac{\partial}{\partial t} P(t;\mathbf{l}) = \frac{\partial^2}{\partial l_\alpha \partial l_\beta} D_{\alpha\beta}(\mathbf{l}) P(t;\mathbf{l}), \quad P(0;\mathbf{l}) = \delta(\mathbf{l} - \mathbf{l}_0), \tag{11.26}$$

where

$$D_{\alpha\beta}(\mathbf{l}) = 2 \left[B_{\alpha\beta}^{\mathrm{eff}}(0) - B_{\alpha\beta}^{\mathrm{eff}}(\mathbf{l}) \right]$$

is the structure matrix of vector field $\mathbf{u}(\mathbf{r}, t)$, and $\mathbf{l}_0$ is the initial distance between the particles.

In the general case, Eq. (11.26) hardly can be solved analytically. However, if the initial distance between particles l_0 is sufficiently small, namely, if $l_0 \ll l_{\mathrm{cor}}$, where l_{cor} is the spatial correlation radius of the velocity field $\mathbf{u}(\mathbf{r}, t)$, we can expand functions $D_{\alpha\beta}(\mathbf{l})$ in the Taylor series to obtain in the first approximation

$$D_{\alpha\beta}(\mathbf{l}) = - \left. \frac{\partial^2 B_{\alpha\beta}^{\mathrm{eff}}(\mathbf{l})}{\partial l_i \partial l_j} \right|_{\mathbf{l}=0} l_i l_j.$$

The use of representation (11.12)–(11.15) simplifies the diffusion tensor reducing it to the form

$$D_{\alpha\beta}(\mathbf{l}) = \frac{1}{d(d+2)} \left[\left(D^{\mathrm{s}}(d+1) + D^{\mathrm{p}} \right) \delta_{\alpha\beta} l^2 - 2 \left(D^{\mathrm{s}} - D^{\mathrm{p}} \right) l_\alpha l_\beta \right], \tag{11.27}$$

where d is the dimension of space.

Substituting now Eq. (11.27) in Eq. (11.26), multiplying both sides of the resulting equation by l^n, and integrating over $\mathbf{l}$, we obtain the closed equation

$$\frac{d}{dt} \ln \langle l^n(t) \rangle$$
$$= \frac{1}{d(d+2)} \left[\left(D^{\mathrm{s}}(d+1) + D^{\mathrm{p}} \right) n \left(d + n - 2 \right) - 2 \left(D^{\mathrm{s}} - D^{\mathrm{p}} \right) n(n - 1) \right],$$

whose solution shows the exponential growth of all moment functions ($n = 1, 2, ...$) in time. In this case, the probability distribution of random process $l(t)/l_0$ will be logarithmic-normal. As a consequence, the typical realization curve of the distance between two particles will be the exponential function of time

$$l^*(t) = l_0 \exp\left\{ \frac{1}{d(d+2)} \left\{ D^{\mathrm{s}} d \left(d - 1 \right) - D^{\mathrm{p}} \left(4 - d \right) \right\} t \right\}. \tag{11.28}$$

It appears that this expression in the two-dimensional case ($d = 2$)

$$l^*(t) = l_0 \exp\left\{ \frac{1}{4} \left(D^{\mathrm{s}} - D^{\mathrm{p}} \right) t \right\}$$

significantly depends of the sign of the difference $\left(D^{\mathrm{s}} - D^{\mathrm{p}} \right)$. In particular, for the divergence-free velocity field ($D^{\mathrm{p}} = 0$), we have the exponentially increasing typical realization curve, which means that particle scatter is exponentially fast for small distances between them. This result is valid for times

$$\frac{1}{4} D^{\mathrm{s}} t \ll \ln \left(\frac{l_{\mathrm{cor}}}{l_0} \right),$$

for which expansion (11.27) holds. In another limiting case of the potential velocity field ($D^{\mathrm{s}} = 0$), the typical realization curve is the exponentially decreasing curve, which means that particles tend to join. In view of the fact that liquid particles themselves are compressed during this process, we arrive at the conclusion that particles must form *clusters*, i.e., compact particle concentration zones located merely in rarefied regions, which agrees with the evolution of the realization (see Fig. 1.1*b*, page 12) obtained by simulating the behavior of the initially homogeneous particle distribution in random potential velocity field (though, for drastically other statistical model of the velocity field). This means that the phenomenon of clustering by itself is independent of the model of the velocity field, although statistical parameters characterizing this phenomenon surely depend on this model.

In the three-dimensional case ($d = 3$) Eq. (11.28) grades into

$$l^{*}(t) = l_0 \exp\left\{ \frac{1}{15}\left(6D^{\mathrm{s}} - D^{\mathrm{p}}\right) t \right\},$$

and typical realization curve will exponentially decay in time under the condition

$$D^{\mathrm{p}} > 6D^{\mathrm{s}},$$

which is stronger than that in the two-dimensional case.

In the one-dimensional case

$$l^{*}(t) = l_0 e^{-D^{\mathrm{p}} t},$$

and typical realization curve always decays in time.

11.2　Diffusion of passive tracer concentration in random velocity field

First of all we note that, in the case of the delta-correlated random velocity field, linear equation (11.1) in the absence of mean flow allows a relatively simple passage to the closed equations for both buoyant tracer average concentration and its higher multipoint correlation functions. For example, averaging Eq. (11.1), using the Furutsu–Novikov formula (8.10), and the expression for variational derivative

$$\frac{\delta\rho(\mathbf{r}, t)}{\delta u_\alpha(\mathbf{r}', t - 0)} = -\frac{\partial}{\partial r_\alpha}\delta(\mathbf{r} - \mathbf{r}')\rho(\mathbf{r}, t)$$

following from Eq. (11.1), we obtain the equation for the average tracer concentration

$$\frac{\partial}{\partial t}\langle\rho(\mathbf{r}, t)\rangle = (D_0 + \mu)\,\Delta\,\langle\rho(\mathbf{r}, t)\rangle. \tag{11.29}$$

Under the condition $D_0 \gg \mu$ ($\mu \ll \sigma_u^2 l_{\mathrm{cor}}^2$), where σ_u^2 is the variance of the random velocity field and l_{cor} is the correlation radius of this field, Eq. (11.29) coincides with the equation for the probability distribution of particle coordinate (11.20); consequently, diffusion coefficient D_0 characterizes only the scales of the region in which tracer is concentrated as a whole and provides no information on the local structure of concentration realization, which is similar to the diffusion in the divergence-free random velocity field.

In a similar way, we obtain that the spatial correlation function of the concentration field

$$\Gamma(\mathbf{r}_1, \mathbf{r}_2, t) = \langle\rho(\mathbf{r}_1, t)\rho(\mathbf{r}_2, t)\rangle$$

satisfies the equation

$$\frac{\partial}{\partial t}\Gamma(\mathbf{r}_1,\mathbf{r}_2,t) = \left[\frac{\partial^2}{\partial r_{1i}\partial r_{1j}} + \frac{\partial^2}{\partial r_{2i}\partial r_{2j}}\right]\left[B_{ij}^{\mathrm{eff}}(0) + \mu\delta_{ij}\right]\Gamma(\mathbf{r}_1,\mathbf{r}_2,t)$$
$$+ \frac{\partial^2}{\partial r_{1i}\partial r_{2j}}B_{ij}^{\mathrm{eff}}(\mathbf{r}_1 - \mathbf{r}_2)\Gamma(\mathbf{r}_1,\mathbf{r}_2,t)$$

coinciding with the equation for the two-particle probability density in the absence of the molecular diffusion ($\mu = 0$).

In the special case of uniform initial distribution of the concentration field ($\rho_0(\mathbf{r}) = \rho_0 = \mathrm{const}$), random field $\rho(\mathbf{r},t)$ will be the homogeneous isotropic random field. In this case $\langle\rho(\mathbf{r},t)\rangle = \rho_0$, and the equation for the correlation function becomes simpler and assumes the form

$$\frac{\partial}{\partial t}\Gamma(\mathbf{r},t) = 2\mu\Delta\Gamma(\mathbf{r},t) + \frac{\partial^2}{\partial r_i\partial r_j}D_{ij}(\mathbf{r})\Gamma(\mathbf{r},t), \quad \Gamma(\mathbf{r},0) = \rho_0^2,$$

where $\mathbf{r} = \mathbf{r}_1 - \mathbf{r}_2$ and

$$D_{ij}(\mathbf{r}) = 2\left[B_{ij}^{\mathrm{eff}}(0) - B_{ij}^{\mathrm{eff}}(\mathbf{r})\right]$$

is the structure matrix of vector field $\mathbf{u}(\mathbf{r},t)$. In the absence of molecular diffusion, this equation coincides with the equation for the probability density of relative diffusion of two particles.

Correlation function $\Gamma(\mathbf{r},t)$ will now depend on the modulus of vector $\mathbf{r}$ ($\Gamma(\mathbf{r},t) = \Gamma(r,t)$) and, as a function of variables $\{r,t\}$, will satisfy the equation

$$\frac{\partial}{\partial t}\Gamma(r,t) = \frac{1}{r^{d-1}}\frac{\partial}{\partial r}r^{d-1}\left[\frac{\partial D_{ii}^{\mathrm{p}}(r,t)}{\partial r} + \left(2\mu + \frac{r_i r_j}{r^2}D_{ij}(\mathbf{r})\right)\frac{\partial}{\partial r}\right]\Gamma(r,t),$$

where d is the dimension of space, as earlier. This equation has the steady-state solution $\Gamma(r) = \Gamma(r,\infty)$ [43, 44]:

$$\Gamma(r) = \rho_0^2\exp\left\{\int\limits_r^\infty dr'\frac{\partial D_{ii}^{\mathrm{p}}(r')/\partial r'}{2\mu + r_i' r_j' D_{i,j}(\mathbf{r}')/r'^2}\right\},$$

which corresponds to the boundary condition

$$\Gamma(\infty) = \rho_0^2.$$

To describe the local behavior of tracer realizations in random velocity field, we need the probability distribution of tracer concentration, which can be obtained only in the absence of molecular diffusion.

To describe the concentration field in the Eulerian description, we introduce the indicator function

$$\varphi(t,\mathbf{r};\rho) = \delta(\rho(t,\mathbf{r}) - \rho), \tag{11.30}$$

which is similar to function (11.16) and is localized on surface $\rho(\mathbf{r},t) = \rho = \mathrm{const}$ in the three-dimensional case or on a contour in the two-dimensional case. The Liouville equation for this function coincides in form with Eq. (3.9)

$$\left(\frac{\partial}{\partial t} + \mathbf{U}(\mathbf{r},t)\frac{\partial}{\partial \mathbf{r}}\right)\Phi(t,\mathbf{r};\rho) = \frac{\partial\mathbf{U}(\mathbf{r},t)}{\partial\mathbf{r}}\frac{\partial}{\partial\rho}\left[\rho\Phi(t,\mathbf{r};\rho)\right], \tag{11.31}$$

so that the on-point probability density of the solution to dynamic equation (11.2) coincides in this case with the ensemble averaged indicator function $P(t, \mathbf{r}; \rho) = \langle \delta(\rho(\mathbf{r}, t) - \rho) \rangle$.

Averaging Eq. (11.31) over an ensemble of realizations of field $\mathbf{u}(\mathbf{r}, t)$, using the Furutsu–Novikov formula (8.10), and the equality

$$\frac{\delta}{\delta u_j(\mathbf{r}', t - 0)} \Phi(t, \mathbf{r}; \rho) = - \left[\delta(\mathbf{r} - \mathbf{r}') \frac{\partial}{\partial r_j} + \frac{\partial \delta(\mathbf{r} - \mathbf{r}')}{\partial r_j} \right] \Phi(t, \mathbf{r}; \rho),$$

we obtain the equation for probability density of the concentration field

$$\left(\frac{\partial}{\partial t} - D_0 \Delta \right) P(t, \mathbf{r}; \rho) = D^{\mathrm{p}} \frac{\partial^2}{\partial \rho^2} \rho^2 P(t, \mathbf{r}; \rho), \quad P(0, \mathbf{r}; \rho) = \delta(\rho_0(\mathbf{r}) - \rho). \tag{11.32}$$

Note that Eq. (11.32) can also be derived by using the following relationship between the one-point probability density of concentration field in the Eulerian representation and the one-time probability density in the Lagrangian description

$$P(t, \mathbf{r}; \rho) = \int d\mathbf{r}_0 \int\limits_0^\infty j \, dj \, P(t; \mathbf{r}, \rho, j | \mathbf{r}_0). \tag{11.33}$$

From Eq. (11.32) follows in particular that moment functions of the concentration field satisfy the equation

$$\left(\frac{\partial}{\partial t} - D_0 \Delta \right) \langle \rho^n(\mathbf{r}, t) \rangle = D^{\mathrm{p}} n(n - 1) \langle \rho^n(\mathbf{r}, t) \rangle, \quad \langle \rho^n(\mathbf{r}, 0) \rangle = \rho_0^n(\mathbf{r}).$$

Its solution can be represented in the form

$$\langle \rho^n(\mathbf{r}, t) \rangle = e^{n(n-1)\tau} \int d\mathbf{r}' P(t; \mathbf{r}|\mathbf{r}') \rho_0^n(\mathbf{r}'),$$

where function $P(t; \mathbf{r}|\mathbf{r}')$ is defined by Eq. (11.20).

For example, in the case of uniform initial tracer concentration ($\rho_0(\mathbf{r}) = \rho_0 = \mathrm{const}$), the tracer concentration is the lognormal quantity whose probability distribution is independent of $\mathbf{r}$; the corresponding probability density and integral distribution function are as follows

$$P(t; \rho) = \frac{1}{2\rho\sqrt{\pi\tau}} \exp\left\{ -\frac{\ln^2(\rho e^\tau / \rho_0)}{4\tau} \right\}, \quad F(t; \rho) = \Phi\left(\frac{\ln(\rho e^\tau / \rho_0)}{2\sqrt{\tau}} \right), \tag{11.34}$$

where $\Phi(z)$ is the error function,

$$\Phi(z) = \frac{1}{\sqrt{2\pi}} \int\limits_{-\infty}^z dy \exp\left\{ -\frac{y^2}{2} \right\},$$

and we use the dimensionless time $\tau = D^{\mathrm{p}} t$. In this case, all moment functions beginning from the second one appear the exponentially increasing functions of time

$$\langle \rho(\mathbf{r}, t) \rangle = \rho_0, \quad \langle \rho^n(\mathbf{r}, t) \rangle = \rho_0^n e^{n(n-1)\tau},$$

and the typical realization curve of the concentration field exponentially decays with time at any fixed spatial point

$$\rho^*(t) = \rho_0 e^{-\tau},$$

which is evidence of cluster behavior of medium density fluctuations in arbitrary divergent flows. The Eulerian concentration statistics at any fixed point is formed due to concentration fluctuations about this curve.

Even the above discussion of the one-point probability density of tracer concentration in the Eulerian representation revealed several regularities concerning the temporal behavior of concentration field realizations at fixed spatial points. Now we show that this distribution additionally allows us to reveal certain features in the space-time structure of concentration field realizations.

For simplicity, we content ourselves with the two-dimensional case. As was mentioned earlier, the analysis of level lines defined by the equality

$$\rho(\mathbf{r}, t) = \rho = \text{const}$$

can give important data on the spatial behavior of realizations, in particular, on different functionals of the concentration field, such as total area $S(t, \rho)$ of regions where $\rho(\mathbf{r}, t) > \rho$ and total tracer mass within these regions $M(t, \rho)$. Average values of these functionals can be expressed in terms of the one-point probability density:

$$\langle S(t, \rho) \rangle = \int_{\rho}^{\infty} d\tilde{\rho} \int d\mathbf{r}\, P(t, \mathbf{r}; \tilde{\rho}), \quad \langle M(t, \rho) \rangle = \int_{\rho}^{\infty} \tilde{\rho}\, d\tilde{\rho} \int d\mathbf{r}\, P(t, \mathbf{r}; \tilde{\rho}). \tag{11.35}$$

Substituting the solution to Eq. (11.32) in these expressions and performing some rearrangement, we easily obtain explicit expressions for these quantities

$$\langle S(t, \rho) \rangle = \int d\mathbf{r}\, \Phi \left(\frac{1}{2\sqrt{\tau}} \ln \left(\frac{\rho_0(\mathbf{r}) e^{-\tau}}{\rho} \right) \right),$$

$$\langle M(t, \rho) \rangle = \int d\mathbf{r}\, \rho_0(\mathbf{r})\, \Phi \left(\frac{1}{2\sqrt{\tau}} \ln \left(\frac{\rho_0(\mathbf{r}) e^{\tau}}{\rho} \right) \right). \tag{11.36}$$

These expressions show in particular that, for $\tau \gg 1$, the average area of regions where concentration exceeds level ρ decreases in time according to the law

$$\langle S(t, \rho) \rangle \approx \frac{1}{\sqrt{\pi \tau \rho}} e^{-\tau/4} \int d\mathbf{r} \sqrt{\rho_0(\mathbf{r})}, \tag{11.37}$$

while the average tracer mass within these regions

$$\langle M(t, \rho) \rangle \approx M_0 - \sqrt{\frac{\rho}{\pi \tau}} e^{-\tau/4} \int d\mathbf{r} \sqrt{\rho_0(\mathbf{r})} \tag{11.38}$$

monotonously tends to the total mass

$$M_0 = \int d\mathbf{r} \rho_0(\mathbf{r}).$$

This is an additional evidence in favor of the above conclusion that tracer particles tend to join in clusters, i.e., in compact regions of enhanced concentration surrounded with rarefied regions.

We illustrate the dynamics of cluster formation by the example of the initially uniform distribution of buoyant tracer over the plane, in which case $\rho_0(\mathbf{r}) = \rho_0 = \text{const}$. In this case, the average specific area (per unit surface area) of regions within which $\rho(\mathbf{r}, t) > \rho$ is

$$s(t, \rho) = \int_{\rho}^{\infty} d\tilde{\rho} P(t; \tilde{\rho}) = \Phi \left(\frac{\ln \left(\rho_0 e^{-\tau} / \rho \right)}{2\sqrt{\tau}} \right), \tag{11.39}$$

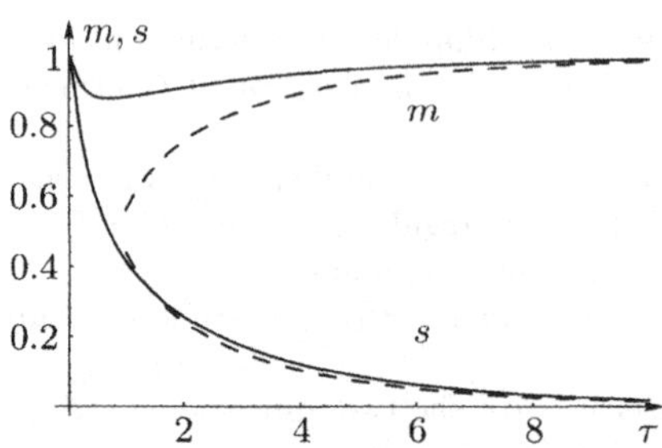

Figure 11.1: Cluster formation dynamics for $\rho/\rho_0 = 0.5$.

where $P(t; \rho)$ is the solution to Eq. (11.32) independent of $\mathbf{r}$ (i.e., function (11.34)), and average specific tracer mass (per unit surface area) within these regions is given by the expression

$$m(t, \rho) = \frac{1}{\rho_0} \int\limits_\rho^\infty \tilde{\rho} d\tilde{\rho} P(t; \tilde{\rho}) = \Phi\left(\frac{\ln(\rho_0 e^\tau/\rho)}{2\sqrt{\tau}}\right). \tag{11.40}$$

From Eqs. (11.39) and (11.40) follows that, for sufficiently large times, average specific area of such regions decreases according to the law

$$s(t, \rho) = \Phi(-\sqrt{\tau}/2) \approx \frac{1}{\sqrt{\pi\tau}} e^{-\tau/4} \tag{11.41}$$

irrespective of ratio ρ/ρ_0; at the same time, these regions concentrate almost all tracer mass

$$m(t, \rho) = \Phi(\sqrt{\tau}/2) \approx 1 - \frac{1}{\sqrt{\pi\tau}} e^{-\tau/4}. \tag{11.42}$$

Nevertheless, the time-dependent behavior of the formation of cluster structure essentially depends on ratio ρ/ρ_0. If $\rho/\rho_0 < 1$, then $s(0, \rho) = 1$ and $m(0, \rho) = 1$ at the initial instant. Then, in view of the fact that particles of buoyant tracer initially tend to scatter, there appear small areas within which $\rho(\mathbf{r}, t) < \rho$ and which concentrate only insignificant portion of the total mass. These regions rapidly grow with time and their mass flows into cluster region relatively quickly approaching asymptotic expressions (11.41) and (11.42) (Fig. 11.1). Note that $s(t^*, \rho) = 1/2$ at instant $\tau^* = \ln(\rho/\rho_0)$.

In the opposite, more interesting case $\rho/\rho_0 > 1$, we have $s(0, \rho) = 0$ and $m(0, \rho) = 0$ at the initial instant. In view of initial scatter of particles, there appear small cluster regions within which $\rho(\mathbf{r}, t) > \rho$; these regions remain at first almost invariable in time and intensively absorb a significant portion of total mass. With time, the area of these regions begins to decrease and the mass within them begins to increase according to asymptotic expressions (11.41) and (11.42) (Fig. 11.2a and Fig. 11.2b).

As we mentioned earlier, a more detailed description of the tracer concentration field in random velocity field requires that spatial gradient $\mathbf{p}(\mathbf{r}, t) = \nabla\rho(\mathbf{r}, t)$ (and, generally, higher-order derivatives) was additionally included in the consideration.

The concentration gradient satisfies dynamic equation (1.31); consequently, the extended indicator function

$$\varphi(t, \mathbf{r}; \rho, \mathbf{p}) = \delta\left(\rho(\mathbf{r}, t) - \rho\right)\delta\left(\mathbf{p}(\mathbf{r}, t) - \mathbf{p}\right)$$

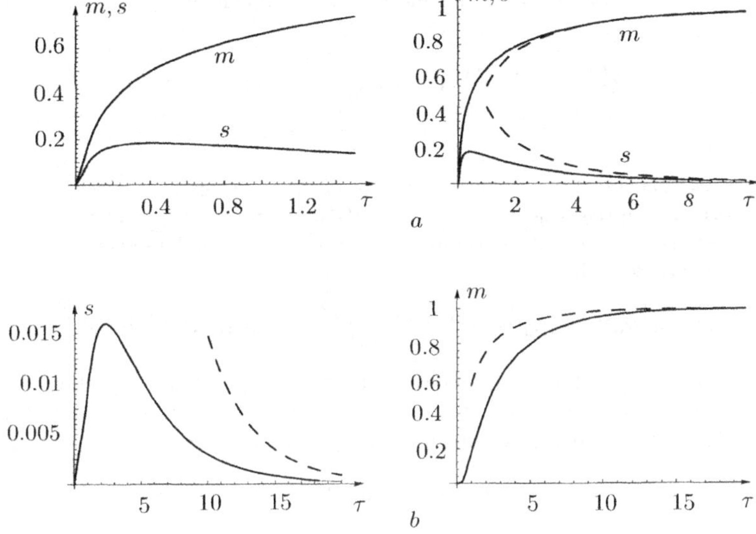

Figure 11.2: Cluster formation dynamics for (*a*) $\rho/\rho_0 = 1.5$ and (*b*) $\rho/\rho_0 = 10$.

for divergence-free velocity field satisfies Eq. (3.11). Averaging now Eq. (3.11) over an ensemble of velocity field realizations in the approximation of the delta-correlated velocity field, we obtain the equation for the one-point joint probability density of the concentration and its gradient $P(t, \mathbf{r}; \rho, \mathbf{p}) = \langle \varphi(t, \mathbf{r}; \rho, \mathbf{p}) \rangle$ (this probability density is a function of the space-time point $(\mathbf{r}, t)$)

$$\frac{\partial}{\partial t} P(t, \mathbf{r}; \rho, \mathbf{p}) = D_0 \Delta P(t, \mathbf{r}; \rho, \mathbf{p})$$

$$+ \frac{1}{d(d+2)} D^{\mathrm{s}} \left((d+1)\frac{\partial^2}{\partial \mathbf{p}^2}\mathbf{p}^2 - 2\frac{\partial^2}{\partial p_k \partial p_l}p_k p_l \right) P(t, \mathbf{r}; \rho, \mathbf{p}),$$

$$P(0, \mathbf{r}; \rho, \mathbf{p})\delta\left(\rho_0(\mathbf{r}) - \rho\right)\delta\left(\mathbf{p}_0(\mathbf{r}) - \mathbf{p}\right). \tag{11.43}$$

In view of the fact that the random velocity field is divergence-free, we can represent the solution to Eq. (11.43) in the form

$$P(t, \mathbf{r}; \rho, \mathbf{p}) = \int d\mathbf{r}_0 P(t, \mathbf{r}|\mathbf{r}_0) P(t, \mathbf{p}|\mathbf{r}_0), \tag{11.44}$$

where $P(t, \mathbf{r}|\mathbf{r}_0)$ and $P(t, \mathbf{p}|\mathbf{r}_0)$ are the corresponding Lagrangian probability densities of particle position and its gradient. The first density is given by Eq. (11.20), and the second satisfies the equation

$$\frac{\partial}{\partial t} P(t, \mathbf{p}|\mathbf{r}_0) = \frac{1}{d(d+2)} D^{\mathrm{s}} \left((d+1)\frac{\partial^2}{\partial \mathbf{p}^2}\mathbf{p}^2 - 2\frac{\partial^2}{\partial p_k \partial p_l}p_k p_l \right) P(t, \mathbf{p}|\mathbf{r}_0). \tag{11.45}$$

From Eq. (11.45) follows that average tracer concentration gradient is invariant

$$\langle \mathbf{p}(\mathbf{r}, t) \rangle = \mathbf{p}_0(\mathbf{r}_0).$$

As regards the moment functions of the concentration gradient modulus, they satisfy the equations

$$\frac{\partial}{\partial t}\left\langle p^n(t|\mathbf{r}_0)\right\rangle = \frac{n(d+n)(d-1)}{d(d+2)}D^s\left\langle p^n(t|\mathbf{r}_0)\right\rangle, \quad \left\langle p^n(0|\mathbf{r}_0)\right\rangle = p_0^n(\mathbf{r}_0) \qquad (11.46)$$

that follow from (11.45).

Consequently, concentration gradient modulus in the Lagrangian representation is the logarithmic-normal quantity whose typical realization curve and all moment functions increase exponentially in time. In particular, the first and second moments in the two-dimensional case are given by the equalities

$$\langle|\mathbf{p}(t|\mathbf{r}_0)|\rangle = |\mathbf{p}_0(\mathbf{r}_0)|e^{3D^s t/8}, \quad \left\langle \mathbf{p}^2(t|\mathbf{r}_0)\right\rangle = \mathbf{p}_0^2(\mathbf{r}_0)e^{D^s t}. \qquad (11.47)$$

In addition, from Eq. (11.43) with allowance for Eq. (4.30) follows that the average total length of contour $\rho(\mathbf{r},t) = \rho = \mathrm{const}$ (remember that we deal with the two-dimensional case) also exponentially increases in time according to the law

$$\langle l(t,\rho)\rangle = l_0 e^{D^s t},$$

where l_0 is the initial length of the contour. Remind that, in the case of the divergence-free velocity field, the number of contours remains unchanged; the contours cannot appear and disappear in the medium, they only evolve in time depending on their spatial distribution at the initial instant.

Thus, the initially smooth tracer distribution becomes with time increasingly inhomogeneous in space; its spatial gradients sharpen and level lines acquire the fractal behavior. We observed such a pattern in Fig. 1.1a, page 12 that shows simulated results (for quite other model of velocity field fluctuations). This means that the above general behavioral characteristics only slightly depend on the fluctuation model.

Above, we studied statistical characteristics of the solution to Eq. (11.2) in the Lagrangian and Eulerian representations and showed that both particle dynamics and Eulerian concentration field show clustering if the velocity field has the nonzero potential component. Along with dynamic equation (11.2), there is certain interest to the equation describing transfer of nonconservative tracer

$$\left(\frac{\partial}{\partial t} + \mathbf{U}(\mathbf{r},t)\frac{\partial}{\partial \mathbf{r}}\right)\rho(\mathbf{r},t) = 0, \quad \rho(\mathbf{r},0) = \rho_0(\mathbf{r}).$$

In this case, particle dynamics in the Lagrangian representation is described by the equation coinciding with Eq. (11.3); consequently, particles are clustered. However, in the Eulerian representation, no clustering occurs. In this case, as in the case of divergence-free velocity field, average number of contours, average area of regions within which $\rho(\mathbf{r},t) > \rho$, and average tracer mass $\int dS\rho(\mathbf{r},t)$ within these regions remain invariant.

Above, we considered the simplest statistical problems on tracer diffusion in random velocity field in the absence of regular flows and molecular diffusion. Moreover, our statistical description used the approximation of random delta-correlated (in time) field. All unaccounted factors act beginning from certain time, so that the above results hold only during the initial stage of diffusion. Furthermore, these factors can give rise to new physical effects. In this section, we outline these additional problems for the divergence-free (noncompressible) velocity field.

11.3 Effect of molecular diffusion

As we mentioned earlier, in the presence of velocity field fluctuations, the initially smooth distribution of tracer becomes increasingly inhomogeneous with time, there appear changes within increasingly shorter scales, and concentration spatial gradients sharpen. In actuality, molecular diffusion smooths these processes, so that the mentioned behavior of tracer concentration holds only for limited temporal intervals.

In the presence of molecular diffusion, the tracer diffusion is described in terms of the second-order partial differential equation (11.1) that do not allow deriving the equation for the one-point probability density. As we mentioned earlier, the only closed equations that can be derived in the general case of divergent hydrodynamic flow are the equations for average concentration field and spacial correlation function.

Problems in which mean concentration gradient assumes nonzero values allow a more complete analysis. This case corresponds to solving Eq. (11.1) with the following initial conditions (here, we again content ourselves with the two-dimensional case)

$$\rho_0(\mathbf{r}) = \mathbf{G}\mathbf{r}, \quad \mathbf{p}_0(\mathbf{r}) = \mathbf{G}.$$

Substituting the concentration field in the form

$$\rho(\mathbf{r}, t) = \mathbf{G}\mathbf{r} + \widetilde{\rho}(\mathbf{r}, t),$$

we obtain the equation for fluctuating portion $\widetilde{\rho}(\mathbf{r}, t)$ of the concentration field

$$\left(\frac{\partial}{\partial t} + \frac{\partial}{\partial \mathbf{r}} \mathbf{u}(\mathbf{r}, t) \right) \widetilde{\rho}(\mathbf{r}, t) = -\mathbf{G}\mathbf{u}(\mathbf{r}, t) + \mu \Delta \widetilde{\rho}(\mathbf{r}, t), \quad \widetilde{\rho}(\mathbf{r}, 0) = 0. \tag{11.48}$$

Unlike the problems discussed earlier, the solution to this problem is the random field statistically homogeneous in space, so that all one-point statistical averages are independent of $\mathbf{r}$ and steady-state probability densities exist for $t \to \infty$ for both concentration field and its gradient.

In this case, Eq. (11.48) easily yields the equation

$$\frac{d}{dt} \langle \widetilde{\rho}^n(\mathbf{r}, t) \rangle = n(n-1)D_0\mathbf{G}^2 \left\langle \widetilde{\rho}^{n-2}(\mathbf{r}, t) \right\rangle - \mu n(n-1) \left\langle \widetilde{\rho}^{n-2}(\mathbf{r}, t)\widetilde{\mathbf{p}}^2(\mathbf{r}, t) \right\rangle, \tag{11.49}$$

where

$$\widetilde{\mathbf{p}}(\mathbf{r}, t) = \frac{\partial}{\partial \mathbf{r}} \widetilde{\rho}(\mathbf{r}, t) = \mathbf{p}(\mathbf{r}, t) - \mathbf{G}.$$

In the steady-state regime (for $t \to \infty$), we obtain from Eq. (11.49) that

$$\left\langle \widetilde{\rho}^{n-2}(\mathbf{r}, t)\widetilde{\mathbf{p}}^2(\mathbf{r}, t) \right\rangle = \frac{D_0\mathbf{G}^2}{\mu} \left\langle \widetilde{\rho}^{n-2}(\mathbf{r}, t) \right\rangle. \tag{11.50}$$

For $n = 2$ in particular, we obtain the expression for the variance of fluctuations of the concentration gradient

$$\lim_{t \to \infty} \left\langle \widetilde{\mathbf{p}}^2(\mathbf{r}, t) \right\rangle = \frac{D_0\mathbf{G}^2}{\mu}. \tag{11.51}$$

Rewrite now Eq. (11.49) in the form

$$\frac{d}{dt} \langle \widetilde{\rho}^n(\mathbf{r}, t) \rangle = n(n-1)D_0\mathbf{G}^2 \left\langle f(\mathbf{r}, t)\widetilde{\rho}^{n-2}(\mathbf{r}, t) \right\rangle, \tag{11.52}$$

where

$$f(\mathbf{r}, t) = 1 - \frac{\mu}{D_0 \mathbf{G}^2} \widetilde{\mathbf{p}}^2(\mathbf{r}, t).$$

Consequently, the variance of concentration is given by the expression ($\langle \tilde{\rho}(\mathbf{r}, t) \rangle = 0$)

$$\left\langle \tilde{\rho}^2(\mathbf{r}, t) \right\rangle = 2 D_0 \mathbf{G}^2 \int\limits_0^t d\tau \, \langle f(\mathbf{r}, \tau) \rangle. \tag{11.53}$$

Note that the correlation function of field $\tilde{\rho}(\mathbf{r}, t)$,

$$\Gamma(\mathbf{r}, t) = \langle \tilde{\rho}(\mathbf{r}_1, t) \tilde{\rho}(\mathbf{r}_2, t) \rangle, \quad \mathbf{r} = \mathbf{r}_1 - \mathbf{r}_2,$$

satisfies the equation

$$\frac{\partial}{\partial t} \Gamma(\mathbf{r}, t) = 2 G_i G_j B_{ij}^{\mathrm{eff}}(\mathbf{r}) + 2 \left[B_{ij}^{\mathrm{eff}}(0) - B_{ij}^{\mathrm{eff}}(\mathbf{r}) + \mu \delta_{ij} \right] \frac{\partial^2}{\partial r_i \partial r_j} \Gamma(\mathbf{r}, t)$$

that follows from Eq. (11.48); consequently, its steady-state limit

$$\Gamma(\mathbf{r}) = \lim_{t \to \infty} \Gamma(\mathbf{r}, t)$$

satisfies the equation

$$G_i G_j B_{ij}^{\mathrm{eff}}(\mathbf{r}) = - \left[B_{ij}^{\mathrm{eff}}(0) - B_{ij}^{\mathrm{eff}}(\mathbf{r}) + \mu \delta_{ij} \right] \frac{\partial^2}{\partial r_i \partial r_j} \Gamma(\mathbf{r}). \tag{11.54}$$

Setting $\mathbf{r} = 0$ in this equation and taking into account Eqs. (11.14) and (11.15), we arrive at equality (11.51). If we differentiate Eq. (11.54) two times with respect to $\mathbf{r}$ and then set $\mathbf{r} = 0$, we obtain the equality

$$\mu^2 \left\langle (\Delta \tilde{\rho}(\mathbf{r}, t))^2 \right\rangle = \frac{1}{2} D_2^s (D_0 + \mu) \mathbf{G}^2. \tag{11.55}$$

Exact equalities (11.51) and (11.55) can appear practicable for testing different numerical schemes and checking simulated results. However, the calculation of the steady-state limit $\langle \tilde{\rho}^2(\mathbf{r}, t) \rangle$ for $t \to \infty$ requires the knowledge of the time-dependent behavior of the second moment $\langle \widetilde{\mathbf{p}}^2(\mathbf{r}, t) \rangle$, which can be obtained only if molecular diffusion is absent. In this case, the probability density of the concentration gradient satisfies Eq. (11.43); in the problem under consideration, this equation assumes the form

$$\frac{\partial}{\partial t} P(t, \mathbf{r}; \mathbf{p}) = \frac{1}{8} D^{\mathrm{s}} \left(3 \frac{\partial^2}{\partial \mathbf{p}^2} \mathbf{p}^2 - 2 \frac{\partial^2}{\partial p_k \partial p_l} p_k p_l \right) P(t, \mathbf{r}; \mathbf{p}),$$

$$P(0, \mathbf{r}; \mathbf{p}) = \delta (\mathbf{p} - \mathbf{G}). \tag{11.56}$$

Consequently, according to Eq. (11.47), we have

$$\left\langle |\widetilde{\mathbf{p}}(\mathbf{r}, t)|^2 \right\rangle = \mathbf{G}^2 \left\{ e^{D^{\mathrm{s}} t} - 1 \right\}. \tag{11.57}$$

The exact formula (11.51) and Eq. (11.57) give a possibility of estimating the time required for quantity $\langle \widetilde{\mathbf{p}}^2(\mathbf{r}, t) \rangle$ to approach the steady-state regime for $t \to \infty$; namely,

$$D^{\mathrm{s}} T_0 \sim \ln \left(\frac{D_0 + \mu}{\mu} \right).$$

As a consequence, we obtain from Eq. (11.53) the following estimate of the steady-state variance of concentration field fluctuations

$$\lim_{t \to \infty} \left\langle \tilde{\rho}^2(\mathbf{r}, t) \right\rangle \sim 2 \frac{D_0}{D^{\mathrm{s}}} \mathbf{G}^2 \ln \left(\frac{D_0 + \mu}{\mu} \right).$$

Taking into account the fact that $D_0 \sim \sigma_u^2 \tau_0$ and $D_0/D^{\mathrm{s}} \sim l_0^2$ (σ_u^2 is the variance of velocity field fluctuations and τ_0 and l_0 are this field temporal and spatial correlation radii, respectively), we see that time T_0, in view of its logarithmic dependence on molecular diffusion coefficient μ, can be not very long

$$T_0 \approx \frac{l_0^2}{\sigma_u^2 \tau_0} \ln \left(\frac{\sigma_u^2 \tau_0}{\mu} \right),$$

and quantity

$$\left\langle \tilde{\rho}^2 \right\rangle \sim \mathbf{G}^2 l_0^2 \ln \left(\frac{\sigma_u^2 \tau_0}{\mu} \right) \quad \text{for} \quad \mu \ll \sigma_u^2 \tau_0.$$

Problems

Problem 63 *Derive the Fokker–Planck equation for particle position in random divergence-free velocity field with allowance for the two-dimensional plane-parallel average flow* $\mathbf{u}_0(\mathbf{r}, t) = v(y)\mathbf{l}$, *where* $\mathbf{r} = (x, y)$, $\mathbf{l} = (1, 0)$.

Instruction. Particle motion is described by the stochastic system of equations

$$\frac{d}{dt} x(t) = v(y) + u_1(\mathbf{r}, t), \quad \frac{d}{dt} y(t) = u_2(\mathbf{r}, t).$$

Solution.

$$\left(\frac{\partial}{\partial t} + v(y) \frac{\partial}{\partial x} \right) P(t; \mathbf{r}) = D_0 \Delta P(t; \mathbf{r}). \tag{11.58}$$

Problem 64 *Show that Eq. (11.58) is statistically equivalent to a particle whose dynamics is described by the equations*

$$\frac{d}{dt} x(t) = v(y) + u_1(t), \quad \frac{d}{dt} y(t) = u_2(t), \tag{11.59}$$

where $u_i(t)$, $i = 1, 2$ *are the statistically independent Gaussian delta-correlated processes with the characteristics*

$$\langle \mathbf{u}(t) \rangle = 0, \quad \langle u_i(t) u_j(t') \rangle = 2\delta_{ij} D_0 \delta(t - t').$$

Problem 65 *Integrate Eqs. (11.59) and evaluate statistical characteristics of particle position in the case of linear shear* $v(y) = \alpha y$.

Solution.

$$y(t) = y_0 + w_2(t), \quad x(t) = x_0 + w_1(t) + \int_0^t d\tau\, v\left(y + w_2(\tau)\right), \tag{11.60}$$

where $w_i(t) = \int_0^t d\tau u_i(\tau)$ are independent Wiener processes with the characteristics

$$\langle \mathbf{w}(t) \rangle = 0, \quad \langle w_i(t) w_j(t') \rangle = 2D_0 \delta_{ij} \min\{t, t'\}.$$

In all cases, coordinate $y(t)$ has the Gaussian probability density with the parameters

$$\langle y(t) \rangle = y_0, \quad \left\langle y^2(t) \right\rangle = y_0^2 + 2D_0 t.$$

For the linear shear, Eqs. (11.60) correspond to the joint Gaussian probability density with the parameters

$$\begin{aligned}
\langle y(t) \rangle &= y_0, \quad \langle x(t) \rangle = x_0 + \alpha y_0 t, \\
\sigma_{xx}^2(t) &= \left\langle (x(t) - \langle x(t) \rangle)^2 \right\rangle = 2D_0 t \left(1 + \alpha t + \frac{1}{3} \alpha^2 t^2 \right), \\
\sigma_{yy}^2(t) &= \left\langle (y(t) - \langle y(t) \rangle)^2 \right\rangle = 2D_0 t, \\
\sigma_{xy}^2(t) &= \left\langle (x(t) - \langle x(t) \rangle)(y(t) - \langle y(t) \rangle) \right\rangle = 2D_0 t (1 + \alpha t).
\end{aligned}$$

Problem 66 *Derive the Fokker–Planck equation for the gradient of particle concentration in divergence-free random velocity field in the case of the two-dimensional average linear shear* $\mathbf{u}_0(\mathbf{r}, t) = \alpha y \mathbf{l}$, *where* $\mathbf{r} = (x, y)$, $\mathbf{l} = (1, 0)$.

Instruction. In the Lagrangian description, the particle concentration gradient in divergence-free velocity field satisfies the equations

$$\begin{aligned}
\frac{d}{dt} p_x(t|\mathbf{r}_0) &= -p_k(t|\mathbf{r}_0) \frac{\partial u_k(\mathbf{r}, t)}{\partial x}, \\
\frac{d}{dt} p_y(t|\mathbf{r}_0) &= -\alpha p_x(t|\mathbf{r}_0) - p_k(t|\mathbf{r}_0) \frac{\partial u_k(\mathbf{r}, t)}{\partial y}, \\
\mathbf{p}(0|\mathbf{r}_0) &= \mathbf{p}_0(\mathbf{r}_0) = \boldsymbol{\nabla} \rho_0(\mathbf{r}_0).
\end{aligned}$$

Solution.

$$\frac{\partial}{\partial t} P(t, \mathbf{p}|\mathbf{r}_0) = \left\{ \alpha p_x \frac{\partial}{\partial p_y} + \frac{1}{8} D^s \left(3 \frac{\partial^2}{\partial \mathbf{p}^2} \mathbf{p}^2 - 2 \frac{\partial^2}{\partial p_k \partial p_l} p_k p_l \right) \right\} P(t, \mathbf{p}|\mathbf{r}_0),$$
$$P(0, \mathbf{p}|\mathbf{r}_0) = \delta(\mathbf{p} - \mathbf{p}_0(\mathbf{r}_0)). \tag{11.61}$$

Problem 67 *Starting from Eq. (11.61), study the behavior of statistical characteristics of particle concentration gradient in divergence-free velocity field in the case of the two-dimensional average linear shear* $\mathbf{u}_0(\mathbf{r}, t) = \alpha y \mathbf{l}$, *where* $\mathbf{r} = (x, y)$, $\mathbf{l} = (1, 0)$.

Solution. Average gradient of tracer concentration field coincides with the problem solution for absent fluctuations of the velocity field

$$\langle p_x(t) \rangle = p_x(0), \quad \langle p_y(t) \rangle = p_y(0) - \alpha p_x(0) t.$$

The second moments of the gradient satisfy a closed system of equations with the cubic characteristic equation in parameter λ

$$\left(\lambda + \frac{1}{2} D^s \right)^2 (\lambda - D^s) = \frac{3}{2} \alpha^2 D^s$$

whose roots essentially depend on parameter α/D^s. For $\alpha/D^s \ll 1$, the approximate expressions for the roots are as follows

$$\lambda_1 = D^s + \frac{2\alpha^2}{3D^s}, \quad \lambda_2 = -\frac{1}{2}D^s + i|\alpha|, \quad \lambda_3 = -\frac{1}{2}D^s - i|\alpha|.$$

Consequently, random factor completely determines the problem solution for times $D^s t \gg 2$. This means that the effects of velocity field fluctuations predominate the effects of the weak gradient of linear shear. In the opposite limit $\alpha/D^s \gg 1$, the approximate expressions for the roots have the form

$$\lambda_1 = \left(\frac{3}{2}\alpha^2 D^s\right)^{1/3}, \quad \lambda_2 = \left(\frac{3}{2}\alpha^2 D^s\right)^{1/3} e^{i(2/3)\pi}, \quad \lambda_2 = \left(\frac{3}{2}\alpha^2 D^s\right)^{1/3} e^{-i(2/3)\pi}.$$

Because real parts of λ_2 and λ_3 are negative, the asymptotic solution of the system for second moments has for $\left(\frac{3}{2}\alpha^2 D^s\right)^{1/3} t \gg 1$ the form

$$\left\langle p^2(t) \right\rangle \sim \exp\left\{ \left(\frac{3}{2}\alpha^2 D_2^s\right)^{1/3} t \right\},$$

from which follows that even small fluctuations of velocity field appear the governing factor in the case of shears characterized by strong gradients.

Problem 68 *Derive the equation for the joint Eulerian probability density of the concentration and concentration gradient fields in random divergence-free velocity field in the case of the two-dimensional average linear shear* $\mathbf{u}_0(\mathbf{r}, t) = v(y)\mathbf{l}$, *where* $\mathbf{r} = (x, y)$, $\mathbf{l} = (1, 0)$.

Solution.

$$\frac{\partial}{\partial t} P(t, \mathbf{r}; \rho, \mathbf{p}) = \left[-\alpha y \frac{\partial}{\partial x} + D_0 \Delta \right] P(t, \mathbf{r}; \rho, \mathbf{p})$$
$$+ \left\{ \alpha p_x \frac{\partial}{\partial p_y} + \frac{1}{8} D^s \left(3\frac{\partial^2}{\partial \mathbf{p}^2} \mathbf{p}^2 - 2\frac{\partial^2}{\partial p_k \partial p_l} p_k p_l \right) \right\} P(t, \mathbf{r}; \rho, \mathbf{p}).$$

The solution to this equation can be represented in the form of integral (11.44), where Lagrangian probability densities $P(t, \mathbf{r}|\mathbf{r}_0)$ and $P(t, \mathbf{p}|\mathbf{r}_0)$ satisfy Eqs. (11.58) and (11.61).

Chapter 12

Wave localization in randomly layered media

The problem on plane wave propagation in layered media is formulated in terms of the one-dimensional boundary-value problem. It attracts attention of many researchers because it is much simpler in comparison with the corresponding two- and three-dimensional problems and provides a deep insight into wave propagation in random media. In view of the fact that the one-dimensional problem allows the exact asymptotic solution, we can use it to trace the effect of different models, medium parameters, and boundary conditions on statistical characteristics of the wave field.

The problem in the one-dimensional statement was given in Part 1.

Let the layer of inhomogeneous medium occupies the portion of space $L_0 < x < L$. The unit-amplitude plane wave is incident on this layer from region $x > L$. The wave field in the inhomogeneous layer satisfies the Helmholtz equation

$$\frac{d^2}{dx^2} u(x) + k^2(x) u(x) = 0, \tag{12.1}$$

where

$$k^2(x) = k^2[1 + \varepsilon(x)]$$

and function $\varepsilon(x)$ describes the inhomogeneities of the medium. In the simplest case of unmatched boundary, we assume that $k(x) = k$, i.e., $\varepsilon(x) = 0$ outside the layer and

$$\varepsilon(x) = \varepsilon_1(x) + i\gamma$$

inside the layer, where $\varepsilon_1(x)$ is the real part responsible for wave scattering in the medium and $\gamma \ll 1$ is the imaginary part responsible for wave absorption in the medium.

The boundary conditions for Eq. (12.1) are formulated as the continuity of function $u(x)$ and derivative $\frac{d}{dx} u(x)$ at layer boundaries; these conditions can we written in the form

$$\left(1 + \frac{i}{k}\frac{d}{dx}\right) u(x)\Bigg|_{x=L} = 2, \qquad \left(1 - \frac{i}{k}\frac{d}{dx}\right) u(x)\Bigg|_{x=L_0} = 0. \tag{12.2}$$

For $x < L$, from Eq. (12.1) follows the equality

$$k\gamma I(x) = \frac{d}{dx} S(x), \tag{12.3}$$

where $S(x)$ is the energy flux density,

$$S(x) = \frac{i}{2k} \left[u(x) \frac{d}{dx} u^*(x) - u^*(x) \frac{d}{dx} u(x) \right],$$

and $I(x)$ is the wave field intensity, $I(x) = |u(x)|^2$. In addition,

$$S(L) = 1 - |R_L|^2, \quad S(L_0) = |T_L|^2,$$

where R_L is the complex reflection coefficient from the medium layer and T_L is the complex transmission coefficient of the wave. Integrating Eq. (12.3) over the inhomogeneous layer, we obtain the equality

$$|R_L|^2 + |T_L|^2 + k\gamma \int\limits_{L_0}^{L} dx\, I(x) = 1. \tag{12.4}$$

If the medium causes no wave attenuation ($\gamma = 0$), then conservation of the energy flux density is expressed by the equality

$$|R_L|^2 + |T_L|^2 = 1.$$

The imbedding method provides a possibility of reformulating boundary-value problem (12.1), (12.2) in terms of the dynamic initial value problem with respect to parameter L (geometric position of the right-hand boundary of the layer) by considering the solution of the problem as a function of this parameter. For example, reflection coefficient R_L satisfies the Riccati equation

$$\frac{d}{dL} R_L = 2ikR_L + \frac{ik}{2}\varepsilon(L)\left(1 + R_L\right)^2, \quad R_{L_0} = 0, \tag{12.5}$$

and the wave field in medium layer $u(x) \equiv u(x; L)$ satisfies the linear equation

$$\frac{\partial}{\partial L} u(x; L) = iku(x; L) + \frac{ik}{2}\varepsilon(L)\left(1 + R_L\right)u(x; L), \quad u(x; x) = 1 + R_x. \tag{12.6}$$

From Eqs. (12.5) and (12.6) follow the equations for the squared modulus of the reflection coefficient $W_L = |R_L|^2$ and the wave field intensity $I(x; L) = |u(x; L)|^2$

$$\frac{d}{dL} W_L = -\frac{k\gamma}{2}\left[4W_L + (R_L + R_L^*)(1 + W_L)\right] - \frac{ik}{2}\varepsilon_1(L)(R_L - R_L^*)(1 - W_L),$$
$$W_{L_0} = 0,$$
$$\frac{\partial}{\partial L} I(x; L) = -\frac{k\gamma}{2}(2 + R_L + R_L^*)I(x; L) - \frac{ik}{2}\varepsilon_1(L)(R_L - R_L^*)I(x; L),$$
$$I(x; x) = |1 + R_x|^2. \tag{12.7}$$

In the case of non-absorptive medium, Eq. (12.7) can be integrated in the analytic form; the resulting wave field intensity inside the inhomogeneous layer is explicitly expressed in terms of the layer reflection coefficient.

$$I(x; L) = \frac{|1 + R_x|^2(1 - W_L)}{1 - W_x}. \tag{12.8}$$

Similarly, the field of the point source located in the layer of random medium is described in terms of the boundary-value problem for Green's function of the Helmholtz equation

$$\frac{d^2}{dx^2}G(x;x_0) + k^2[1 + \varepsilon(x)]G(x;x_0) = 2ik\delta(x - x_0),$$

$$\left(\frac{d}{dx} + ik\right)G(x;x_0)\Big|_{x=L_0} = 0, \quad \left(\frac{d}{dx} - ik\right)G(x;x_0)\Big|_{x=L} = 0. \tag{12.9}$$

Note that the source at the layer boundary $x_0 = L$ corresponds to boundary-value problem (12.1), (12.2) on wave incidence on the layer, i.e.,

$$G(x;L) = u(x;L).$$

The solution of boundary-value problem (12.9) has the structure

$$G(x;x_0) = G(x_0;x_0)\begin{cases} \exp\left[ik\int\limits_x^{x_0}\psi_1(\xi)\,d\xi\right], & x_0 \geq x, \\[2ex] \exp\left[ik\int\limits_{x_0}^x\psi_2(\xi)\,d\xi\right], & x_0 \leq x, \end{cases} \tag{12.10}$$

where the field at the source location, by virtue of the derivative gap condition

$$\frac{dG(x;x_0)}{dx}\Big|_{x=x_0+0} - \frac{dG(x;x_0)}{dx}\Big|_{x=x_0-0} = 2ik,$$

is determined by the formula

$$G(x_0;x_0) = \frac{2}{\psi_1(x_0) + \psi_2(x_0)}$$

and functions $\psi_i(x)$ satisfy the Riccati equations

$$\begin{aligned} \frac{d}{dx}\psi_1 &= ik\left[\psi_1^2 - 1 - \varepsilon(x)\right], & \psi_1(L_0) &= 1, \\ \frac{d}{dx}\psi_2 &= -ik\left[\psi_2^2 - 1 - \varepsilon(x)\right], & \psi_2(L) &= 1. \end{aligned} \tag{12.11}$$

Introduce new functions $R_i(x)$ related to functions $\psi_i(x)$ by the formula

$$\psi_i(x) = \frac{1 - R_i(x)}{1 + R_i(x)}, \quad i = 1,2.$$

With these functions, the wavefield in region $x < x_0$ can be written in the form

$$G(x;x_0) = \frac{[1 + R_1(x_0)][1 + R_2(x_0)]}{1 - R_1(x_0)R_2(x_0)}\exp\left[ik\int\limits_x^{x_0}d\xi\frac{1 - R_1(\xi)}{1 + R_1(\xi)}\right], \tag{12.12}$$

where parameter $R_1(L) = R_L$ is the reflection coefficient of the plane wave incident on the layer from region $x > L$. In a similar way, quantity $R_2(x_0)$ is the reflection coefficient of the wave incident on the medium layer (x_0, L) from the homogeneous half-space $x < x_0$ (i.e., from region with $\varepsilon = 0$).

Problems with perfectly reflecting boundaries (at which either $G(x; x_0)$ or $\frac{d}{dx}G(x; x_0)$ vanishes) are of great interest for applications. Indeed, in the latter case, we have $R_2(x_0) = 1$ for the source located at this boundary; consequently,

$$G_{\text{ref}}(x; x_0) = \frac{2}{1 - R_1(x_0)} \exp\left[ik \int\limits_x^{x_0} d\xi \frac{1 - R_1(\xi)}{1 + R_1(\xi)}\right], \quad x \le x_0. \tag{12.13}$$

In addition, the expression for wave field intensity $I(x; x_0) = |G(x; x_0)|^2$ follows from Eq. (12.9)

$$k\gamma I(x; x_0) = \frac{d}{dx} S(x; x_0), \tag{12.14}$$

for $x < x_0$, where energy flux density $S(x; x_0)$ is given by the expression

$$S(x; x_0) = \frac{i}{2k}\left[G(x; x_0)\frac{d}{dx}G^*(x; x_0) - G^*(x; x_0)\frac{d}{dx}G(x; x_0)\right].$$

Using Eq. (12.12), we can represent $S(x; x_0)$ in the form $(x \le x_0)$

$$S(x; x_0) = S(x_0; x_0) \exp\left[-k\gamma \int\limits_x^{x_0} d\xi \frac{|1 + R_1(\xi)|^2}{1 - |R_1(\xi)|^2}\right],$$

where the energy flux density at the point of source location is

$$S(x_0; x_0) = \frac{\left[1 - |R_1(x_0)|^2\right]|1 + R_2(x_0)|^2}{|1 - R_1(x_0) R_2(x_0)|^2}. \tag{12.15}$$

Below, our concern will be with statistical problems on waves incident on random half-space $(L_0 \to -\infty)$ and source-generated waves in infinite space $(L_0 \to -\infty, L \to \infty)$ for sufficiently small absorption $(\gamma \to 0)$. One can see from Eq. (12.14) that these limit processes are not commutable in the general case. Indeed, if $\gamma = 0$, then energy flux density $S(x; x_0)$ is conserved in the whole half-space $x < x_0$. However, integrating Eq. (12.14) over half-space $x < x_0$ in the case of small but finite absorption, we obtain the restriction on the energy confined in this half-space

$$k\gamma \int\limits_{-\infty}^{x_0} dx I(x; x_0) = S(x_0; x_0) = \frac{\left[1 - |R_1(x_0)|^2\right]|1 + R_2(x_0)|^2}{|1 - R_1(x_0) R_2(x_0)|^2}. \tag{12.16}$$

Three simple statistical problems are of interest:
- wave incidence on medium layer (of finite and infinite thickness);
- wave source in the medium layer or infinite medium;
- effect of boundaries on statistical characteristics of the wave field.

All these problems can be exhaustively solved in the analytic form. One can easily simulate these problems numerically and compare the simulated and analytic results.

We will assume that $\varepsilon_1(x)$ is the Gaussian delta-correlated random process with the parameters

$$\langle\varepsilon_1(L)\rangle = 0, \quad \langle\varepsilon_1(L)\varepsilon_1(L')\rangle = B_\varepsilon(L - L') = 2\sigma_\varepsilon^2 l_0 \delta(L - L'), \tag{12.17}$$

where $\sigma_\varepsilon^2 \ll 1$ is the variance and l_0 is the correlation radius of random function $\varepsilon_1(L)$. This approximation means that asymptotic limit process $l_0 \to 0$ in the exact problem solution with a finite correlation radius l_0 must give the result coinciding with the solution to the statistical problem with parameters (12.17).

In view of smallness of parameter σ_ε^2, all statistical effects can be divided into two types, local and accumulated due to multiple wave reflections in the medium. Our concern will be with the latter.

The statement of boundary-value wave problems in terms of the imbedding method clearly shows that two types of wave field characteristics are of immediate interest. The first type of characteristics deals with quantities, such as values of the wave field at layer boundaries (reflection and transmission coefficients R_L and T_L), field at the point of source location $G(x_0; x_0)$, and energy flux density at the point of source location $S(x_0; x_0)$. The second type of characteristics deals with statistical characteristics of wave field intensity in the medium layer, which is the subject matter of the statistical theory of radiative transfer.

12.1 Statistics of scattered field at layer boundaries

12.1.1 Reflection and transmission coefficients

Complex reflection coefficient of wave reflection from a medium layer satisfies the closes Riccati equation (12.5).

Represent reflection coefficient in the form

$$R_L = \sqrt{W_L}\, e^{i\phi_L},$$

where $\sqrt{W_L}$ is the modulus and ϕ_L is the phase of the reflection coefficient. Then, starting from Eq. (12.5), we obtain the system of equations for squared modulus of the reflection coefficient $W_L = \rho_L^2 = |R_L|^2$ and its phase

$$\frac{d}{dL} W_L = -2k\gamma W_L + k\varepsilon_1(L)\sqrt{W_L}\,(1 - W_L)\sin\phi_L, \quad W_{L_0} = 0,$$

$$\frac{d}{dL}\phi_L = 2k + k\varepsilon_1(L)\left\{1 + \frac{1 + W_L}{2\sqrt{W_L}}\cos\phi_L\right\}, \quad \phi_{L_0} = 0. \tag{12.18}$$

Fast functions producing only little contribution to accumulated effects are omitted in the dissipative terms of system (12.18) (cf. with Eq (12.7)).

Introduce the indicator function

$$\varphi(L; W) = \delta(W_L - W)$$

that satisfies the Liouville equation

$$\frac{\partial}{\partial L}\varphi(L; W) = 2k\gamma\frac{\partial}{\partial W}\left\{W\varphi(L; W)\right\}$$

$$-k\varepsilon_1(L)\frac{\partial}{\partial W}\left\{\sqrt{W}\,(1 - W)\sin\phi_L\varphi(L; W)\right\}, \quad \varphi(L_0; W) = \delta(W - 1). \tag{12.19}$$

Averaging this equation over an ensemble of realizations of function $\varepsilon_1(L)$ and using the Furutsu Novikov formula (8.10) and expression (12.17), we obtain the unclosed equation for probability density $P(L; W) = \langle\varphi(L; W)\rangle$

$$\frac{\partial}{\partial L}P(L; W) = 2k\gamma\frac{\partial}{\partial W}\left\{W P(L; W)\right\}$$

$$-k^2\sigma_\varepsilon^2 l_0\frac{\partial}{\partial W}(1 - W)\left\langle\left[\sqrt{W}\cos\phi_L + \frac{1}{2}(1 + W)\cos^2\phi_L\right]\varphi(L; W)\right\rangle$$

$$+k^2\sigma_\varepsilon^2 l_0\frac{\partial}{\partial W}\left\{\sqrt{W}\,(1 - W)\frac{\partial}{\partial W}\left[\sqrt{W}\,(1 - W)\left\langle\sin^2\phi_L\varphi(L; W)\right\rangle\right]\right\}.$$

In view of the fact that the phase of the reflection coefficient

$$\phi_L = k(L - L_0) + \tilde{\phi}_L,$$

rapidly varies within distances about the wavelength, we can additionally average this equation over fast oscillations, which will be valid under the natural restriction $k/D \gg 1$. Thus we arrive at the Fokker–Planck equation

$$\frac{\partial}{\partial L} P(L; W) = 2k\gamma \frac{\partial}{\partial W} W P(L; W) - 2D \frac{\partial}{\partial W} W (1 - W) P(L; W)$$

$$+ D \frac{\partial}{\partial W} W (1 - W)^2 \frac{\partial}{\partial W} P(L; W), \quad P(L_0, W) = \delta (W - 1) \tag{12.20}$$

with the diffusion coefficient

$$D = \frac{k^2 \sigma_\varepsilon^2 l_0}{2}.$$

Representation of quantity W_L in the form

$$W_L = \frac{u_L - 1}{u_L + 1}, \quad u_L = \frac{1 + W_L}{1 - W_L}, \quad u_L \geq 1. \tag{12.21}$$

appears more convenient in some cases. Quantity u_L satisfies the stochastic system of equations

$$\frac{d}{dL} u_L = -k\gamma \left(u_L^2 - 1 \right) + k\varepsilon_1(L) \sqrt{u_L^2 - 1} \sin \phi_L, \quad u_{L_0} = 1,$$

$$\frac{d}{dL} \phi_L = 2k + k\varepsilon_1(L) \left\{ 1 + \frac{u_L}{\sqrt{u_L^2 - 1}} \cos \phi_L \right\}, \quad \phi_{L_0} = 0,$$

and we obtain that probability density $P(L; u) = \langle \delta(u_L - u) \rangle$ of random quantity u_L satisfies the Fokker–Planck equation

$$\frac{\partial}{\partial L} P(L; u) = k\gamma \frac{\partial}{\partial u} \left(u^2 - 1 \right) P(L; u) + D \frac{\partial}{\partial u} \left(u^2 - 1 \right) \frac{\partial}{\partial u} P(L; u). \tag{12.22}$$

Note that quantity inverse to the diffusion coefficient defines the natural spatial scale related to medium inhomogeneities and is usually called the *localization length*

$$l_{\text{loc}} = 1/D.$$

In further analysis of wave field statistics, we will see that this quantity determines the scale of the *dynamic wave localization* in separate wave field realizations, although the *statistical localization* related to statistical characteristics of the wave field may not occur in some cases.

Nondissipative medium

If the medium is non-absorptive (i.e., if $\gamma = 0$), then Eq. (12.22) for the dimensionless layer thickness $\eta = D(L - L_0)$ assumes the form

$$\frac{\partial}{\partial \eta} P(\eta; u) = \frac{\partial}{\partial u} \left(u^2 - 1 \right) \frac{\partial}{\partial u} P(\eta; u). \tag{12.23}$$

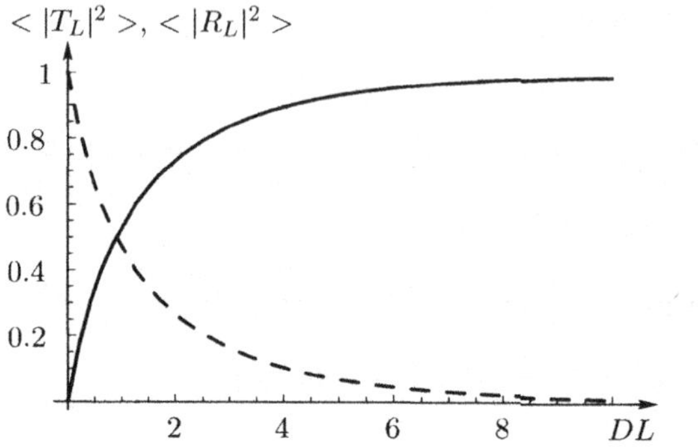

Figure 12.1: Quantities $\langle |R_L|^2 \rangle$ and $\langle |T_L|^2 \rangle$ versus layer thickness.

The solution to this equation can be easily obtained using the integral *Meller–Fock transform*. This solution has the form

$$P(\eta, u) = \int\limits_0^\infty d\mu\, \mu \tanh(\pi\mu) \exp\left\{ -\left(\mu^2 + \frac{1}{4}\right)\eta \right\} P_{-\frac{1}{2}+i\mu}(u), \qquad (12.24)$$

where $P_{-1/2+i\mu}(x)$ is the first-order *Legendre function (conal function)*.

In view of the formula

$$\int\limits_1^\infty \frac{dx}{(1+x)^n} P_{-\frac{1}{2}+i\mu}(x) = \frac{\pi}{\cosh(\mu\pi)} K_n(\mu),$$

where

$$K_{n+1}(\mu) = \frac{1}{2n}\left[\mu^2 + \left(n - \frac{1}{2}\right)^2 \right] K_n(\mu), \quad K_1(\mu) = 1,$$

representation (12.24) offers a possibility of calculating statistical characteristics of reflection and transmission coefficients $W_L = |R_L|^2$ and $|T_L|^2 = 1 - |R_L|^2 = 2/(1 + u_L)$; in particular, we obtain the expression for the moments of the transmission coefficient squared modulus

$$\left\langle |T_L|^{2n} \right\rangle = 2^n \pi \int\limits_0^\infty d\mu \frac{\mu \sinh(\mu\pi)}{\cosh^2(\mu\pi)} K_n(\mu) e^{-(\mu^2+1/4)\eta}. \qquad (12.25)$$

Figure 12.1 shows coefficients $\langle W_L \rangle = \langle |R_L|^2 \rangle$ and $\langle |T_L|^2 \rangle = 1 - \langle |R_L|^2 \rangle$ as functions of layer thickness.

For sufficiently thick layers, namely, if $\eta = D(L - L_0) \gg 1$, Eq. (12.25) yields the asymptotic formula for the moments of the reflection coefficient squared modulus

$$\left\langle |T_L|^{2n} \right\rangle \approx \frac{[(2n-3)!!]^2 \pi^2 \sqrt{\pi}}{2^{2n-1}(n-1)!} \frac{1}{\eta\sqrt{\eta}} e^{-\eta/4}.$$

As may be seen, all moments of the reflection coefficient modulus $|T_L|$ vary with layer thickness according to the universal law (only the numerical factor is changed).

The fact that all moments of quantity $|T_L|$ tend to zero with increasing layer thickness means that $|R_L| \to 1$ with a probability equal to unity, i.e., *the half-space of randomly layered nondissipative medium completely reflects the incident wave*. It is clear that this phenomenon is independent of the statistical model of medium and the condition of applicability of the description based on the additional averaging over fast oscillations related to the reflection coefficient phase.

In the approximation of the delta-correlated random process $\varepsilon_1(L)$, random processes W_L and u_L are obviously the Markovian processes with respect to parameter L. It is obvious that the transition probability density

$$p(u, L|u', L') = \langle \delta(u_L - u|u_{L'} = u') \rangle$$

also satisfies in this case Eq. (12.23), i.e.,

$$\frac{\partial}{\partial L} p(u, L|u', L') = D \frac{\partial}{\partial u} \left(u^2 - 1 \right) \frac{\partial}{\partial u} p(u, L|u', L')$$

with the initial condition

$$p(u, L''|u', L') = \delta(u - u').$$

The corresponding solution has the form, i.e.,

$$p(u, L|u', L') = \int\limits_{0}^{\infty} d\mu\, \mu \tanh(\pi\mu) e^{-D\left(\mu^2 + 1/4\right)(L - L')} P_{-\frac{1}{2}+i\mu}(u) P_{-\frac{1}{2}+i\mu}(u'). \tag{12.26}$$

At $L' = L_0$ and $u' = 1$, expression (12.26) grades into the one-point probability density (12.24).

Dissipative medium

In the case of absorptive medium, Eqs. (12.20) and (12.22) cannot be solved analytically for the layer of finite thickness. Nevertheless, in the limit of half-space ($L_0 \to -\infty$), quantities W_L and u_L have the steady-state probability density independent of L and satisfying the equations

$$2\left(\beta - 1 + W\right) P(W) + (1 - W)^2 \frac{d}{dW} P(W) = 0, \quad 0 < W < 1,$$

$$\beta P(u) + \frac{d}{du} P(u) = 0, \quad u > 1, \tag{12.27}$$

where $\beta = k\gamma/D$ is the dimensionless absorption coefficient.

Solutions to Eqs. (12.27) have the form

$$P(W) = \frac{2\beta}{(1 - W)^2} \exp\left\{ -\frac{2\beta W}{1 - W} \right\}, \quad P(u) = \beta e^{-\beta(u-1)}, \tag{12.28}$$

and Fig. 12.2 shows function $P(W)$ for different values of parameter β.

The physical meaning of probability density (12.28) is obvious. It describes the statistics of the reflection coefficient of the random layer sufficiently thick for the incident wave could not reach its end because of dynamic absorption in the medium.

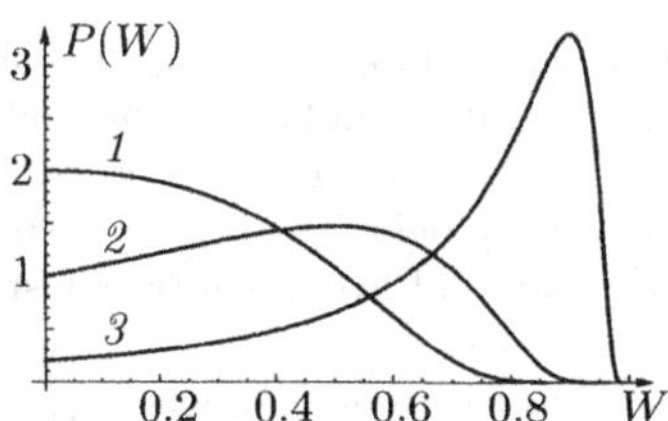

Figure 12.2: Probability density of squared reflection coefficient modulus $P(W)$. Curves *1* to *3* correspond to $\beta = 1$, 0.5, and 0.1, respectively.

Using distributions (12.28), we can calculate all moments of quantity $W_L = |R_L|^2$. For example, we have for the average square of reflection coefficient modulus

$$\langle W \rangle = \int_0^1 dW \, W P(W) = \int_1^\infty du \frac{u-1}{u+1} P(u) = 1 + 2\beta e^{2\beta} \, \mathrm{Ei}(-2\beta),$$

where $\mathrm{Ei}(-x) = -\int_x^\infty \frac{dt}{t} e^{-t} \; (x > 0)$ is the *integral exponent*. Using asymptotic expansions of function $\mathrm{Ei}(-x)$

$$\mathrm{Ei}(-x) = \begin{cases} \ln x & (x \ll 1), \\ -e^{-x} \frac{1}{x}\left(1 - \frac{1}{x}\right) & (x \gg 1), \end{cases}$$

we obtain the asymptotic expansions of quantity $\langle W \rangle = \langle |R_L|^2 \rangle$

$$\langle W \rangle \approx \begin{cases} 1 - 2\beta \ln(1/\beta), & \beta \ll 1, \\ 1/2\beta, & \beta \gg 1. \end{cases} \tag{12.29}$$

To determine higher moments of quantity $W_L = |R_L|^2$, we multiply the first equation in (12.27) by W^n and integrate the result over W from 0 to 1. As a result, we obtain recurrence equation

$$n \left\langle W^{n+1} \right\rangle - 2(\beta + n) \left\langle W^n \right\rangle + n \left\langle W^{n-1} \right\rangle = 0 \quad (n = 1, 2, ...). \tag{12.30}$$

Using this equation, we can recursively calculate all higher moments. For example, we have for $n = 1$

$$\left\langle W^2 \right\rangle = 2(\beta + 1) \left\langle W \right\rangle - 1.$$

The steady-state probability distribution can be obtained not only by limiting process $L_0 \to -\infty$, but also $L \to \infty$. Equation (12.20) was solved numerically for two values $\beta = 1.0$ and $\beta = 0.08$ for different initial conditions. Figure 12.3 shows moments $\langle W_L \rangle$, $\langle W_L^2 \rangle$ calculated from the obtained solutions versus dimensionless layer thickness $\eta = D(L - L_0)$. The curves show that the probability distribution approaches its steady-state behavior relatively rapidly ($\eta \sim 1.5$) for $\beta \geq 1$ and much slower ($\eta \geq 5$) for strongly stochastic problem with $\beta = 0.08$.

Note that, for the problem under consideration, energy flux density and wave field intensity at layer boundary $x = L$ can be expressed in terms of the reflection coefficient. Consequently, we have for $\beta \ll 1$

$$\langle S(L, L) \rangle = 1 - \langle W_L \rangle = 2\beta \ln(1/\beta), \quad \langle I(L, L) \rangle = 1 + \langle W_L \rangle = 2. \tag{12.31}$$

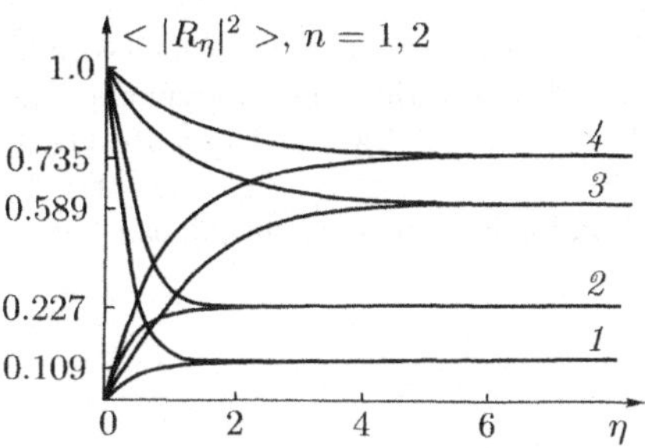

Figure 12.3: Statistical characteristics of quantity $W_L = |R_L|^2$. Curves *1* and *2* show the second and first moments at $\beta = 1$, and curves *3* and *4* show the second and first moments at $\beta = 0.08$.

Taking into account that $|T_L| = 0$ in the case of random half-space and using Eq. (12.4), we obtain that the wave field energy contained in this half-space

$$E = D \int\limits_{-\infty}^{L} dx I(x; L),$$

has the probability distribution

$$P(E) = \beta P(W)_{W=(1-\beta E)} = \frac{2}{E^2} \exp\left\{-\frac{2}{E}(1 - \beta E)\right\} \theta(1 - \beta E), \qquad (12.32)$$

so that we have, in particular,

$$\langle E \rangle = 2\ln(1/\beta) \qquad (12.33)$$

for $\beta \ll 1$.

Note that probability distribution (12.32) allows the limit process to $\beta = 0$; as a result, we obtain the limiting probability density

$$P(E) = \frac{2}{E^2} \exp\left\{-\frac{2}{E}\right\} \qquad (12.34)$$

that decays according to the power law for large energies E. The corresponding integral distribution function has the form

$$F(E) = \exp\left\{-\frac{2}{E}\right\}.$$

A consequence of Eq. (12.34) is the fact that all moments of the total wave energy appear infinite. Nevertheless, the total energy in separate wave field realizations can be limited to arbitrary value with a finite probability.

12.1.2 Source inside the layer of a medium

If the source of plane waves is located inside the medium layer, the wave field and energy flux density at the point of source location are given by Eqs. (12.13) and (12.15). Quantities

$R_1(x_0)$ and $R_2(x_0)$ are statistically independent within the framework of the model of the delta-correlated fluctuations of $\varepsilon_1(x)$, because they satisfy dynamic equations (12.11) for nonoverlapping space portions. In the case of the infinite space ($L_0 \to -\infty, L \to \infty$), probability densities of quantities $R_1(x_0)$ and $R_2(x_0)$ are given by Eq. (12.28); as a result, average intensity of the wave field and average energy flux density at the point of source location are given by the expressions

$$\langle I(x_0; x_0)\rangle = 1 + \frac{1}{\beta}, \quad \langle S(x_0; x_0)\rangle = 1. \tag{12.35}$$

The infinite increase of the average intensity at the point of source location for $\beta \to 0$ is evidence of wave energy accumulation in a randomly layered medium; at the same time, average energy flux density at the point of source location is independent of medium parameter fluctuations and coincides with energy flux density in free space.

For the source located at perfectly reflecting boundary $x_0 = L$, we obtain from Eqs. (12.13) and (12.15)

$$\langle I_{\text{ref}}(L; L)\rangle = 4\left(1 + \frac{2}{\beta}\right), \quad \langle S_{\text{ref}}(L; L)\rangle = 4, \tag{12.36}$$

i.e., average energy flux density of the source located at the reflecting boundary is also independent of medium parameter fluctuations and coincides with energy flux density in free space.

Note the singularity of the above formulas (12.35), (12.36) for $\beta \to 0$, which shows that absorption (even arbitrarily small) serves the regularizing factor in the problem on the point source.

Using Eq. (12.16), we can obtain the probability distribution of wave field energy in the half-space

$$E = D \int\limits_{-\infty}^{x_0} dx I(x; x_0).$$

In particular, for the source located at reflecting boundary, we obtain the expression

$$P_{\text{ref}}(E) = \sqrt{\frac{2}{\pi}} \frac{1}{E\sqrt{E}} \exp\left\{-\frac{2}{E}\left(1 - \frac{\beta E}{4}\right)^2\right\},$$

that allows limiting process $\beta \to 0$, which is similar to the case of the wave incidence on the half-space of random medium.

12.1.3 Statistical energy localization

In view of Eq. (12.16), the obtained results related to wave field at fixed spatial points (at layer boundaries and at the point of source location) offer a possibility of making certain general conclusions about the behavior of the wave field average intensity inside the random medium.

For example, Eq. (12.16) yields the expression for average energy contained in the half-space $(-\infty, x)$

$$\langle E\rangle = D \int\limits_{-\infty}^{x_0} dx \langle I(x; x_0)\rangle = \frac{1}{\beta}\langle S(x_0; x_0)\rangle. \tag{12.37}$$

In the case of the plane wave ($x_0 = L$) incident on the half-space $x \leq L$, Eqs. (12.31) and (12.37) yield, for $\beta \ll 1$, the expressions

$$\langle E \rangle = 2\ln(1/\beta), \quad \langle I(L; L \rangle = 2. \qquad (12.38)$$

Consequently, the space segment of length

$$Dl_\beta \cong \ln(1/\beta),$$

concentrates the most portion of average energy, which means that there occurs the *wave field statistical localization* caused by wave absorption. Note that, in the absence of medium parameter fluctuations, energy localization occurs on scales about absorption length $Dl_{\mathrm{abs}} \cong 1/\beta$. However, we have $l_{\mathrm{abs}} \gg l_\beta$ for $\beta \ll 1$. If $\beta \to 0$, then $l_\beta \to \infty$, and statistical localization of the wave filed disappears in the limiting case of non-absorptive medium.

In the case of the source in unbounded space, we have

$$\langle E \rangle = \frac{1}{\beta}, \quad \langle I(x_0; x_0) \rangle = 1 + \frac{1}{\beta},$$

and average energy localization is characterized, as distinct from the foregoing case, by spatial scale $D|x - x_0| \cong 1$ for $\beta \to 0$.

In a similar way, we have for the source located at reflecting boundary

$$\langle E \rangle = \frac{4}{\beta}, \quad \langle I_{\mathrm{ref}}(L; L) \rangle = 4\left(1 + \frac{2}{\beta}\right),$$

from which follows that average energy localization is characterized by the half spatial scale $D(L - x)| \cong 1/2$ for $\beta \to 0$.

In the considered problems, wave field average energy essentially depends on parameter β and tends to infinity for $\beta \to 0$. However, this is the case only for average quantities. In our further analysis of the wave field in random medium, we will show that the field is localized in separate realization due to the *dynamic localization* even in non-absorptive media.

12.2   Statistical theory of radiative transfer

Now, we dwell on the statistical description of a wave field in random medium (statistical theory of radiative transfer). We consider two problems of which the first concerns the wave incident on the medium layer and the second concerns the waves generated by a source located in the medium.

12.2.1   Normal wave incidence on the layer of random media

In the general case of absorptive medium, the wave field is described by the boundary-value problem (12.1), (12.2). We introduce complex opposite waves

$$u(x) = u_1(x) + u_2(x), \quad \frac{d}{dx}u(x) = -ik[u_1(x) - u_2(x)],$$

related to the wave field through the relationships

$$
\begin{aligned}
u_1(x) &= \frac{1}{2}\left[1 + \frac{i}{k}\frac{d}{dx}\right] u(x), \quad u_1(L) = 1, \\
u_2(x) &= \frac{1}{2}\left[1 - \frac{i}{k}\frac{d}{dx}\right] u(x), \quad u_2(L_0) = 0,
\end{aligned}
$$

and, consequently, the boundary-value problem (12.1), (12.2) can be rewritten as

$$
\begin{aligned}
\left(\frac{d}{dx} + ik\right) u_1(x) &= -\frac{ik}{2}\varepsilon(x)\left[u_1(x) + u_2(x)\right], \quad u_1(L) = 1, \\
\left(\frac{d}{dx} - ik\right) u_2(x) &= -\frac{ik}{2}\varepsilon(x)\left[u_1(x) + u_2(x)\right], \quad u_2(L_0) = 0.
\end{aligned}
$$

The wave field as a function of parameter L satisfies imbedding equation (12.6). It is obvious that the opposite waves will also satisfy Eq. (12.6), but with different initial conditions:

$$
\begin{aligned}
\frac{\partial}{\partial L} u_1(x; L) &= ik\left(1 + \frac{1}{2}\varepsilon(L)\left(1 + R_L\right)\right) u_1(x; L), \quad u_1(x; x) = 1, \\
\frac{\partial}{\partial L} u_2(x; L) &= ik\left(1 + \frac{1}{2}\varepsilon(L)\left(1 + R_L\right)\right) u_2(x; L), \quad u_2(x; x) = R_x,
\end{aligned}
$$

where reflection coefficient R_L satisfies Eq. (12.5).

Introduce now opposite wave intensities $W_1(x; L) = |u_1(x; L)|^2$ and $W_2(x; L) = |u_2(x; L)|^2$ satisfying the equations

$$
\begin{aligned}
\frac{\partial}{\partial L} W_1(x; L) &= -k\gamma W_1(x; L) + \frac{ik}{2}\varepsilon(L)\left(R_L - R_L^*\right) W_1(x; L), \\
\frac{\partial}{\partial L} W_2(x; L) &= -k\gamma W_2(x; L) + \frac{ik}{2}\varepsilon(L)\left(R_L - R_L^*\right) W_2(x; L), \\
W_1(x; x) &= 1, \qquad W_2(x; x) = |R_x|^2.
\end{aligned} \tag{12.39}
$$

Quantity $W_L = |R_L|^2$ appeared in the initial condition of Eq. (12.39) satisfies Eq. (12.7), or the equation

$$
\frac{d}{dL} W_L = -2k\gamma W_L - \frac{ik}{2}\varepsilon_1(L)\left(R_L - R_L^*\right)\left(1 - W_L\right), \quad W_{L_0} = 0. \tag{12.40}
$$

In Eqs. (12.39) and (12.40), we omitted dissipative terms producing no contribution in accumulated effects.

As earlier, we will assume that $\varepsilon_1(x)$ is the Gaussian delta-correlated process with correlation function (12.17). In view of the fact that Eqs. (12.39), (12.40) are the first-order equations with initial conditions, we can use the standard procedure of deriving the Fokker–Planck equation for the joint probability density of quantities $W_1(x; L), W_2(x; L)$, and W_L

$$
P(x; L; W_1, W_2, W) = \langle \delta(W_1(x; L) - W_1)\delta(W_2(x; L) - W_2)\delta(W_L - W) \rangle.
$$

As a result, we obtain the Fokker–Planck equation

$$\frac{\partial}{\partial L}P(x;L;W_1,W_2,W)$$

$$= k\gamma\left(\frac{\partial}{\partial W_1}W_1 + \frac{\partial}{\partial W_2}W_2 + 2\frac{\partial}{\partial W}W\right)P(x;L;W_1,W_2,W)$$

$$+D\left[\frac{\partial}{\partial W_1}W_1 + \frac{\partial}{\partial W_2}W_2 - \frac{\partial}{\partial W}(1-W)\right]P(x;L;W_1,W_2,W)$$

$$+D\left[\frac{\partial}{\partial W_1}W_1 + \frac{\partial}{\partial W_2}W_2 - \frac{\partial}{\partial W}(1-W)\right]^2 W P(x;L;W_1,W_2) \qquad (12.41)$$

with the initial condition

$$P(x;x;W_1,W_2,W) = \delta(W_1 - 1)\delta(W_2 - W)P(x;W),$$

where function $P(L;W)$ is the probability density of reflection coefficient squared modulus W_L, which satisfies Eq. (12.20). As earlier, the diffusion coefficient in Eq. (12.41) is $D = k^2\sigma_\varepsilon^2 l_0/2$. Deriving this equation, we used an additional averaging over fast oscillations $(u(x) \sim e^{\pm ikx})$ that appear in the solution of the problem with $\varepsilon = 0$.

In view of the fact that Eqs. (12.39) are linear in $W_n(x;L)$, we can introduce the generating function of moments of opposite wave intensities

$$Q(x;L;\mu,\lambda,W) = \int\limits_0^1 dW_1 \int\limits_0^1 dW_2\, W_1^{\mu-\lambda}W_2^\lambda P(x;L;W_1,W_2,W), \qquad (12.42)$$

which satisfies the simpler equation

$$\frac{\partial}{\partial L}Q(x;L;\mu,\lambda,W) = -k\gamma\left(\mu - 2\frac{\partial}{\partial W}W\right)Q(x;L;\mu,\lambda,W)$$

$$-D\left[\mu + \frac{\partial}{\partial W}(1-W)\right]Q(x;L;\mu,\lambda,W)$$

$$+\left[\mu - \frac{\partial}{\partial W}(1-W)\right]^2 W Q(x;L;\mu,\lambda,W) \qquad (12.43)$$

with the initial condition

$$Q(x;x;\mu,\lambda,W) = W^\lambda P(x;W).$$

With function $Q(x;L;\mu,\lambda,W)$, we can determine the moment functions of opposite wave intensities by the formula

$$\left\langle W_1^{\mu-\lambda}(x;L)W_2^\lambda(x;L)\right\rangle = \int\limits_0^1 dW\, Q(x;L;\mu,\lambda,W). \qquad (12.44)$$

Equation (12.43) describes statistics of the wave field in medium layer $L_0 \leq x \leq L$. In particular, if we set $x = L_0$, it describes the transmission coefficient of the wave.

In the limiting case of the half-space ($L_0 \to -\infty$), Eq. (12.43) grades into the equation

$$\frac{\partial}{\partial\xi}Q(\xi;\mu,\lambda,W) = -\beta\left(\mu - 2\frac{\partial}{\partial W}W\right)Q(\xi;\mu,\lambda,W)$$

$$-\left[\mu + \frac{\partial}{\partial W}(1-W)\right]Q(\xi;\mu,\lambda,W) + \left[\mu - \frac{\partial}{\partial W}(1-W)\right]^2 W Q(\xi;\mu,\lambda,W),$$

$$Q(0;\mu,\lambda,W) = W^\lambda P(W), \qquad (12.45)$$

where $\xi = D(L - x) > 0$ is the dimensionless distance, and steady-state (independent of L) probability density of the reflection coefficient modulus $P(W)$ is given by Eq. (12.28). In this case, Eq. (12.44) assumes the form

$$\left\langle W_1^{\mu-\lambda}(\xi)W_2^{\lambda}(\xi)\right\rangle = \int_0^1 dW Q(\xi; \mu, \lambda, W). \tag{12.46}$$

Further discussion will be more convenient if we consider separately the cases of absorptive (dissipative) and non-absorptive (nondissipative) random medium.

Nondissipative medium (stochastic wave parametric resonance and dynamic wave localization)

Let us now discuss the equations for moments of opposite wave intensities in non-absorptive medium, i.e., to Eq. (12.45) at $\beta = 0$ in the limit of the half-space ($L_0 \to -\infty$). In this case, $W_L = 1$ with a probability equal to unity, and the solution to Eq. (12.45) has the form

$$Q(x, L; \mu, \lambda, W) = \delta(W - 1)e^{D\lambda(\lambda-1)(L-x)},$$

so that

$$\left\langle W_1^{\lambda-\mu}(x; L)W_2^{\mu}(x; L)\right\rangle = e^{D\lambda(\lambda-1)(L-x)}.$$

In view of arbitrariness of parameters λ and μ, this means that

$$W_1(x; L) = W_2(x; L) = W(x; L)$$

with a probability equal to unity and quantity $W(x; L)$ has the lognormal probability density. In addition, the mean value of this quantity is equal to unity, and its higher moments beginning from the second one exponentially increase with the distance in medium

$$\langle W(x; L)\rangle = 1, \quad \langle W^n(x; L)\rangle = e^{Dn(n-1)(L-x)}, \quad n = 2, 3, \ldots.$$

Note that wave field intensity $I(x; L)$ has in this case the form

$$I(x; L) = 2W(x; L)\left(1 + \cos\phi_x\right), \tag{12.47}$$

where ϕ_L is the phase of the reflection coefficient.

In view of the lognormal probability distribution, the typical realization curve of function $W(x; L)$ is the curve exponentially decaying with distance in the medium

$$W^*(x; L) = e^{-D(L-x)}. \tag{12.48}$$

For example, realizations of function $W(x; L)$ satisfy with a probability of $1/2$ the inequality

$$W(x; L) < 4e^{-D(L-x)/2}$$

within the whole of the half-space.

In physics of disordered systems, the exponential decay of typical realization curve (12.48) with increasing $\xi = D(L - x)$ is usually identified with the property of *dynamic localization*, and quantity

$$l_{\mathrm{loc}} = \frac{1}{D}$$

is usually called the *localization length*. Here,

$$l_{\text{loc}}^{-1} = -\frac{\partial}{\partial L} \langle \varkappa(x; L) \rangle ,$$

where

$$\varkappa(x; L) = \ln W(x; L).$$

Physically, the lognormal property of the wave field intensity $W(x; L)$ implies the existence of large spikes relative typical realization curve (12.48) towards both large and small intensities. This result agrees with the example of simulations given in Chapter 1 (see Fig. 1.6, page 19). However, these spikes of intensity contain only small energy, because random area below curve $W(x; L)$,

$$S(L) = D \int\limits_{-\infty}^{L} dx W(x; L),$$

has, in accordance with the lognormal probability distribution attribute, the steady-state (independent of L) probability density

$$P(S) = \frac{1}{S^2} \exp\left\{ -\frac{1}{S} \right\}$$

that coincides with the distribution of total energy of the wave field in the half-space (12.34) if we set $E = 2S$. This means that the term dependent on fast phase oscillations of reflection coefficient in Eq. (12.47) only slightly contributes to total energy.

Thus, the knowledge of the one-point probability density provides an insight into the evolution of separate realizations of wave field intensity in the whole space and allows estimating the parameters of this evolution in terms of statistical characteristics of fluctuating medium.

Dissipative medium

In the presence of a finite (even arbitrary small) absorption in the medium occupying the half-space, the exponential growth of moment functions must cease and give place to attenuation. If $\beta \gg 1$ (i.e., if the effect of absorption is great in comparison with the effect of diffusion), then

$$P(W) = 2\beta e^{-2\beta W},$$

and, as can be easily seen from Eq. (12.45), opposite wave intensities $W_1(x; L)$ and $W_2(x; L)$ appear statistically independent, i.e., uncorrelated. In this case,

$$\langle W_1(\xi) \rangle = \exp\left\{ -\beta\xi \left(1 + \frac{1}{\beta} \right) \right\}, \quad \langle W_2(\xi) \rangle = \frac{1}{2\beta} \exp\left\{ -\beta\xi \left(1 + \frac{1}{\beta} \right) \right\}.$$

Figures. 12.4–12.6 show the examples of moment functions of random processes obtained by numerical solution of Eq. (12.45) and calculation of quadrature (12.46) for different values of parameter β. Different figures mark the curves corresponding to different values of parameter β. Figure 12.4 shows average intensities of the transmitted and reflected waves. The curves monotonically decrease with increasing ξ. Figure 12.5 shows the corresponding curves for second moments. We see that $\langle W_1^2(0) \rangle = 1$ and $\langle W_2^2(0) \rangle = \langle |R_L|^4 \rangle$ at $\xi = 0$. For $\beta < 1$, the curves as functions of ξ become nonmonotonic; the moments

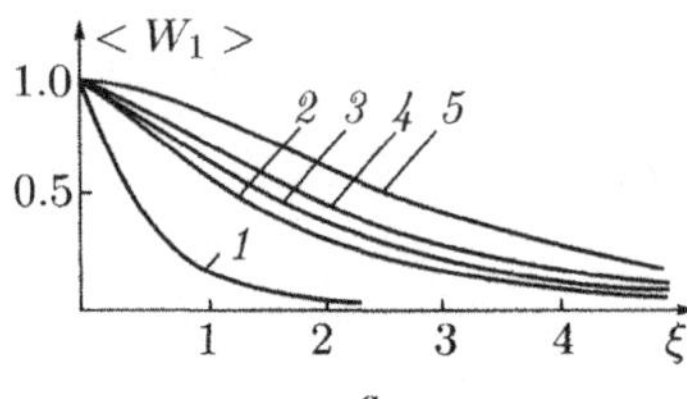
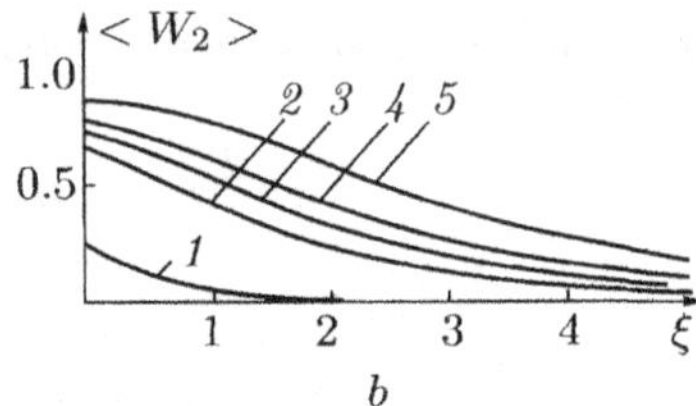

Figure 12.4: Distribution of wave field average intensity along the medium; (a) the transmitted wave and (b) the reflected wave. Curves 1 to 5 correspond to parameter $\beta = 1$, 0.1, 0.06, 0.04 and 0.02, respectively.

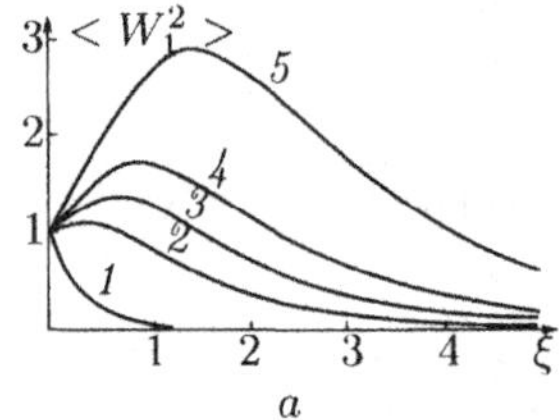
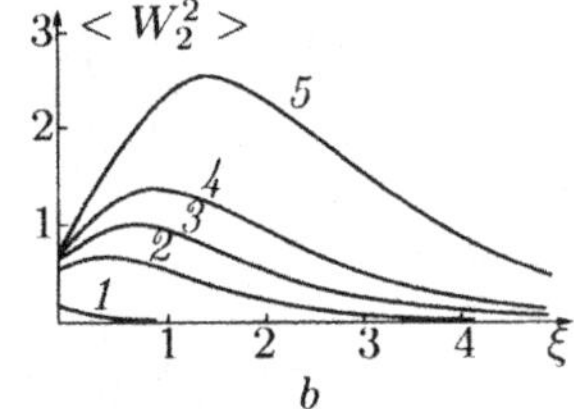

Figure 12.5: Distribution of the second moment of wave field intensity along the medium; (a) the transmitted wave and (b) the reflected wave. Curves 1 to 5 correspond to parameter $\beta = 1$, 0.1, 0.06, 0.04 and 0.02, respectively.

first increase, then pass the maximum, and finally monotonically decay. With decreasing parameter β, the position of the maximum moves to the right and the maximum value increases. The fact that moments of intensity behave in the layer as exponentially increasing functions is evidence of the *phenomenon of stochastic wave parametric resonance*, which is similar to the ordinary parametric resonance. The only difference consists in the fact that values of intensity moments at layer boundary are asymptotically predetermined; as a result, the wave field intensity exponentially increases inside the layer and its maximum occurs approximately in the middle of the layer.

Figure 12.6 shows curves for mutual correlation of intensities of the transmitted and reflected waves $\langle \Delta W_1(\xi) \Delta W_2(\xi) \rangle$ (here, $\Delta W_n(\xi) = W_n(\xi) - \langle W_n(\xi) \rangle$). For $\beta \geq 1$, this correlation disappears. For $\beta < 1$, the correlation is strong, and wave division into opposite waves appears physically senseless, but mathematically useful technique. For $\beta \geq 1$, such a division is justified in view of the lack of mutual correlation.

As was shown earlier, in the case of the half-space of random medium with $\beta = 0$, all wave field moments beginning from the second one exponentially increase with the distance the wave travels in the medium. It is clear, that problem solution for small β ($\beta \ll 1$) must show the singular behavior in β in order to vanish the solution for sufficiently long distances.

One can show that, in asymptotic limit $\beta \ll 1$, intensities of the opposite waves are equal with a probability equal to unity, and the solution for small distances from the boundary coincides with the solution corresponding to the stochastic parametric resonance.

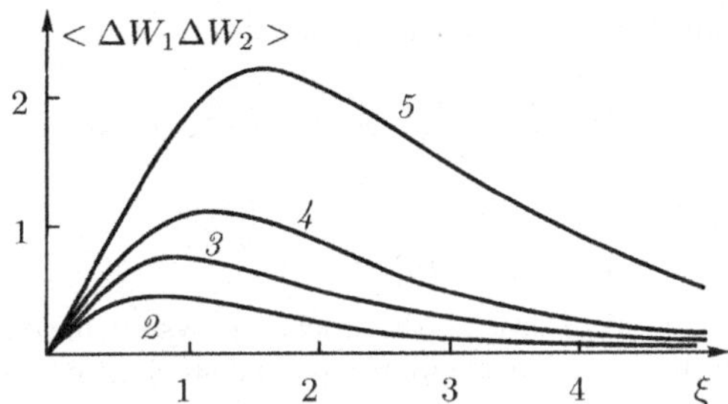

Figure 12.6: Correlation between the intensities of transmitted and reflected waves. Curves *1* to *5* correspond to parameter $\beta = 1$, 0.1, 0.06, 0.04 and 0.02, respectively.

For sufficiently great distances ξ, namely

$$\xi \gg 4 \left(n - \frac{1}{2} \right) \ln \left(\frac{n}{\beta} \right),$$

quantities $\langle W^n(\xi) \rangle$ are characterized by the universal spatial localization behavior

$$\langle W^n(\xi) \rangle \cong A_n \frac{1}{\beta^{n-1/2}} \ln \left(\frac{1}{\beta} \right) \frac{1}{\xi \sqrt{\xi}} e^{-\xi/4},$$

which coincides to a numerical factor with the asymptotic behavior of moments of the transmission coefficient of a wave passed through the layer of thickness ξ in the case $\beta = 0$.

Thus, there are three regions, where moments of opposite wave intensities show essentially different behavior. In the first region (it corresponds to the stochastic wave parametric resonance), the moments exponentially increase with the distance in medium and wave absorption plays only insignificant role. In the second region, absorption plays the most important role, because namely absorption ceases the exponential growth of moments. In the third region, the decrease of moment functions of opposite wave intensities is independent of absorption. The boundaries of these regions depend on parameter β and tend to infinity for $\beta \to 0$.

12.2.2 Plane wave source located in random medium

Consider now the asymptotic solution of the problem on the plane wave source in infinite space ($L_0 \to -\infty$, $L \to \infty$) under the condition $\beta \to 0$. In this case, it appears convenient to calculate the average wave field intensity in region $x < x_0$ using relationships (12.14) and (12.15)

$$\beta \langle I(x; x_0) \rangle = \frac{1}{D} \frac{\partial}{\partial x} \langle S(x; x_0) \rangle = \frac{1}{D} \frac{\partial}{\partial x} \langle \psi(x; x_0) \rangle,$$

where quantity

$$\psi(x; x_0) = \exp \left\{ -\beta D \int_x^{x_0} d\xi \frac{|1 + R_\xi|^2}{1 - |R_\xi|^2} \right\}$$

satisfies, as a function of parameter x_0, the stochastic equation

$$\frac{\partial}{\partial x_0}\psi(x;x_0) = -\beta D\frac{|1+R_{x_0}|^2}{1-|R_{x_0}|^2}\psi(x;x_0), \quad \psi(x;x) = 1.$$

Introduce function

$$\Phi(x;x_0;u) = \psi(x;x_0)\delta(u_{x_0} - u), \tag{12.49}$$

where function $u_L = (1+W_L)/(1-W_L)$ satisfies the stochastic system of equations (12.21). Differentiating Eq. (12.49) with respect to x_0, we obtain the stochastic equation

$$\frac{\partial}{\partial x_0}\Phi(x;x_0;u) = -\beta D\left\{u + \sqrt{u^2-1}\cos\phi_{x_0}\right\}\Phi(x;x_0;u)$$
$$+\beta D\frac{\partial}{\partial u}\left\{\left(u^2-1\right)\Phi(x;x_0;u)\right\} - k\varepsilon_1(x_0)\frac{\partial}{\partial u}\left\{\sqrt{u^2-1}\sin\phi_{x_0}\Phi(x;x_0;u)\right\}. \tag{12.50}$$

Average now Eq. (12.50) over an ensemble of realizations of random process $\varepsilon_1(x_0)$ assuming it, as earlier, the Gaussian process delta-correlated in x_0. Using the Furutsu–Novikov formula (8.10), the following expression for the variational derivatives

$$\frac{\delta\Phi(x;x_0;u)}{\delta\varepsilon_1(x_0)} = -k\frac{\partial}{\partial u}\left\{\sqrt{u^2-1}\sin\phi_{x_0}\Phi(x;x_0;u)\right\},$$

$$\frac{\delta\phi_{x_0}}{\delta\varepsilon_1(x_0)} = k\left[1 + \frac{u_{x_0}}{\sqrt{u_{x_0}^2-1}}\right]\cos\phi_{x_0},$$

and additionally averaging over fast oscillations (over the phase of the reflection coefficient), we obtain that function

$$\Phi(\xi;u) = \langle\Phi(x;x_0;u)\rangle = \langle\psi(x;x_0)\delta(u_{x_0} - u)\rangle,$$

where $\xi = D|x - x_0|$, satisfies the equation

$$\frac{\partial}{\partial\xi}\Phi(\xi;u) = -\beta u\Phi(\xi;u) +$$
$$+\beta\frac{\partial}{\partial u}\left(u^2-1\right)\Phi(\xi;u) + \frac{\partial}{\partial u}\left(u^2-1\right)\frac{\partial}{\partial u}\Phi(\xi;u),$$
$$\Phi(0;u) = P(u) = \beta e^{-\beta(u-1)}. \tag{12.51}$$

The average intensity can now be represented in the form

$$\beta\langle I(x;x_0)\rangle = -\frac{\partial}{\partial\xi}\int_1^\infty du\Phi(\xi;u) = \beta\int_1^\infty duu\Phi(\xi;u).$$

Equation (12.51) allows limiting process $\beta \to 0$. As a result, we obtain a simpler equation

$$\frac{\partial}{\partial\xi}\tilde{\Phi}(\xi;u) = -u\tilde{\Phi}(\xi;u) + \frac{\partial}{\partial u}u^2\tilde{\Phi}(\xi;u) + \frac{\partial}{\partial u}u^2\frac{\partial}{\partial u}\tilde{\Phi}(\xi;u),$$
$$\tilde{\Phi}(0;u) = e^{-u}. \tag{12.52}$$

Consequently, localization of average intensity in space is described by the quadrature

$$\Phi_{\text{loc}}(\xi) = \int\limits_{1}^{\infty} duu\tilde{\Phi}(\xi; u),$$

where

$$\Phi_{\text{loc}}(\xi) = \lim_{\beta \to 0} \beta \left\langle I(x; x_0) \right\rangle = \lim_{\beta \to 0} \frac{\left\langle I(x; x_0) \right\rangle}{\left\langle I(x_0; x_0) \right\rangle}.$$

Thus, the average intensity of the wave field generated by the point source has for $\beta \ll 1$ the following asymptotic behavior

$$\left\langle I(x; x_0) \right\rangle = \frac{1}{\beta} \Phi_{\text{loc}}(\xi).$$

Equation (12.52) can be easily solved with the use of the Kantorovich–Lebedev transform; as a result, we obtain the expression for the *localization curve*

$$\Phi_{\text{loc}}(\xi) = 2\pi \int\limits_{0}^{\infty} d\tau\, \tau \left(\tau^2 + \frac{1}{4} \right) \frac{\sinh(\pi\tau)}{\cosh^2(\pi\tau)} e^{-\left(\tau^2 + \frac{1}{4}\right)\xi}. \tag{12.53}$$

For small distances ξ, the localization curve decays according to relatively fast law

$$\Phi_{\text{loc}}(\xi) \approx e^{-2\xi}. \tag{12.54}$$

For great distances ξ (namely, for $\xi \gg \pi^2$), it decays significantly slower, according to the universal law

$$\Phi_{\text{loc}}(\xi) \approx \frac{\pi^2 \sqrt{\pi}}{8} \frac{1}{\xi\sqrt{\xi}} e^{-\xi/4}, \tag{12.55}$$

but for all that

$$\int\limits_{0}^{\infty} d\xi \Phi_{\text{loc}}(\xi) = 1.$$

Function (12.53) is given in Fig. 12.7, where asymptotic curves (12.54) and (12.55) are also shown for comparison purposes.

A similar situation occurs in the case of the plane wave source located at the reflecting boundary. In this case, we obtain the expression

$$\lim_{\beta \to 0} \frac{\left\langle I_{\text{ref}}(x; L) \right\rangle}{\left\langle I_{\text{ref}}(L; L) \right\rangle} = \frac{1}{2} \Phi_{\text{loc}}(\xi), \quad \xi = D(L - x). \tag{12.56}$$

12.3   Numerical simulation

The above theory rests on two simplifications—on using the delta-correlated approximation of function $\varepsilon_1(x)$ (or the diffusion approximation) and eliminating slow (within the scale of a wavelength) variations of statistical characteristics by averaging over fast oscillations. Averaging over fast oscillations is validated only for statistical characteristics of the reflection coefficient in the case of random medium occupying a half-space. For statistical characteristics of the wave field intensity in medium, the corresponding validation appears very difficult if at all possible (this method is merely physical than mathematical).

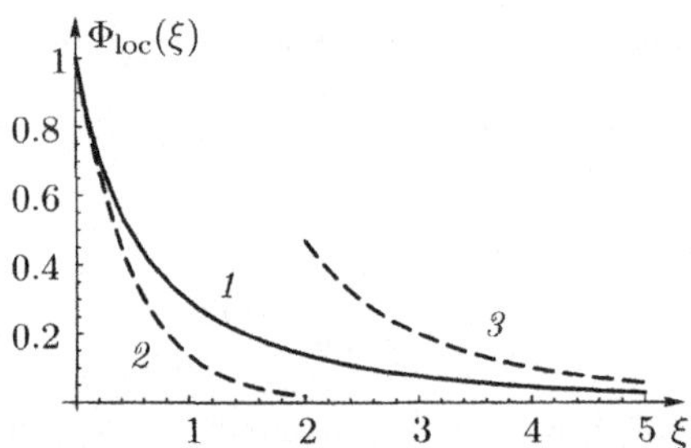

Figure 12.7: Localization curve for a source in infinite space (12.53) (curve *1*). Curves *2* and *3* correspond to asymptotic expressions for small and large distances from the source.

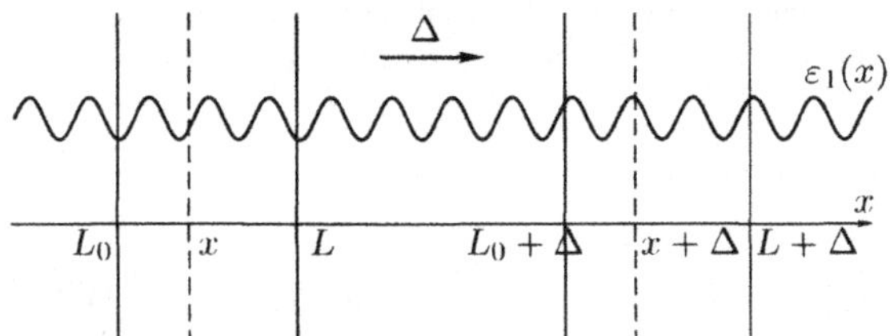

Figure 12.8: Averaging over parameter Δ by the procedure based on ergodicity of imbedding equations for a half-space of random medium.

Numerical simulation of the exact problem offers a possibility of both verifying these simplifications and obtaining results concerning more difficult situations for which no analytic results exists.

In principle, such numerical simulation could be performed by way of multiply solving the problem for different realizations of medium parameters followed by averaging the obtained solutions over an ensemble of realizations. However, such an approach is not very practicable because it requires a vast body of realizations of medium parameters. Moreover, it is unsuitable for real physical problems, such as wave propagation in Earth's atmosphere and ocean, where only a single realization is usually available. A more practicable approach is based on the ergodic property of boundary-value problem solutions with respect to the displacement of the problem along the single realization of function $\varepsilon_1(x)$ defined along the half-axis (L_0, ∞) (see Fig. 12.8). This approach assumes that statistical characteristics are calculated by the formula

$$\langle F(L_0; x, x_0; L) \rangle = \lim_{\delta \to \infty} F_\delta(L_0; x, x_0; L),$$

where

$$F_\delta(L_0; x, x_0; L) = \frac{1}{\delta} \int_0^\delta d\Delta F(L_0 + \Delta; x + \Delta, x_0 + \Delta; L + \Delta).$$

In the limit of a half-space $(L_0 \to -\infty)$, statistical characteristics are independent of L_0, and, consequently, the problem have ergodic property with respect to the position

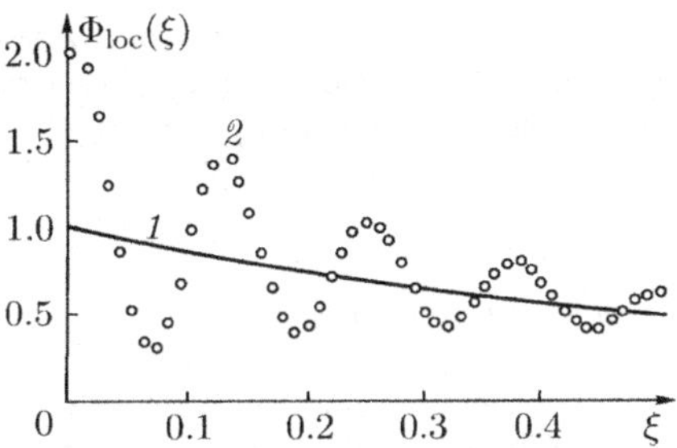

Figure 12.9: Average intensity $2\langle I_{\mathrm{ref}}(x;L)\rangle / \langle I_{\mathrm{ref}}(L;L)\rangle$ of the field of a source located at reflecting boundary ($\beta = 0.08$). Curve *1* shows the localization curve (12.53) and circles *2* show the simulated result.

of the right-hand layer boundary L (simultaneously, parameter L is the variable of the imbedding method), because this position is identified in this case with the displacement parameter. As a result, having solved the imbedding equation for the sole realization of medium parameters, we simultaneously obtain all desired statistical characteristics of this solution by using the obvious formula

$$\langle F(x, x_0; L)\rangle = \frac{1}{\delta} \int\limits_0^\delta d\xi\, F\left(\xi, \xi + x_0 - x; \xi + (L - x_0) + (x_0 - x)\right)$$

for sufficiently large interval $(0, \delta)$. This approach offers a possibility of calculating even the wave statistical characteristics that cannot be obtained within the framework of current statistical theory, and this calculation requires no additional simplifications.

In the case of the layer of finite thickness, the problem is not ergodic with respect to parameter L. However, the corresponding solution can be expressed in terms of two independent solutions of the problem on the half-space [2] and, consequently, it can be reduced to the case ergodic with respect to L.

As an example let us discuss the case of the point source located at reflecting boundary $x_0 = L$ with boundary condition $dG(x, x_0; L)/dx|_{x=L} = 0$. Figure 12.9 shows the quantity $\langle I_{\mathrm{ref}}(x, x_0)\rangle$ simulated with $\beta = 0.08$ and $k/D = 25$. In region $\xi = D(L - x) < 0.3$, one can see oscillations of period $T = 0.13$. For larger ξ, simulated results agree well with localization curve (12.56).

Problems

Problem 69 *Derive the probability distribution of the reflection coefficient phase in the problem on a plane wave incident on the half-space of nondissipative medium [45].*

Instruction. In this case, reflection coefficient has the form $R_L = \exp\{i\phi_L\}$, where phase ϕ_L satisfies the imbedding equation following from Eq. (12.18)

$$\frac{d}{dL}\phi_L = 2k + k\varepsilon_1(L)\left(1 + \cos\phi_L\right), \quad \phi_{L_0} = 0. \tag{12.57}$$

The solution to Eq. (12.57) is defined over the whole axis ϕ_L $(-\infty, \infty)$. A more practicable characteristic is the probability distribution in interval $(-\pi, \pi)$, which must be independent of L in the case of the half-space. To obtain such a distribution, it is convenient to consider singular function $z_L = \tan(\phi_L/2)$ instead of phase ϕ_L. The corresponding dynamic equation has the form

$$\frac{d}{dL} z_L = k\left(1 + z_L^2\right) + k\varepsilon_1(L), \quad z_{L_0} = 0.$$

Solution. Assuming that $\varepsilon_1(L)$ is the Gaussian delta-correlated random function with the parameters given in Eq. (12.17), we obtain that probability density $P(L, z) = \langle \delta(z_L - z) \rangle$ defined over the whole axis $(-\infty, \infty)$ satisfies the Fokker–Planck equation

$$\frac{\partial}{\partial L} P(L, z) = -k\frac{\partial}{\partial z}\left(1 + z_L^2\right) P(L, z) + 2D\frac{\partial^2}{\partial z^2} P(L, z).$$

For the random half-space ($L_0 \to -\infty$), the steady-state (independent of L) solution to the Fokker–Planck equation can be obtained as $P(z) = \lim_{L_0 \to -\infty} P(L, z)$ and satisfies the equation

$$-\kappa\frac{d}{dz}\left(1 + z^2\right) P(z) + \frac{d^2}{dz^2} P(z) = 0, \tag{12.58}$$

where

$$\kappa = \frac{k}{2D}, \quad D = \frac{k^2 \sigma_\varepsilon^2 l_0}{2}.$$

The solution to Eq. (12.58) corresponding to constant probability flux density has the form

$$P(z) = J(\kappa) \int_z^\infty d\xi \exp\left\{-\kappa\xi\left[1 + \frac{\xi^3}{3} + z(z + \xi)\right]\right\},$$

where $J(\kappa)$ is the steady-state probability flux density,

$$J^{-1}(\kappa) = \sqrt{\frac{\pi}{\kappa}} \int_0^\infty \xi^{-1/2} d\xi \exp\left\{-\kappa\left(\xi + \frac{\xi^3}{12}\right)\right\}.$$

The corresponding probability distribution of the wave phase over interval $(-\pi, \pi)$ is as follows

$$P(\phi) = \left.\frac{1 + z^2}{2} P(z)\right|_{z=\tan(\phi/2)}.$$

For $\kappa \gg 1$, we have asymptotically

$$P(z) = \frac{1}{\pi(1 + z^2)},$$

which corresponds to the uniform distribution of the reflection coefficient phase

$$P(\phi) = \frac{1}{2\pi}, \quad -\pi < \phi < \pi.$$

In the opposite limiting case $\kappa \ll 1$, we obtain

$$P(z) = \kappa^{1/3}\left(\frac{3}{4}\right)^{1/6} \frac{1}{\sqrt{\pi}\Gamma(1/6)} \Gamma\left(\frac{1}{3}, \frac{\kappa z^3}{3}\right),$$

where $\Gamma(\mu, z)$ is the *incomplete gamma function*. From this expression follows that

$$P(z) = \kappa^{1/3} \left(\frac{3}{4}\right)^{1/6} \frac{1}{\sqrt{\pi}\,\Gamma(1/6)} \left(\frac{3}{\kappa z^3}\right)^{2/3}$$

for $\kappa|z|^3 \gg 3$ and $|z| \to \infty$.

Problem 70 *In the diffusion approximation, derive the equation for the reflection coefficient probability density.*

Solution. The equation for the reflection coefficient probability density has the form of the Fokker–Planck equation (12.20) with the diffusion coefficient

$$D(k, l_0) = \frac{k^2}{4} \int\limits_{-\infty}^{\infty} d\xi\, B_\varepsilon(\xi) \cos(2k\xi) = \frac{k^2}{4} \Phi_\varepsilon(2k),$$

where $\Phi_\varepsilon(q) = \int\limits_{-\infty}^{\infty} d\xi\, B_\varepsilon(\xi) e^{iq\xi}$ is the spectral function of random process $\varepsilon_1(x)$. Applicability range of this equation is given by the conditions

$$D(k, l_0)l_0 \ll 1, \quad \alpha = \frac{k}{D(k, l_0)} \gg 1.$$

Problem 71 *Derive the equation for the reflection coefficient probability density in the case of matched boundary.*

Instruction. In the case of matched boundary, reflection coefficient satisfies the Riccati equation

$$\frac{d}{dL} R_L = 2ikR_L - k\gamma R_L + \frac{\xi(L)}{2}\left(1 - R_L^2\right), \quad R_{L_0} = 0,$$

where $\xi(L) = \varepsilon_1'(L)$. For the Gaussian process with correlation function $B_\varepsilon(x)$, random process $\xi(x)$ is also the Gaussian process with correlation function

$$B_\xi(x - x') = \langle \xi(x)\xi(x')\rangle = -\frac{\partial^2}{\partial x^2} B_\varepsilon(x - x').$$

Solution. The equation for the reflection coefficient probability density has the form of the Fokker–Planck equation (12.20) with the diffusion coefficient

$$D(k, l_0) = \frac{k^2}{4} \Phi_\varepsilon(2k).$$

Bibliography

1. Klyatskin, V. I., 1986, Metod Pogruzheniya v Teorii Rasprostraneniya Voln (The Imbedding Method in Wave Propagation Theory). — Moscow: Nauka (in Russian).

2. Klyatskin, V. I., 1994, The imbedding method in statistical boundary-value wave problems / Progress in Optics. **XXXIII**, 1 – 128, ed. E. Wolf. — Amsterdam: North-Holland.

3. Novikov, E. A., 1964, Functionals and the random force method in turbulence theory, Zh. Eksper. Teor. Phys. **47**(5), 1919 – 1926 [Sov. Phys. JETP **20**(5), 1290 – 1294, 1965].

4. Klyatskin, V. I., 1975, Statisticheskoe Opisanie Dinamicheskikh Sistem s Fluktuiruyushchimi Parametrami (Statistical Description of Dynamical Systems with Fluctuating Parameters). — Moscow: Nauka (in Russian).

5. Klyatskin, V. I., 1980, Stochasticheskie Uravneniya i Volny v Sluchaino-Neodnorodnoi Srede (Stochastic Equations and Waves in Randomly Inhomogeneous Medium). — Moscow: Nauka (in Russian).

6. Klyatskin, V. I., 1985, Ondes et Équations Stochastiques dans les milieus Aléatoirement non Homogènes. — Besançon Cedex: Les Éditions de Physique.

7. Klyatskin, V. I. and A. I. Saichev, 1997, Statistical theory of the diffusion of a passive tracer in a random velocity field, Zh. Éxper. Teor. Fiz. **111**(4), 1297 – 1313 [JETP **84**(4), 716 – 724, 1997].

8. Klyatskin, V. I. and A. I. Saichev, 1992, Statistical and dynamical localization of plane waves in randomly layered media, Uspekhi Fiz. Nauk **162**(3), 161 – 194 [Soviet Physics Usp. **35**(3), 231 – 247, 1992].

9. Klyatskin, V. I. and D. Gurarie, 1999, Coherent phenomena in stochastic dynamical systems, Uspekhi Fiz. Nauk **169**(2), 171 – 207 [Physics - Uspekhi **42**(2), 165 – 198, 1999].

10. Nicolis, G. and I. Prigogin, 1989, Exploring Complexity, an Introduction. — NY: W. H. Freeman and Company.

11. Isichenko, M. B., 1992, Percolation, statistical topography, and transport in random media // Rev. Modern Phys. **64**(4), 961 – 1043.

12. Klyatskin, V. I., 2000, Stochastic transport of passive tracer by random flows, Izvestiya AN, Fiz. Atm. i okeana **36**(2), 177 – 201, [Atmospheric and Oceanic Physics **36**(2), 757 – 765, 2000].

13. Klyatskin, V. I. and K. V. Koshel', 2000, The simplest example of the development of a cluster-structured passive tracer field in random flows, Uspekhi Fiz. Nauk **170**(7), 771 – 778 [Physics - Uspekhi **43**(7), 717 – 723, 2000].

14. Klyatskin, V. I., 2001, Stochasticheskie Uravneniya Glazami Fizika (Osnovnye Idei, Tochnye Resultaty i Asimptoticheskie Priblizheniya) (Stochastic Equations through the Eye of the Physicist (Basic Ideas, Exact Results, and Asymptotic Approximations)). — Moscow: Fizmatlit (in Russian).

15. Koshel', K. V. and O. V. Aleksandrova, 1999, Some resullts of numerical modeling of diffusion of passive scalar in randon velocity field, Izvestiya AN, Fiz. Atm. i okeana **35**(5), 638 – 648 [Izvestiya, Atmospheric and Oceanic Physics **35**, 1999].

16. Zirbel, C. L. and E. Çinlar, 1996, Mass transport by Brownian motion / In: Stochastic Models in Geosystems, eds. S. A. Molchanov and W. A. Woyczynski, IMA Volumes in Math. and its Appl., **85**, 459 – 492. — NY: Springer-Verlag.

17. Mesinger, F. and Y. Mintz, 1970, Numerical simulation of the 1970 – 1971 Eole experiment, Numerical simulation of weather and climate, Technical Report No. 4. — Los Angeles: Dep. Meteorology. Univ. of California.

18. Mesinger, F. and Y. Mintz, 1970, Numerical simulation of the clustering of the constant-volume balloons in the global domain, Numerical simulation of weather and climate, Technical Report No. 5. — Los Angeles: Dep. Meteorology, Univ. of California.

19. Yaroshchuk, I. O., 1986, Chislennoe Modelirovanie Rasprostraneniya Ploskikh Voln v Sluchaino Sloistych Lineinych i Nelineinych Sredach (Numerical Modeling of Plane Wave Propagation in Randomly Layered Linear and Nonlinear Media), Ph.D thesis. — Vladivostok: Pacific Oceanological Institute.

20. Maxey, M. R., 1987, The Gravitational Settling of Aerosol Particles in Homogeneous turbulence and random flow field, J. Fluid Mech. **174**, 441 – 465.

21. Lamb, H., 1932, Hydrodinamics, Sixth Edition. — New York: Dover Publications.

22. Klyatskin, V. I. and T. Elperin, 2002, Clustering of the Low-Inertial Particle Number Density Field in Random Divergence-Free Hydrodynamic Flows, Zh. Éxper. Teor. Fiz. **122**(2), 327 – 340 [JETP **95**(2), 328 – 340, 2002].

23. Rytov, S. M., Yu. A. Kravtsov and V. I. Tatarskii, 1977–1978, Vvedenie v Statisticheskuyu Radiofiziku (Introduction to Statistical Radiophysics). — Moscow: Nauka (in Russian) [Principles of Statistical Radiophysics, v. **1–4**. — Berlin: Springer-Verlag, 1987–1989].

24. Tatarskii, V. I., 1967, Rasprostranenie Voln v Turbulentnoi Atmosfere (Wave Propagation in Turbulent Atmosphere). — Moscow: Nauka (in Russian) [The Effects of the Turbulent Atmosphere on Wave Propagation. — Springfield, Va.: National Technical Information Service, 1977].

25. Monin, A. S. and A. M. Yaglom, 1965, 1967, Statisticheskaya Gidromekhanika (Statistical Hydromechanics), **1, 2**. — Moscow: Nauka (in Russian) [Statistical Fluid Mechanics. — Cambridge Massachusetts: MIT Press, 1971, 1975].

26. Ambartsumyan, V. A., 1943, Diffuse reflection of light by a foggy medium, Comptes Rendus (Doklady) de l'USSR.**38**(8),229–232.

27. Ambartsumyan, V. A., 1944, On the problem of diffuse reflection of light, Journal of Physics USSR. **8**(1), 65.

28. Ambartsumyan, V. A., 1989, On the principle of invariance and its some applications, in: Principle of Invariance and its Applications, 9 – 18, eds. M. A. Mnatsakanyan, H. V. Pickichyan. — Yerevan: Armenian SSR Acad. of Sciences.

29. Casti, J. and R. Kalaba, 1973, Imbedding Methods in Applied Mathematics. — Reading, MA: Addison-Wesley.

30. Kagiwada, H. H. and R. Kalaba, 1974, Integral Equations Via Imbedding Methods. — Reading, MA: Addison-Wesley.

31. Bellman, R. and G. M. Wing, 1992, An Introduction to Invariant Imbedding / Classics in Applied Mathematics **8**. — Philadelphia: SIAM.

32. Golberg, M. A., 1975, Invariant imbedding and Riccati transformations, Appl. Math. and Com. **1**(1), 1 – 24.

33. Furutsu, K., 1963, On the statistical theory of electromagnetic waves in a fluctuating medium, J. Res. NBS. **D-67**, 303.

34. Shapiro, V. E. and V. M. Loginov, 1978, 'Formulae of differentiation' and their use for solving stochastic equations, Physica.**91A**, 563 – 574.

35. Shapiro, V. E. and V. M. Loginov, 1983, Dinamicheskie systemy pod vozdeistviem sluchainykh vliyanii (Dynamical systems under random influences). — Novosibirsk: Nauka (in Russian).

36. Bourret, R. C., U. Frish and A. Pouquet,1973, Brownian motion of harmonical oscillator with stochastic frequency, Physica **A 65**(2), 303 – 320.

37. Klyatskin, V. I. and T. Elperin, 2002, Diffusion of Low-Inertia Particles in a Field of Random Forces and the Kramers Problem, Izvestiya AN, Fiz. Atm. i okeana **38**(6), 817 – 823 [Atmospheric and Oceanic Physics **38**(6), 725 – 731, 2002].

38. Tatarskii, V. I., 1969, Light propagation in a medium with random index refraction inhomogeneities in the Markov process approximation, Zh. Éxper. Teor. Fiz. **56**(6), 2106 – 2117 [Sov. Phys. JETP **29**(6), 1133 – 1138, 1969].

39. Klyatskin, V. I., 1971, Space-time description of stationary and homogeneous turbulence, Izvestiya AN SSSR, Mekhanika Zhidkosti i Gaza No 4, 120 – 127 [Fluid Dynamics **6**(4), 655 – 661, 1971].

40. Kuzovlev, Yu. E. and G. N. Bochkov, 1977, Operator methods of analysing stochastic non-gaussian processes and systems, Izv. Vuzov., Radiofizika **20**(10), 1505 – 1515 [Radiophys. & Quantum Electron. **20**(10), 1036 – 1044, 1977].

41. Saichev, A. I. and M. M. Slavinskii, 1985, Equations for moment functions of waves propagating in random inhomogeneous media with elongated inhomogeneities, Izv. Vuzov., Radiofizika **28**(1), 75 – 83 [Radiophys. & Quantum Electron. **28**(1), 55 – 61, 1985].

42. Virovlyanskii, A. L., A. I. Saichev and M. M. Slavinskii, 1985, Moment functions of waves propagation in waveguides with elongated random inhomogeneities of the refractive index, Izv. Vuzov., Radiofizika **28** (9), 1149 – 1159 [Radiophys. & Quantum Electron. **28**(9), 794 – 803, 1985].

43. Balkovsky, E., G. Falkovich and A. Fouxon, 2000, Clustering of inertial particles in turbulent flows, http://arxiv.org/abs/chao-dyn/9912027.

44. Balkovsky, E., G. Falkovich and A. Fouxon, 2001, Intermittent Distribution of Inertial Particles in Turbulent Flows, Phys. Rev. Letters **86**(13), 2790 – 2793.

45. Guzev, M. A., V. I. Klyatskin and G. V. Popov, 1992, Phase fluctuations and localization length in layered randomly inhomogeneous media, Waves in Random Media **2**(2), 117   123.

Index

Made in the USA
Monee, IL
07 July 2026